Nonlinear Ordinary Differential Equations

Martin Hermann · Masoud Saravi

Nonlinear Ordinary Differential Equations

Analytical Approximation and Numerical Methods

 Springer

Martin Hermann
Department of Numerical Mathematics
Friedrich Schiller University
Jena
Germany

Masoud Saravi
Department of Science
Shomal University
Amol
Iran

ISBN 978-81-322-3845-4   ISBN 978-81-322-2812-7   (eBook)
DOI 10.1007/978-81-322-2812-7

To our wives,
Gudrun and Mahnaz

and our children,
Alexander, Sina and Sayeh

Preface

Nonlinear ordinary differential equations (ODEs) play a fundamental role in scientific modeling, technical and economic processes as well as in inner mathematical questions. Examples include the complicated calculation of the trajectories of space gliders and airplanes, forecast calculations, spread of AIDS, and the automation of the parking maneuver of road vehicles on the basis of mechatronic systems.

However, the mathematical analysis of nonlinear ODEs is much more complicated than the study of linear ones. The complexity of nonlinear problems can be illustrated by the following comparison: if the amount of linear problems is set into relation to the size of the earth then the amount of nonlinear problems corresponds to the size of our universe. Therefore, the special type of the nonlinearity must be taken into consideration if a nonlinear problem is theoretically and/or numerically analyzed.

One of the greatest difficulties of nonlinear problems is that, in general, the combination of known solutions is not again a solution. In linear problems, for example, a family of linearly independent solutions can be used to construct the general solution through the superposition principle.

For second- and higher order nonlinear ODEs (more generally speaking, systems of equations) the solutions can seldom be represented in closed form, though implicit solutions and solutions involving non-elementary integrals are encountered. It is therefore important for the applications to have analytical and numerical techniques at hand that can be used to compute approximate solutions.

The first part of this book presents some analytical approximation methods which have been approved in practice. They are particularly suitable for nonlinear lower order ODEs.

High-dimensional systems of nonlinear ODEs are very often found in the applications. The solutions of these systems can only be approximated by appropriate numerical methods. Therefore, in the second part of this book, we deal with numerical methods for the solution of systems of first-order nonlinear (parameter-depending) ODEs, which are subjected to two-point boundary conditions. It is precisely this focus on analytical as well as numerical methods that makes this book particularly attractive.

The book is intended as a primary text for courses in the theory of ODEs and the numerical treatment of ODEs for graduate students. It is assumed that the reader has a basic knowledge of functional analysis, in particular analytical methods for nonlinear operator equations, and numerical analysis. Moreover, the reader should have substantial knowledge of mathematical techniques for solving initial and boundary value problems in linear ODEs as they are represented in our book [63], *A First Course in Ordinary Differential Equations—Analytical and Numerical Methods* (Springer India, 2014). Physicists, chemists, biologists, computer scientists, and engineers whose work involves the solution of nonlinear ODEs will also find the book useful both as a reference and as a tool for self-studying.

In Chaps. 1–4, we have avoided the traditional definition–theorem–proof format. Instead, the mathematical background is described by means of a variety of problems, examples, and exercises ranging from the elementary to the challenging problems. In Chap. 5, we deviate from this principle since the corresponding mathematical proofs are constructive and can be used as the basis for the development of appropriate numerical methods.

The book has been prepared within the scope of a German–Iranian research project on mathematical methods for nonlinear ODEs, which was started in 2012.

We now outline the contents of the book. In Chap. 1, we present the basic theoretical principles of nonlinear ODEs. The existence and uniqueness statements for the solutions of the corresponding initial value problems (IVPs) are given. To help the reader understand the structure and the methodology of nonlinear ODEs, we present a brief review of elementary methods for finding analytical solutions of selected problems. Some worked examples focusing on the analytical solution of ODEs from the applications are used to demonstrate mathematical techniques, which may be familiar to the reader. In Chap. 2, the analytical approximation methods are introduced. Special attention is paid to the variational iteration method and the Adomian decomposition method. For smooth nonlinear problems both methods generate series, which converge rapidly to an exact solution. These approximation methods are applied to some nonlinear ODEs, which play an important role in the pure and applied mathematics. In Chap. 3, another four analytical approximations methods are considered, namely the perturbation method, the energy balance method, the Hamiltonian approach, and the homotopy perturbation method. These methods have proved to be powerful tools for approximating solutions of strongly nonlinear ODEs. They are applied in many areas of applied physics and engineering. The accuracy of the approximated solutions is demonstrated by a series of examples which are based on strongly nonlinear ODEs.

The numerical part of the book begins with Chap. 4. Here, numerical methods for nonlinear boundary value problems (BVPs) of systems of first-order ODEs are studied. Starting from the shooting methods for linear BVPs [63], we present the simple shooting method, the method of complementary functions, the multiple shooting method, and a version of the stabilized march method for nonlinear BVPs. A MATLAB code for the multiple shooting method is provided such that the reader can perform his own numerical experiments. Chapter 5 is devoted to nonlinear parametrized BVPs since in the applications the majority of nonlinear problems depend on external parameters. The first step in the treatment of such BVPs is their formulation as a nonlinear operator equation $T(y, \lambda) = 0$ in Banach spaces, and to apply the Riesz–Schauder theory. Under the assumption that $T_y(z_0)$ is a Fredholm operator with index zero, where $z_0 \equiv (y_0, \lambda_0)$ is a solution of the operator equation, the concept of isolated solutions and singular solutions (turning points and bifurcation points) is discussed. Appropriate extended systems are presented to compute the singular points with standard techniques. In each case, the extended operator equation is transformed back into an extended nonlinear BVP. Since the operator equation is bad conditioned in the neighborhood of singular points, we use the technique of transformed systems to determine solutions near to singularities. In addition to the one-parameter problem, we study the operator equation under the influence of initial imperfections, which can be described by a second (perturbation) parameter. Feasible path-following methods are required for the generation of bifurcation diagrams of parametrized BVPs. We present some methods that follow simple solution curves. The chapter is concluded with the description of two real-life problems.

At this point it is our pleasure to thank all those colleagues who have particularly helped us with the preparation of this book. Our first thanks go to Dr. Dieter Kaiser, without whose high motivation, deep MATLAB knowledge, and intensive computational assistance this book could never have appeared. Moreover, our thanks also go to Dr. Andreas Merker who read through earlier drafts or sections of the manuscript and made important suggestions or improvements. We gratefully acknowledge the financial support provided by the Ernst Abbe Foundation Jena for our German–Iranian joint work.

It has been a pleasure working with the Springer India publication staff, in particular with Shamim Ahmad and Ms. Shruti Raj.

Finally, the authors wish to emphasize that any helpful suggestion or comment for improving this book will be warmly appreciated and can be sent by e-mail to: martin.hermann@uni-jena.de and masoud@saravi.info.

Martin Hermann
Masoud Saravi

Contents

About the Authors

Martin Hermann is professor of numerical mathematics at the Friedrich Schiller University (FSU) Jena, Germany. His activities and research interests include scientific computing and numerical analysis of nonlinear parameter-dependent ODEs. He is also the founder of the Interdisciplinary Centre for Scientific Computing (1999), where scientists of different faculties at the FSU Jena work together in the fields of applied mathematics, computer science, and applications. Since 2003, he has headed an international collaborative project with the Institute of Mathematics at the National Academy of Sciences Kiev (Ukraine), studying, for example, the sloshing of liquids in tanks. Since 2003, Dr. Hermann has been a curator at the Collegium Europaeum Jenense of the FSU Jena (CEJ) and the first chairman of the Friends of the CEJ. In addition to his professional activities, he volunteers in various organizations and associations. In German-speaking countries, his books *Numerical Mathematics and Numerical Treatment of ODEs: Initial and Boundary Value Problems* count among the standard works on numerical analysis. He has also published over 70 articles in refereed journals.

Masoud Saravi is professor of mathematics at the Shomal University, Iran. His research interests include the numerical solution of ODEs, partial differential equations (PDEs), integral equations, differential algebraic equations (DAE), and spectral methods. In addition to publishing several papers with his German colleagues, Dr. Saravi has published more than 15 successful titles on mathematics. The immense popularity of his books is a reflection of his more than 20 years of educational experience, and a result of his accessible writing style, as well as a broad coverage of well laid-out and easy-to-follow subjects. He has recently retired from Azad University and currently cooperates with Shomal University and is working together with the Numerical Analysis Group and the Faculty of Mathematics and Computer Sciences of FSU Jena (Germany). He started off his academic studies at the UK's Dudley Technical College before receiving his first degree in mathematics and statistics from the Polytechnic of North London, and his advanced degree in numerical analysis from Brunel University. After obtaining his M.Phil. in Applied Mathematics from Iran's Amir Kabir University, he completed his Ph.D. in Numerical Analysis on solutions of ODEs and DAEs by using spectral methods at the UK's Open University.

Chapter 1
A Brief Review of Elementary Analytical Methods for Solving Nonlinear ODEs

1.1 Introduction

Nonlinear ordinary differential equations (ODEs) are encountered in various fields of mathematics, physics, mechanics, chemistry, biology, economics, and numerous applications. Exact solutions to those equations play an important role in the proper understanding of qualitative features of many phenomena and processes in various areas of natural science. However, even if there exists a solution, only for a few ODEs it is possible to determine this exact solution in closed form. Although some methods have been proposed in the literature to determine analytical (closed form) solutions to special ODEs, there is not a general analytical approach to evaluate analytical solutions. Surveys of the literature with numerous references, and useful bibliographies, have been given in [5, 67, 89, 106], and more recently in [50].

In this chapter, we present the basic theoretical principles of nonlinear ODEs. The existence and uniqueness statements for the solutions of the corresponding initial value problems (IVPs) are given. To enable the reader to understand the structure and the methodology of nonlinear ODEs, we present a brief review of elementary methods for finding analytical solutions of selected problems. Some worked examples focusing on the analytical solution of ODEs from the applications are used to demonstrate mathematical techniques, which may be familiar to the reader.

1.2 Analytical Solution of First-Order Nonlinear ODEs

The general form of a first-order ODE is

$$f(x, y, y') = 0. \tag{1.1}$$

Since (1.1) is order one, we may guess that the solution (if it exists) contains a constant c. That is, the family of solution will be of the form

© Springer India 2016
M. Hermann and M. Saravi, *Nonlinear Ordinary Differential Equations*,
DOI 10.1007/978-81-322-2812-7_1

$$F(x, y, c) = 0. \tag{1.2}$$

Moreover, if (1.1) can be written as $y' = g(x, y)$, then we search a family of solutions $y = G(x, c)$. We mention, that the constant c can be specified by an additional condition.

As we know, a linear equation has only one general solution whereas a nonlinear equation may also have a singular solution. Consider the IVP

$$y' = f(x, y), \quad y(x_0) = y_0. \tag{1.3}$$

This equation may have no solution or one or more than one solution. For example, the IVP

$$(y')^2 + y^4 = 0, \quad y(0) = 1,$$

has no real solution. The IVP

$$y' - 2y = 0, \quad y(0) = 1,$$

has one solution $y(x) = e^{2x}$ and the IVP

$$y' = \sqrt{y}, \quad y(0) = 0,$$

has more than one solution, namely, $y(x) \equiv 0$ and $y(x) = \dfrac{1}{4}x^2$. This leads to the following two fundamental questions:

1. Does there exist a solution of the IVP (1.3)?
2. If a solution exists, is this solution unique?

The answers to these questions are provided by the following existence and uniqueness theorems.

Theorem 1.1 (Peano's existence theorem)
Let $R(a, b)$ be a rectangle in the x-y plane, and assume that the point (x_0, y_0) is inside it such that

$$R(a, b) \equiv \{(x, y) : |x - x_0| < a, \quad |y - y_0| < b\}.$$

If $f(x, y)$ is continuous and $|f(x, y)| < M$ at all points $(x, y) \in R$, then the IVP (1.3) has at least one solution $y(x)$, which is defined for all x in the interval $|x - x_0| < c$, where $c = \min\{a, b/M\}$.

Proof See, e.g., the monograph [46]. ∎

Theorem 1.2 (Uniqueness theorem of Picard and Lindelöf)
Let the assumptions of Theorem 1.1 be satisfied. If in addition $\frac{\partial f}{\partial y}(x, y)$ is continuous and bounded for all points $(x, y) \in R$, then the IVP (1.3) has a unique solution $y(x)$, which is defined for all x in the interval $|x - x_0| < c$.

Proof See, e.g., the monograph [46].   ∎

Moreover, we have

Theorem 1.3 (Picard iteration)
Let the assumptions of Theorem 1.2 be satisfied. Then, the IVP (1.3) is equivalent to Picard's iteration method

$$y_n(x) = y(x_0) + \int_{x_0}^{x} f(t, y_{n-1}(t))dt, \quad n = 1, 2, 3, \ldots \qquad (1.4)$$

That is, the sequence $y_0, y_1, \ldots$ converges to $y(x)$.

Proof See, e.g., the monograph [46].   ∎

We note that in this chapter our intention is not to solve a nonlinear first-order ODE by Picard's iteration method, but to present Theorem 1.3 just for discussing the existence and uniqueness theorems.

Now, as before, let $c \equiv \min\{a, b/M\}$ and consider the interval $I : |x - x_0| \le c$ and a smaller rectangle

$$T \equiv \{(x, y) : |x - x_0| \le c, \quad |y - y_0| \le b\}.$$

By the mean value theorem of differential calculus, one can prove that if $\partial f/\partial y$ is continuous in the smaller rectangle $T \subset R(a, b)$, then there exists a positive constant K such that

$$|f(x, y_1) - f(x, y_2)| \le K|y_1 - y_2|, \quad (x, y_1), (x, y_2) \in T. \qquad (1.5)$$

Definition 1.4 If the function $f(x, y)$ satisfies the inequality (1.5) for all (x, y_1), $(x, y_2) \in T$, it is said that $f(x, y)$ satisfies a uniform *Lipschitz condition* w.r.t. y in T and K is called *Lipschitz constant*.   □

Remark 1.5 We can conclude that if $\partial f/\partial y$ is continuous in T, then $f(x, y)$ satisfies a uniform Lipschitz condition. On the contrary, it may $f(x, y)$ satisfy a uniform Lipschitz condition, but $\partial f/\partial y$ is not continuous in T. For example, $f(x, y) = |x|$ satisfies a Lipschitz condition in a rectangle, which contains $(0, 0)$, but $\partial f/\partial y$ is not continuous.   □

The following result can be easily proved.

Proposition 1.6 *If on a closed bounded region $R(a, b)$ the function $f(x, y)$ is a continuous function of its arguments and satisfies a Lipschitz condition, then there exists a unique solution $y(x)$ of (1.3), which is defined on an interval I. Furthermore, the sequence of functions $\{y_n(x)\}_1^\infty$ defined by (1.4) converges to $y(x)$ on I.*

As we pointed out it may be difficult to find closed form solutions whenever we deal with nonlinear problems. We are going to introduce some analytical methods, known as elementary methods, which are based on integration. But, first of all, let us write (1.1) in the form

$$M(x, y)dx + N(x, y)dy = 0. \tag{1.6}$$

This equation can be solved in the following cases.

(i) *Separable equations*
If (1.6) can be written in the form $M(x)dx + N(y)dy = 0$, we say that the variables are separable, and the solution can be obtained by integrating both sides with respect to x, i.e.

$$\int M(x)dx + \int N(y(x))\frac{dy}{dx} \cdot dx = c.$$

Then, for simplicity one can write

$$\int M(x)dx + \int N(y)dy = c.$$

Example 1.7 Solve the IVP $(y + 1)e^x y' + xy^2 = 0, \quad y(0) = 1$.

Solution. The given equation can be written as

$$\left(1 + \frac{1}{y}\right) dy = -xe^{-x}dx.$$

Integration yields

$$y + \ln(y) = (x + 1)e^{-x} + c.$$

Using $y(0) = 1$, we obtain $c = 0$. Hence $y + \ln(y) = (x + 1)e^{-x}$.   □

(ii) *Homogeneous equations*
We call $f(x, y)$ homogeneous of degree k if $f(tx, ty) = t^k f(x, y)$, for some constant $k \neq 0$. If in (1.6), $M(x, y)$, and $N(x, y)$ have the same degree, or if $f(x, y)$ is of degree zero, then the substitution $z = y/x$ will reduce the given equation into a separable form. Let $y = xz$, then $y' = z + xz'$. Substituting this into the equation gives $z + xz' = f(z)$. After separating the variables, we obtain

$$\frac{z'}{f(z) - z} = \frac{1}{x}.$$

Example 1.8 Solve $y' = \dfrac{y(2x^2 - xy + y^2)}{x^2(2x - y)}$.

Solution. Since the degree of

$$f(x, y) = \frac{y(2x^2 - xy + y^2)}{x^2(2x - y)}$$

is zero, the substitution $y = xz$ and $y' = z + xz'$ leads to

$$\left(\frac{2}{z^3} - \frac{1}{z^2}\right) dz = \frac{dx}{x}.$$

Thus,

$$\frac{1}{z} - \frac{1}{z^2} = \ln(|x|) + c.$$

Since $z = y/x$, we have

$$\frac{x}{y} - \frac{x^2}{y^2} = \ln(|x|) + c.$$

$\square$

(iii) *Exact equations*
The total differential of the function $z(x, y)$ is given by

$$dz = z_x dx + z_y dy. \tag{1.7}$$

The next definition is based on the connection between (1.6) and (1.7). We say that (1.6) is *exact*, if $M(x, y)dx + N(x, y)dy$ is the total differential of $z(x, y)$. Then, the ODE can be written as $dz = 0$. Now, we have the following result.

Theorem 1.9 *If $M(x, y)$ and $N(x, y)$ have continuous first partial derivatives over some domain D, then (1.6) is exact if and only if $M_y = N_x$.*

Proof See e.g., [46]. ∎

In the case of an exact equation, it holds $z_x = M$ and $z_y = N$. Define $z(x, y)$ by

$$z(x, y) \equiv \int_x M(x, y)dx + g(y).$$

From this we have

$$z_y = \frac{\partial}{\partial y} \int_x M(x, y)dx + g'(y),$$

and this is equal to $N(x, y)$. Therefore

$$g'(y) = N(x, y) - \frac{\partial}{\partial y} \int_x M(x, y) dx.$$

We can determine $g(y)$ if $g'(y)$ is independent from x, and this is true whenever the derivative of the right-hand side w.r.t. x is zero. Performing this differentiation leads to

$$\frac{\partial}{\partial x} \left[N(x, y) - \frac{\partial}{\partial y} \int_x M(x, y) dx \right] = \frac{\partial N}{\partial x} - \frac{\partial}{\partial x} \frac{\partial}{\partial y} \int_x M(x, y) dx$$

$$= \frac{\partial N}{\partial x} - \frac{\partial M}{\partial y}.$$

For an exact equation, the last expression is zero.

Example 1.10 Find the family of solutions of the differential equation

$$\left(\cos(2x) - 2y \sin(2x) - 2xy^3 \right) dx + \left(\cos(2x) - 3x^2 y^2 \right) dy = 0.$$

Solution. The variables are not separable. We have

$$M(x, y) = \cos(2x) - 2y \sin(2x) - 2xy^3, \quad N(x, y) = \cos(2x) - 3x^2 y^2.$$

Then $M_y = N_x = -2 \sin(2x) - 6xy^2$. Hence the equation is exact. We have $z_x = M = \cos(2x) - 2y \sin(2x) - 2xy^3$. Integration w.r.t. x gives

$$z(x, y) = \frac{\sin(2x)}{2} + y \cos(2x) - x^2 y^3 + g(y).$$

Differentiation w.r.t. y leads to

$$z_y = \cos(2x) - 3x^2 y^2 + g'(y).$$

Now, since $z_y = N(x, y) = \cos(2x) - 3x^2 y^2$, it is

$$g'(y) = 0, \quad \text{i.e. } g(y) = c.$$

Therefore, the family of solutions will be

$$\frac{\sin(2x)}{2} + y \cos(2x) - x^2 y^3 = c.$$

□

(iv) *Non-exact equations*

Let us come back to (1.6) and assume that this equation is not exact. Thus, it holds $M_y \neq N_x$. To determine a solution in that case can be very difficult or impossible, even if there exists a solution. But, sometimes Eq. (1.6) can be transformed into an exact equation by multiplying it with a function $F(x, y) \not\equiv 0$, which is called the *integrating factor* that is usually not unique. If such a function exists then the modified ODE

$$F(x, y)M(x, y) + F(x, y)N(x, y) = 0$$

will be exact. Therefore $(FM)_y = (FN)_x$, and we have

$$M F_y + F M_y = N F_x + F N_x.$$

We can rewrite this formula as

$$\frac{N F_x - M F_y}{F} = M_y - N_x. \tag{1.8}$$

This is a first-order partial differential equation that may have more than one solution. But to compute such a solution will be more difficult than to solve the original Eq. (1.6). Fortunately, if we restrict the integrating factor to be a function of the single variable x or y, then we can find F. Suppose $F = F(x)$, then $F_y = 0$, and (1.8) is reduced to

$$\frac{dF}{F} = \frac{M_y - N_x}{N}.$$

It is not difficult to show that

$$F = \exp\left(\int \frac{M_y - N_x}{N} dx\right).$$

On the other hand, if we have $F = F(y)$, then $F_x = 0$, and it holds

$$F = \exp\left(-\int \frac{M_y - N_x}{M} dy\right).$$

Thus, we can say that $F = F(x)$, when $(M_y - N_x)/N$ is a function of x, and $F = F(y)$, when $(M_y - N_x)/M$ is a function of y.

Example 1.11 Determine a particular solution of the ODE

$$2 \tan(y) \sin(2x)dx + (\sin(y) - \cos(2x))dy = 0.$$

Solution. We have $M_y = 2(1 + \tan^2(y)) \sin(2x)$ and $N_x = 2 \sin(2x)$. Hence, the equation is not exact, but $(M_y - N_x)/M = \tan(y) = g(y)$. Thus

$$F(y) = \exp\left(-\int g(y)dy\right) = \exp\left(\int -\tan(y)dy\right)$$

$$= \exp\left(\ln(\cos(y))\right) = \cos(y).$$

Multiplying the differential equation by $\cos(y)$, we get

$$2\sin(y)\sin(2x)dx + \cos(y)(\sin(y) - \cos(2x))dy = 0.$$

This equation is exact, and it can be shown that its solution is

$$2\sin(y)\cos(2x) + \cos^2(y) = c. \qquad \square$$

Example 1.12 Determine a particular solution of the IVP

$$\cos(x)\big(\sin(x) - \sin(y)\big)dx + \sin(x)\cos(y)dy = 0, \quad y\left(\frac{\pi}{2}\right) = 0.$$

Solution. We have $M_y = -\cos(x)\cos(y)$ and $N_x = \cos(x)\cos(y)$. Hence, the equation is not exact, but

$$\frac{M_y - N_x}{M} = -2\cot(x) = g(x).$$

Thus

$$F(x) = \exp\left(-\int g(x)dx\right) = \exp\left(-2\int \cot(x)dx\right)$$

$$= \exp\left(-2\ln(\sin(x))\right) = \frac{1}{\sin^2(x)}.$$

Multiplying the differential equation by $1/\sin^2(x)$, we get

$$\frac{\cos(x)}{\sin^2(x)}\big(\sin(x) - \sin(y)\big)dx + \frac{\cos(x)}{\sin(x)}dy = 0.$$

This equation is exact. The corresponding family of solutions is

$$\ln(\sin(x)) + \frac{\sin(y)}{\sin(x)} = c.$$

To determine the particular solution, which satisfies the given initial condition, we substitute $x = \pi/2$ and $y = 0$ into this family of solutions, obtaining $c = 0$. Hence, the solution of the IVP is

$$\ln\big(\sin(x)\big) + \frac{\sin(y)}{\sin(x)} = 0. \qquad \square$$

(v) *Reducible equations*
Some nonlinear ODEs can be reduced to linear or separable equations by forming
the dependent variable $z = g(y)$. The most common examples of such equations are
the Bernoulli and the Riccati equation. In the rest of this section, we will consider
these two equations. But first, we remember that the general form of a linear ODE is

$$y' + p(x)y = q(x). \tag{1.9}$$

When $q(x) = 0$, this equation is reduced to a separable equation. Assume $q(x) \neq 0$
and consider (1.9). This equation is not exact but it can easily be shown that it has
the integrating factor

$$F(x) = \exp\left(\int p(x)dx\right).$$

Multiplying the given equation by this factor, we get

$$\exp\left(\int p(x)dx\right)(y' + p(x)y) = \exp\left(\int p(x)dx\right)q(x).$$

This equation is exact. Integrating both sides yields

$$\exp\left(\int p(x)dx\right)y = \int \exp\left(\int p(x)dx\right)q(x)dx + c.$$

Hence,

$$y = \exp\left(-\int p(x)dx\right)\int \exp\left(\int p(x)dx\right)q(x)dx$$
$$+ c\exp\left(-\int p(x)dx\right), \tag{1.10}$$

which is the general solution of (1.9).

(a) *Bernoulli equation*
The general form of a Bernoulli equation is

$$y' + p(x)y = q(x)y^n, \tag{1.11}$$

where n is a real number. Applications of this equation can be seen in the study of
population dynamics and in hydrodynamic stability. When $n = 0$ or $n = 1$, this
equation is reduced to a linear equation. Therefore, we assume $n \neq 0$ or $n \neq 1$.
    If we set $z \equiv y^{1-n}$, then $z' = (1 - n)y'y^{2-n}$. Thus,

$$y' = \frac{1 - n}{y^{2-n}}z'.$$

Substituting these expressions into (1.11), we obtain the following *linear* ODE for the function z

$$z' + (1 - n)p(x)z = (1 - n)q(x). \tag{1.12}$$

Example 1.13 A mass m is accelerated by a time-varying force $\exp(-\alpha t)v^3$, where v is its velocity. There is also a resistive force βv, where β is a constant, owing to its motion through the air. The ODE modeling the motion of the mass is therefore

$$mv' = \exp(-\alpha t)v^3 - \beta v.$$

Find an expression for the velocity v of the mass as a function of time. Given is the initial velocity v_0.

Solution. We write the equation in the form

$$v' + \frac{\beta}{m}v = \frac{1}{m}e^{-\alpha t}v^3.$$

This is a Bernoulli equation with $n = 3$. Thus, $z = v^{1-3} = v^{-2}$, and the derivative is $v' = -z'/(2v^{-3})$. Substituting these expressions into the Bernoulli equation, we obtain

$$z' - \frac{2\beta}{m}z = -\frac{2}{m}e^{-\alpha t}.$$

It is not difficult to show that the general solution of this equation is

$$z = \frac{2e^{-\alpha t}}{m(\alpha + \frac{2\beta}{m})} + c\exp\left(\frac{2\beta}{m}t\right).$$

Since $z = v^{-2}$, we have

$$v = \left[\frac{2e^{-\alpha t}}{m\left(\alpha + \frac{2\beta}{m}\right)} + c\exp\left(\frac{2\beta}{m}t\right)\right]^{-2}.$$

Now, the initial condition $v(0) = v_0$ must be used to find the particular solution. □

Example 1.14 For the two-dimensional projectile motion under the influence of gravity and air resistance, using the assumption that the air resistance is proportional to the square of the velocity, the model equation relating the speed $s(t)$ of the projectile to the angle of inclination $\theta(t)$ is

$$\frac{ds}{d\theta} - \tan(\theta)s = b\sec(\theta)s^3,$$

where b is a constant, which measures the amount of air resistance. Solve this equation.

Solution. The given equation is a Bernoulli equation with $n = 3$. We have $z = s^{1-3} = s^{-2}$, and the derivative is $z' = -2s's^{-3}$. Thus $s' = -z'/(2s^{-3})$. Substituting these expressions into the Bernoulli equation, we get

$$\frac{dz}{d\theta} + 2\tan(\theta)z = -2b\sec(\theta).$$

This is a linear ODE and one can show that its general solution is

$$z = \frac{c - b\left(\sec(\theta)\tan(\theta) + \ln(\sec(\theta) + \tan(\theta))\right)}{\sec^2(\theta)}.$$

Since $z = s^{-2}$, it follows

$$s = \frac{\sec(\theta)}{\sqrt{c - b\left(\sec(\theta)\tan(\theta) + \ln(\sec(\theta) + \tan(\theta))\right)}}.$$

$\square$

(b) *Riccati equation*
The general form of a Riccati equation is

$$y' = p(x) + q(x)y + r(x)y^2. \tag{1.13}$$

This equation appears in numerous physical applications and is closely related to linear homogeneous second-order ODEs. If we substitute

$$y = -\frac{u'(x)}{q(x)u(x)}$$

into (1.13), we obtain

$$u''(x) - \left(\frac{r'(x)}{r(x)} + q(x)\right)u'(x) + p(x)r(x)u(x) = 0. \tag{1.14}$$

Obviously, if in (1.13) it holds $p(x) \equiv 0$, then we have a Bernoulli equation. Therefore, we assume $p(x) \neq 0$. In that case, Eq. (1.13) cannot be solved by elementary methods. But, if a particular solution y_1 is known, then the solution of this equation can be obtained by the substitution $y = y_1 + 1/z$, where z is the general solution of the ODE

$$z' + (q + 2ry_1)z = -r.$$

Example 1.15 Show that $y = \cos(x)$ is a particular solution of the ODE

$$\left(1 - \sin(x)\cos(x)\right)y' - y + \cos(x)y^2 = -\sin(x),$$

then solve this equation.

Solution. We have $y = \cos(x)$, and therefore $y' = -\sin(x)$. Substituting these functions into the equation leads to

$$\left(1 - \sin(x)\cos(x)\right)(-\sin(x)) - \cos(x) + \cos(x)\cos^2(x)$$

$$= -\sin(x) + \sin^2(x)\cos(x) - \cos(x) + \cos(x)\left(1 - \sin^2(x)\right) = -\sin(x).$$

Thus, $y = \cos(x)$ is a particular solution of the ODE. We set $y \equiv \cos(x) + 1/z$. Then, $y' = -\sin(x) - z'/z^2$. Substituting these expressions into the ODE, we obtain

$$\left(1 - \sin(x)\cos(x)\right)\left(-\sin(x) - \frac{z'}{z^2}\right)$$

$$-\left(\cos(x) + \frac{1}{z}\right) + \cos(x)\left(\cos(x) + \frac{1}{z}\right)^2 = -\sin(x).$$

By simplifying this expression, we get

$$\left(1 - \sin(x)\cos(x)\right)z' + \left(1 - 2\cos^2(x)\right)z = \cos(x).$$

Since $\left(1 - \sin(x)\cos(x)\right)' = 1 - 2\cos^2(x)$, the integration of the above equation yields

$$z = \frac{\sin(x) + c}{1 - \sin(x)\cos(x)}.$$

We also have $y = \cos(x) + 1/z$, i.e. $z = 1/(y - \cos(x))$. Therefore,

$$y = \frac{c\cos(x) - 1}{c - \sin(x)}.$$

Remark 1.16 Another choice for $y(x)$ is $y(x) = y_1(x) + z(x)$, where $z(x)$ is the family of solutions of the equation

$$z' + (p - 2qy_1) = qz^2. \qquad \square$$

Example 1.17 Solve the ODE $y' + 2xy = y^2 + x^2 + 1$. A corresponding particular solution is $y_1(x) = x$.

Solution. Let $y = x + z$. Then, $y' = 1 + z'$. Substituting these expressions into the given equation, we get $z' = z^2$. It is not difficult to show that the family of solutions of this equation is $z = 1/(c - x)$. That is, $y = x + 1/(c - x)$. $\square$

1.3 High-Degree First-Order ODEs

So far, we have considered first-order ODEs with degree one. In this section, we briefly discuss equations with a higher degree.

The general form of such an equation is

$$(y')^n + p_1(x, y)(y')^{n-1} + \cdots + p_{n-1}(x, y)y' + p_n(x, y) = 0. \qquad (1.15)$$

Unfortunately, solving (1.15) is extremely difficult, and in the majority of cases impossible, even if a solution exists. However, there are some special cases of (1.15), which can be solved analytically. Here, we will discuss two of these special cases.

(i) *Equations solvable for* y'

Suppose (1.15) can be factorized in the form

$$[y' - f_1(x, y)][y' - f_2(x, y)] \cdots [y' - f_n(x, y)] = 0.$$

We set each factor equal to zero, i.e.,

$$y' - f_i(x, y) = 0, \quad i = 1, 2, \ldots, n,$$

and determine the solutions

$$\phi_i(x, y, c) = 0, \quad i = 1, 2, \ldots, n,$$

of these factors. Now, the family of solutions of (1.15) can be written as

$$\phi_1(x, y, c)\phi_2(x, y, c) \cdots \phi_n(x, y, c) = 0.$$

Example 1.18 Solve the ODE $x^2(y')^2 + xy' - y^2 - y = 0$.

Solution. In terms of y', this equation can be compared with an algebraic equation of degree two. Hence, we can write

$$(xy' - y)(xy' + y + 1) = 0.$$

From first factor we obtain $y - c_1 x = 0$ and from the second one, we get $x(y + 1) - c_2 = 0$. Thus

$$(y - c_1 x)(x(y + 1) - c_2) = 0$$

is the solution. $\square$

(ii) *Equations solvable for x or y*
Suppose we can write (1.15) in the form

$$y = f(x, y').$$ (1.16)

Differentiating w.r.t. x gives

$$y' = F(x, p, p'),$$

where $y' = p$. This equation has the two variables x and p. The corresponding solution can be written in the form

$$\phi(x, y', c) = 0.$$ (1.17)

The elimination of y' from (1.16) and (1.17) gives the desired solution. A similar operation can be done when the independent variable is missing.

The methods presented in (i) and (ii) can be helpful for solving (1.15). However, when $n \geq 3$, we encounter difficulties to factorize (1.15) or to eliminate y'. To recognize this, let us consider the equation

$$y = 3xy' - (y')^3.$$

Differentiation w.r.t. x yields

$$y' = 3y' + 3xy'' - 3y''(y')^2.$$

It follows

$$3(x - p^2)p' = -2p.$$

If we rewrite this equation in the form

$$2p \, dx + 3(x - p^2) \, dp = 0,$$

where $p = y'$, then it is not difficult to show that the equation has the integrating factor $F = \sqrt{p}$. Multiplying this equation by F and integrating term by term, we obtain

$$p\sqrt{p}(x - p^2) = c, \quad \text{i.e. } y'\sqrt{y'}(x - (y')^2) = c.$$

The next step is to eliminate y' from this equation and the given equation. Obviously, this elimination is nearly impossible.

We close this section by considering two special nonlinear ODEs.

(iii) *Clairaut equation*
The general form of a Clairaut equation is

$$y = xy' + f(y').$$

(1.18)

Differentiating w.r.t. x gives

$$\left[x + f'(p)\right]\frac{dp}{dx} = 0,$$

where $y' = p$. We can have $dp/dx = 0$ or $x + f'(p) = 0$. If $dp/dx = 0$, then $p = y' = c$. Now, putting $y' = c$ in (1.18), we obtain

$$y = cx + f(c).$$

This is the family of solutions of the Clairaut equation. One can eliminate y' from (1.18) and $x + f'(y') = 0$. It results in a singular solution because it does not contain c.

Example 1.19 Solve the ODE $y = xy' - 2(y')^3$, and find its singular solution.

Solution. Obviously, this is a Clairaut equation. Hence, its family of solutions is $y = cx - 2c^3$. We have $x + f'(y') = x - 6(y')^2 = 0$. Now, if we eliminate y' from this equation and the given equation, we obtain

$$y = 4\left(\frac{x}{6}\right)^{\frac{3}{2}},$$

which is a singular solution.   □

(iv) *Lagrange equation*
A more general form of a Clairaut equation is given by

$$y = f(y') + xg(y'),$$

(1.19)

and known as Lagrange equation. If we differentiate it w.r.t. x and set $y' \equiv p$, then this equation can be written in the form

$$\frac{dx}{dp} - \frac{g'(p)}{p - g(p)}x = \frac{f'(p)}{p - g(p)}.$$

Obviously, this is a linear equation and can be solved by the formula given in (1.10).

Example 1.20 Solve the equation $y = x(y')^2 - (y')^2$.

Solution. This is a Lagrange equation with $f(y') = -g(y') = (y')^2$. Therefore,

$$\frac{dx}{dp} - \frac{2p}{p - p^2}x = -\frac{2p}{p - p^2}.$$

That is,

$$\frac{dx}{dp} + \frac{2}{p - 1}x = -\frac{2}{p - 1}.$$

The general solution of this equation can be found by Eq. (1.14). We obtain

$$x = 1 + \frac{c}{(p - 1)^2} = 1 + \frac{c}{(y' - 1)^2}.$$

By elimination of y' from this equation and the given equation, we get

$$y = x + c + 2\sqrt{\frac{c}{x - 1}}.$$

What makes the solution of the Lagrange equation tedious is the elimination of y'. In general, this difficulty is the same difficulty as it exists in the case (iii).

1.4   Analytical Solution of Nonlinear ODEs by Reducing the Order

We can simplify a nonlinear ODE to a first-order equation by reducing its order, in particular when one of its dependent or independent variables is missing. We consider the second-order ODE in general form

$$f(x, y, y', y'') = 0, \tag{1.20}$$

and deal with two special cases.

1. Suppose (1.20) does not contain y. Hence, we have

$$f(x, y', y'') = 0. \tag{1.21}$$

   The substitution $y' = p$ can be used to transform this equation into a first-order ODE. Since $y'' = p'$, we get

$$f(x, p, p') = 0.$$

2. Suppose (1.20) does not contain x, i.e., it is an autonomous ODE. So we have

$$f(y, y', y'') = 0. \tag{1.22}$$

Let $y' = p$, then $y'' = p(dp/dy)$. Therefore, (1.22) can be written as

$$f(y, p, p') = 0.$$

This is an equation of first-order and can be solved for p.

Example 1.21 A classical problem of mechanical engineering is the hanging cable problem. The shape of a hanging cable of uniform density is governed by the ODE

$$\frac{d^2 y}{dx^2} = \frac{d}{T} \sqrt{1 + \left(\frac{dy}{dx}\right)^2},$$

where d and T are the constant density and the tension in the cable at its lowest point, respectively. Solve this ODE.

Solution. Set $y' \equiv p$, then $y'' = p'$. Hence, the equation becomes

$$p' = \frac{d}{T} \sqrt{1 + p^2}.$$

Obviously, the variables are separable, and we can write

$$\frac{dp}{\sqrt{1 + p^2}} = \frac{d}{T} dx.$$

Integrating both sides, we obtain

$$\sinh^{-1}(p) = \frac{d}{T} x + c_1, \quad \text{i.e.,} \quad p = y' = \sinh\left(\frac{d}{T} x + c_1\right).$$

One more integration leads to

$$y = \frac{T}{d} \cosh\left(\frac{d}{T} x + c_1\right) + c_2.$$

We mention that, for this example, we could also use the method presented in Case 2. $\square$

Example 1.22 The ODE

$$\frac{d^2 y}{dx^2} = \left[\left(\frac{dy}{dx}\right)^2 + 1\right] \frac{dy}{dx}$$

occurs in the analysis of satellite orbits. Find all solutions of this equation.

Solution. Set $y' \equiv p$, then $y'' = p(dp/dy)$. Substituting these expressions into the ODE, we get

$$p\frac{dp}{dy} = [p^2 + 1]p.$$

Obviously, $p = 0$ or $dp/dy = p^2 + 1$. From the first possibility, we get $y' = 0$, i.e., $y = c$. For the second possibility, we can write $dp/(p^2 + 1) = dy$. Obviously, the variables are separable. Therefore, we integrate both sides, and obtain $\tan^{-1}(p) = y + c_1$, i.e., $y' = \tan(y + c_1)$. We can write

$$y' \cot (y + c_1) = dx.$$

Integrating this equation yields

$$\ln \left(\sin (y + c_1) \right) = x + \ln(c_2), \quad \text{i.e.,} \quad y = \sin^{-1}\left(e^{c_2 x} - c_1\right). \qquad \square$$

Example 1.23 In the theory of the plane jet, which is studied in the hydrodynamics [21], we are encountered with the ODE

$$ay''' + yy'' + (y')^2 = 0, \quad a > 0,$$

which is subjected to the conditions $y(0) = y''(0) = 0$ and $y'(x) = 0$, when $x \to +\infty$. Determine the solution of this problem.

Solution. Let us write

$$\frac{d}{dx}(ay'' + yy') = 0.$$

By integration we get

$$ay'' + yy' = c_1.$$

Using the given initial conditions $y(0) = y''(0) = 0$, we obtain $c_1 = 0$. Hence, the equation becomes

$$ay'' + yy' = 0.$$

One more integration w.r.t. x leads to

$$ay' + \frac{y^2}{2} = c_2.$$

This equation can be written as

$$\frac{2a\,dy}{b^2 - y^2} = dx,$$

with $b^2 = 2c_2$. The variables are separable. By integration, we obtain

$$\frac{2a}{b} \tanh^{-1}\left(\frac{y}{b}\right) = x + c_3.$$

We use the initial condition $y(0) = 0$, and obtain $c_3 = 0$. Thus,

$$\frac{2a}{b} \tanh^{-1}\left(\frac{y}{b}\right) = x, \quad \text{i.e.,} \quad y = b \tanh\left(\frac{bx}{2a}\right). \qquad \square$$

What makes the solution of such equations difficult is the integration. Let us consider a next example.

Example 1.24 The mathematical model for the vibration of the simple pendulum is given by the following ODE

$$\theta'' + \frac{g}{l} \sin(\theta) = 0,$$

where θ denotes the inclination of the string to the downward vertical, g is the gravitational constant and l is the length of pendulum. Solve this equation.

Solution. In absence of an independent variable, we set $\theta' \equiv p$, then $\theta'' = p(dp/d\theta)$. Substituting these expressions into the given equation, we obtain

$$p\frac{dp}{d\theta} + \frac{g}{l} \sin(\theta) = 0.$$

This is a first-order ODE and it can be shown that

$$\frac{p^2}{2} - \frac{g}{l} \cos(\theta) = c_1.$$

Now, if we insert $p = \theta'$ into the above equation, we get

$$\theta' = \pm 2\sqrt{c_1 + \frac{g}{l} \cos(\theta)}.$$

The next integration is impossible. We have to use a numerical method to obtain the solution (see e.g., the monographs [57, 63]). $\qquad \square$

Let us consider a last example, which exhibits another difficulty.

Example 1.25 Solve the ODE $y'' = -\lambda e^y, \quad y(0) = y(1) = 0$.

Solution. This is the well-known *Bratu-Gelfand equation* in 1D planar coordinates. The range of applications of this equation is including the solid fuel ignition model in thermal combustion theory, physical applications arising from chemical reaction theory, radiative heat transfer, nanotechnology as well as the expansion of the universe [63].

Set $y' \equiv p$, then $y'' = p(dp/dy)$. Then, the ODE becomes

$$p\frac{dp}{dy} = -\lambda e^y, \quad \text{i.e.,} \quad p\,dp = -\lambda e^y dy.$$

The integration of the above equation gives

$$\frac{p^2}{2} = -\lambda e^y + c_1.$$

This equation can be written as

$$p = \pm\sqrt{a^2 - 2\lambda e^y},$$

where $a^2 = 2c_1$. Let $a^2 - 2\lambda e^y = z^2$. Then

$$-2\lambda y' e^y = 2z'z, \quad \text{i.e.,} \quad y' = \frac{2z'z}{a^2 - z^2}.$$

The substitution of $y' = p$ into this expression leads to

$$\frac{2z'z}{a^2 - z^2} = \pm z, \quad \text{i.e.} \quad \frac{2dz}{a^2 - z^2} = \pm dx.$$

The integration of both sides gives

$$\frac{2}{a}\tanh^{-1}\left(\frac{z}{a}\right) = c_2 \pm x.$$

It follows

$$z = a\tanh\left(b \pm \frac{a}{2}x\right), \quad b \equiv \frac{a}{2}c_2.$$

Since $a^2 - 2\lambda e^y = z^2$, the solution can be formulated in term of y as

$$\lambda e^y = a^2\left[1 - \tanh^2\left(b \pm \frac{a}{2}x\right)\right] = \frac{a^2}{\cosh^2\left(b \pm \frac{a}{2}x\right)}.$$

It is not difficult to show that

$$y(x) = 2\ln\left(\frac{a}{\sqrt{2\lambda}\cosh\left(b \pm \frac{a}{2}x\right)}\right).$$

Difficulties occur when we apply the boundary conditions. If we use $y(0) = 0$, we come to $a = \sqrt{2\lambda}\cosh(b)$. Then, the second boundary condition $y(1) = 0$ implies $y(x) = 0$, with $\lambda = 0$. This is the trivial solution. For $\lambda = -\pi^2$, the exact solution is

$$y = -2\ln\left(\sqrt{2}\cos\left(\pi\left[\frac{x}{2} - \frac{1}{4}\right]\right)\right).$$

This BVP will be studied in the next chapter by approximation methods.   □

1.5 Transformations of Nonlinear ODEs

In many cases, a change of variables may be useful to solve a nonlinear ODE. Possible choices are $z = g(x)$, $z = h(y)$, or other forms. Although the right choice of the variables is difficult, in the majority of cases the presence of the variables in the given ODE may help to find such a choice.

Example 1.26 Solve the IVP $xe^y y' - e^y = \dfrac{2}{x^2}$, $y(1) = 0$.

Solution. Because of the presence of the function e^y in the ODE, one can guess $z = e^y$. Hence, $z' = e^y y'$. Substituting this into in the ODE and simplifying, we get

$$z' - \frac{z}{x} = \frac{2}{x^2}.$$

This is a linear ODE and it can be shown that its particular solution is

$$z = 2x - \frac{1}{x}.$$

Transforming back the above equation by $z = e^y$, we obtain

$$y = \ln\left(2x - \frac{1}{x}\right)$$

as the solution of the given IVP.   □

Example 1.27 Solve the ODE

$$2y' = (x - y)\cot(x) - [(x - y)\cot(x)]^{-1} + 2.$$

Solution. The presence of $(x - y)$ encourages us to choose $z = x - y$. Hence, $y' = 1 - z'$. If we substitute this into the ODE, we get

$$z' + \frac{1}{2}\cot(x)z = \frac{1}{2}\cot(x)z^{-1}.$$

This is a Bernoulli equation and can easily be solved.   □

Example 1.28 Solve the ODE $x^3 y'' + 2x^2 = (xy' - y)^2$.

Solution. We may write this equation as

$$xy'' + 2 = \left(y' - \frac{y}{x}\right)^2.$$

One can choose $z = y/x$. Then, $y' = z + xz'$ and $y'' = xz'' + 2z'$. Substituting these expressions into the above equation and simplifying the result, yields

$$x^2 z'' + 2xz' + 2 = x^2(z')^2.$$

This equation does not contain the function z. Therefore, we set $z' \equiv p$ and get $z'' = p'$. By substituting these expressions into the equation, we obtain

$$x^2 p' + 2xp + 2 = x^2 p^2,$$

which is a Riccati equation. It can easily be seen that $p_1 = -1/x$ is a particular solution. Hence, we have $z = -\ln(x)$. That is, $y = -x\ln(x)$.

There are several approaches to find the family of solutions. The first one is to set $p \equiv -1/x + 1/z$. With this choice, the family of solutions can be determined as described above. This solution method brings to mind that another choice may be $y = -x\ln(z)$. If we use this change of variable, we come to the following Cauchy-Euler equation

$$x^2 z'' + 2xz' - 2z = 0,$$

which can easily be solved.   □

A list of the exact solutions of some nonlinear ODEs and the corresponding transformations can be found in [96].

Sometimes a group of transformations is required to obtain an ODE, which can be solved exactly.

Example 1.29 Solve the ODE $x^4 y'' + (xy' - y)^3 = 0$.

Solution. If we rewrite the equation as

$$y'' + \left(\frac{1}{x}\right)^4 (xy' - y)^3 = 0,$$

we can set $u \equiv xy' - y$ and $v \equiv 1/x$. With these changes, the equation is transformed into

$$\frac{du}{dv} = -vu^3.$$

Separating the variables, it can easily be shown that

$$u = \frac{1}{\sqrt{v^2 + c_1{}^2}}.$$

Substituting back the expressions for u and v, we obtain

$$xy' - y = \frac{x}{\sqrt{1 + c_1{}^2 x^2}}.$$

This is a first-order linear ODE. It is not difficult to show that

$$y = x\left(c_2 + \ln\left(\frac{x}{c_1 + \sqrt{c_1^2 + x^2}} \right) \right)$$

is the desired solution.   □

Let us now consider a class of ODEs, which is given by

$$y'' + \frac{a}{y}(y')^2 + \frac{b}{x}y' + \frac{c}{x^2}y + dx^r y^s = 0, \tag{1.23}$$

where a, b, c, d and $r \neq -2$, $s \neq 1$ are arbitrary constants (see [96]). A variant of this equation is obtained by setting $y \equiv e^u$ and inserting this expression into (1.23). It results

$$u'' + (a + 1)(u')^2 + \frac{b}{x}u' + \frac{c}{x^2} + dx^r e^{(s-1)u} = 0. \tag{1.24}$$

A large number of nonlinear ODEs, which play an important role in applied physics, can be obtained from (1.23) or (1.24) by a specific choice of the constants. Typical examples are:

1. *Ivey's equation*

$$y'' - \frac{(y')^2}{y} + \frac{2}{x}y' + ky^2 = 0,$$

Occurring in space-charge theory, it is a special case of (1.23) with $a = -1$, $b = 2$, $c = 0$, $d = k$, $r = 0$ and $s = 2$.

2. *Thomas-Fermi's equation*

$$y'' = x^{-1/2}y^{3/2}.$$

This equation is a special case of (1.23) with $a = b = c = 0, d = -1, r = -1/2$ and $s = 3/2$.

3. *Emden-Lane-Fowler equation*

$$y'' + \frac{2}{x}y' = y^m.$$

It is a special case of (1.23) with $a = 0, b = 2, c = 0, d = 1, r = 0$ and $s = m$.

4. *Poisson-Boltzman's equation*

$$y'' + \frac{\alpha}{x}y' = e^y.$$

In case of plane, cylindrical, or spherical symmetry, we have $\alpha = 0, 1, 2$, respectively. This is a special case of (1.24) with $a = -1, b = \alpha, c = 0, d = -1$, $r = 0$, and $s = 2$.

5. *Bratu's equation*

$$y'' = -\lambda e^y.$$

This is a special case of (1.24) with $a = -1, b = c = 0, d = \lambda, r = 0$ and $s = 2$.

A reliable procedure for obtaining a particular solution of such equations is to use the special ansatz $y \equiv ax^n$.

Let us consider the first three equations. When we set $y \equiv ax^n$, we obtain $y' = nax^{n-1}$ and $y'' = an(n-1)x^{n-2}$. Substituting these expressions into the given equations and equating both sides, gives the values of a and n. Let us first have a look at Ivey's equation

$$y'' - \frac{(y')^2}{y} + \frac{2}{x}y' + ky^2 = 0.$$

We set $y \equiv ax^n$ and substitute y, y' and y'' into the equation. It follows

$$n(n-1)ax^{n-2} - an^2x^{n-2} + 2nax^{n-2} + ka^2x^{2n} = 0,$$

that is,

$$n(n-1) - n^2 + 2n + kax^{n+2} = 0.$$

Since this relation must be satisfied for all x, we set $x = 1$. It can easily be seen that a solution of the resulting algebraic equation is $n = -2$ and $a = 2/k$. Thus, a particular solution of Ivey's equation is

$$y = \frac{2}{k x^2}.$$

To compute a solution of the Thomas-Fermi's equation and the Emden-Lane-Fowler's equation, we use the change of variables $y \equiv \ln(u)$ and transform these equations into the form (1.23). Then, we set $u = ax^n$ and proceed as before.

Let us consider the Poisson-Boltzman's equation with $\alpha = 2$, i.e.

$$y'' + \frac{2}{x} y' = e^y.$$

If we set $y \equiv \ln(u)$, we obtain

$$u'' + \frac{2}{x} u' - \frac{(u')^2}{u} = u^2.$$

Now, when we set $u \equiv ax^n$, we have $u' = anx^{n-1}$ and $u'' = an(n-1)x^{n-2}$. The substitution of these expressions into the above ODE yields

$$n(n-1) - n^2 + 2n = ax^{n+2}.$$

As before we can set $x = 1$. A solution of the resulting algebraic equation is $n = a = -2$. Thus, we have $u = -2x^{-2}$, and a particular solution of the Poisson-Boltzman's equation is

$$y = \ln\left(-\frac{2}{x^2}\right). \tag{1.25}$$

Another strategy by which one can attempt to determine a particular solution of a second-order ODE is to reduce the order by a group of transformations. In some cases it is then easier to solve the nonlinear first-order ODE. We will see that this is true for the Poisson-Boltzman's equation. Let us rewrite this equation in the form

$$x^2 y'' + 2xy' = x^2 e^y.$$

When we set $u \equiv xy'$ and $v \equiv x^2 e^y$, the ODE is transformed into the equation

$$\frac{dv}{du} = \frac{v(u+2)}{v-u}.$$

This equation cannot be solved analytically. But if we write

$$(v-u)dv - v(u+2)du = 0,$$

then one may choose $v = u$ and $u = -2$, i.e., $xy' = -2$ and $x^2 e^y = -2$. These two relations lead to

$$y = \ln\left(-\frac{2}{x^2}\right). \tag{1.26}$$

Comparing (1.25) with (1.26), we see that we have determined the same particular solution as before.

Let us come back to the Thomas-Fermi's equation

$$y'' = x^{-1/2} y^{3/2}.$$

Setting $u \equiv x^3 y$ and $v \equiv x^4 y'$, we obtain

$$\frac{dv}{du} = \frac{u^{3/2} + 4v}{v + 3u}.$$

We write

$$\left(u^{3/2} + 4v\right) du - (v + 3u)dv = 0,$$

and demand

$$u^{3/2} + 4v = 0 \quad \text{and} \quad v + 3u = 0.$$

It is not difficult to show that the intersection points are $(144, -532)$ and $(0, 0)$. If we choose $u = 144$ or $v = -532$, we obtain the particular solution

$$y = \frac{144}{x^3}$$

of the Thomas-Fermi's equation.

As we could see above, sometimes it is possible to solve some classes of ODEs without imposing any additional condition. But this is not our goal. The main task in this book is to solve initial or boundary value problems.

Remark 1.30 The first integral method, which is based on the ring theory of commutative algebra, was first proposed by Feng [37] to solve the Benjamin-Bona-Mahony, Gardner and Foam drainage equations. Here, we use a part of the first integral method for the analytical treatment of the Foam drainage equation, without applying the division theorem. First we transform this equation into an ODE by $\xi = x + ct$, which is known as the wave variable. Then, we solve it by elementary methods.   □

Example 1.31 Solve the Foam drainage equation

$$\frac{\partial u}{\partial t} + \frac{\partial}{\partial x}\left(u^2 - \frac{\sqrt{u}}{2}\frac{\partial u}{\partial x}\right) = 0.$$

Solution. First, we write the above equation in the form

$$\frac{\partial u}{\partial t} + 2\frac{\partial u}{\partial x}u - \frac{1}{4\sqrt{u}}\left(\frac{\partial u}{\partial x}\right)^2 - \frac{\sqrt{u}}{2}\frac{\partial^2 u}{\partial x^2} = 0.$$

Using the wave variable $\xi \equiv x + ct$, this partial differential equation can be transformed into the following ODE (see [107])

$$cu' + 2uu' - \frac{(u')^2}{4\sqrt{u}} - \frac{u''\sqrt{u}}{2} = 0,$$

where $(\cdot)'$ denotes the derivative w.r.t. the variable ξ. Assume $u = v^2$, then $u' = 2vv'$ and $u'' = 2((v')^2 + vv'')$. If we substitute these expressions into the ODE, we get

$$2cvv' + 4v^3v' - 2v(v')^2 - v^2v'' = 0.$$

Writing this equation in the form

$$2cvv' + 4v^3v' - (v^2v')' = 0,$$

and then integrating, we obtain

$$cv^2 + v^4 - v^2v' = c_1.$$

If we assume $c_1 = 0$, we get

$$v^2(c + v^2 - v') = 0.$$

Thus, $v^2 = 0$, that is $u = 0$, which is an obvious solution. Moreover, it also applies $v' = c + v^2$, where $v' = dv/d\xi$. By separating the variables, we can write

$$\frac{dv}{c + v^2} = d\xi.$$

A simple integration leads to

$$\frac{1}{\sqrt{c}}\tan\left(\frac{v}{\sqrt{c}}\right) = \xi + c_2.$$

Since $u = v^2$ and $\xi = x + ct$, we get

$$\frac{1}{\sqrt{c}}\tan\left(\sqrt{\frac{u}{c}}\right) = x + ct + c_2.$$

It follows

$$u = c \left(\tan^{-1} \left(\sqrt{c} \, (x + ct + c_2) \right) \right)^2.$$

This is a solution of the Foam drainage equation.

On the other hand, if we assume $c_1 \neq 0$, then it can be shown that

$$\frac{-1}{2b\sqrt{a+b}} \tan^{-1} \left(\sqrt{\frac{u}{a+b}} \right) + \frac{1}{2b\sqrt{a-b}} \tan^{-1} \left(\sqrt{\frac{u}{a-b}} \right) = x + ct + c_2,$$

where

$$a \equiv \frac{c}{2} \text{ and } b \equiv \sqrt{\frac{c^2}{4} + c_2}.$$

This is another solution of the Foam drainage equation in implicit form. □

The Benjamin-Bona-Mahony, Gardner equations can also be solved by this procedure (see [107]). We end this chapter by considering another example.

Example 1.32 Solve the following Burgers' equation

$$\frac{\partial u}{\partial t} + u \frac{\partial u}{\partial x} + \frac{\partial^2 u}{\partial x^2} = 0, \quad u(x, 0) = x.$$

Solution. Set $\xi \equiv x + ct$, then the PDE is transformed into

$$cu' + uu' + u'' = 0,$$

where $(\cdot)'$ denotes the derivative w.r.t. the variable ξ. This is an autonomous ODE. We set $u' \equiv p$. Then, $u'' = p(dp/du)$. Inserting these expressions into the ODE yields

$$cp + up + p \frac{dp}{du} = 0.$$

Obviously, $p = 0$, i.e., $u = a$ is a solution. Moreover, it also applies

$$\frac{dp}{du} = -(c + u).$$

Thus,

$$p = u' = -\frac{1}{2}(c + u)^2 + c_1.$$

Let us assume that $c_1 = 0$. It follows

$$-\frac{du}{(c+u)^2} = \frac{1}{2} d\xi.$$

By integration, we obtain

$$\frac{1}{c+u} = \frac{1}{2}\xi + c_2 = \frac{1}{2}(x + ct) + c_2.$$

Using the initial condition $u(x, 0) = x$, we determine the integration constant c_2 as

$$c_2 = \frac{1}{c+x} - \frac{1}{2}x.$$

This leads to following particular solution

$$u(x, t) = \frac{2(c+x)}{2 + ct(c+x)} - c = \frac{2x - c^2t(c+x)}{2 + ct(c+x)}. \qquad \square$$

1.6 Exercises

Exercise 1.1 Solve the following differential equations:

(1) $\tan(y)dx + (x - \sin^2(y))dy = 0,$

(2) $xdx + \sin^2\left(\frac{y}{x}\right)(ydx - xdy) = 0, \quad y(1) = 0,$

(3) $\left(\sin(\theta) - \cos(\theta)\right)dr + r\left(\sin(\theta) + \cos(\theta)\right)d\theta = 0,$

(4) $y\,(e^x + y)\,dx + (e^x + 2xy)\,dy = 0,$

(5) $2\tan(y)\sin(2x)dx + \left(\sin(y) - \cos(2x)\right)dy = 0,$

(6) $y\left(2x^2 - xy + y^2\right)dx - x^2(2x - y)dy = 0,$

(7) $\left(\cos(2r) - 2\theta\sin(2r) - 2r\theta^3\right)dr + \left(\cos(2r) - 3r^2\theta^2\right)d\theta = 0,$

(8) $y\left(\ln(x) - \ln(y)\right)dx + (y - x\ln(x) + x\ln(y))\,dy = 0, \quad y(1) = 1,$

(9) $x' - 2x\sin(y) = -x\sqrt{x}\sin(y),$

(10) $x^3ydx + \left(3x^4 - y^3\right)dy = 0,$

(11) $y^3\sec^2(x)dx = \left(1 - 2y^2\tan(x)\right)dy,$

(12) $(5x - 3y)dx + (4x + 2y + 1)dy = 0,$

(13) $44(x - 2y + 2)dx - (x - 2y - 3)dy = 0,$

(14) $2y(ydx + 2xdy) + x^3(5ydx + 3xdy) = 0,$

(15) $2x(2ydx + dx + dy) + y^3(3ydx + 5xdy) = 0,$

(16) $y' + x\sin(2y) = 2xe^{-x^2}\cos^2(y),$

(17) $\left(2x^2 + 5xy^3\right)y' + 4xy + 3y^4 = 0,$

(18) $2y^2 - \ln(x) + 3x(y - 2x)y' = 0,$

(19) $xy' - 3y = x^5\sqrt[3]{y},$

(20) $x\,(x - \sec(y)\ln(x))\,y' + x\tan(y) - y\sec(y) = 0.$

Exercise 1.2 Show that by choosing the appropriate change of variables, the following equations lead to equations with separable variables:

(1) $y' = e^{ay} f(ay + \ln(x))$,

(2) $xy' + n = f(x)g(x^n e^y)$,

(3) $y' + nx^{-1} \ln(y) = yf(x)g(x^{2n} \ln(y))$,

(4) $\cos(y)y' + \cot(x)\sin(y) = f(x)g(\sin(x)\sin(y))$,

(5) $\sin(2x)y' + \sin(2y) = \cos^2(y)f(x)g(\tan(x)\tan(y))$.

Exercise 1.3 By introducing the appropriate change of variables, solve the following equations:

(1) $xy' = y + x^{n+1} \sec\left(\dfrac{y}{x}\right)$,

(2) $xy' - y = y^2 - x^4$,

(3) $3y^2 y' = 2\left(y^3 - 4\right) - 5x\left(y^3 - 4\right)^3$,

(4) $x^2 y'' + x(y+1)y' + y = 0$,

(5) $xyy'' + 2x(y')^2 + 3yy' = 0$,

(6) $x(y+x)y'' + x(y')^2 - (y-x)y' = y$,

(7) $\left(xy\sqrt{x^2 - y^2} + x\right)y' = y - x^2\sqrt{x^2 - y^2}$,

(8) $x^2 y'' + xy' + y = y^{-3}$,

(9) $y' = f(x)e^{ay} + g(x)$,

(10) $\cos(x)y'' + (y')^2 = 1$.

Exercise 1.4 Find the family of solutions of the following ODEs by reducing the order of the equation:

(1) $x^4 y'' = y'\left(y' + x^3\right)$,

(2) $yy'' + (y+1)(y')^2 = 0$,

(3) $yy'' + (y')^3 = (y')^2$,

(4) $e^{y'} y'' = x$,

(5) $yy'' = (y')^2 \left(1 - y'\sin(y) - yy'\cos(y)\right)$.

Exercise 1.5 For following nonlinear ODEs, find a particular solution:

(1) $x^2 y'' - (y')^2 + 2y = 0$,

(2) $xy''' + 3y'' = xe^{-y'}$,

(3) $x^2 y'' - 2(y')^3 + 6y = 0$,

(4) $y'' + \dfrac{2}{x}y' = y^m, \quad m \neq 3$,

(5) $y''' - \dfrac{15}{x^2}y' = 3y^2$.

Exercise 1.6 For the following equation, discuss the particular solution for different values of a, b, and c:

$$x^2 y'' + axy' = bx^2 e^{cy}.$$

Exercise 1.7 By using a group of transformations, find the family of solution of the following ODEs:

(1) $x^3 y'' = e^{y/x}$,
(2) $x^2 y'' + xy' - y = x^4 y^2$,
(3) $\tan(y)y' = x^{n-1} \cos(y)$,
(4) $xy'' + y' = x^2 e^{ay} (y')^3$,
(5) $y' - \cot(x) \tan(y) = \sin(x) \sin(y)$.

Chapter 2
Analytical Approximation Methods

2.1 Introduction

As we mentioned in the previous chapter, most of the nonlinear ODEs have no
explicit solutions, i.e., solutions, which are expressible in finite terms. Even if an
explicit solution can be determined, it is often too complicated to analyze the principal
features of this solution. Due to such difficulties, the study of nonlinear mathematical
problems is the most time-consuming and difficult task for researchers dealing with
nonlinear models in the natural sciences, engineering, and scientific computing. With
the increasing interest in the development of nonlinear models, a variety of analytical
asymptotic and approximation techniques have been developed in recent years to
determine *approximate solutions* of partial and ordinary differential equations. Some
of these techniques are the perturbation method, the variational iteration method, the
homotopy perturbation method, the energy balance method, the variational approach
method, the parameter-expansion method, the amplitude-frequency formulation, the
iteration perturbation method, and the Adomian decomposition method.

In this chapter, we present the variational iteration method and the Adomian
decomposition method since these techniques have good convergence characteristics
and can be used to treat strongly nonlinear ODEs.

2.2 The Variational Iteration Method

The variational iteration method (VIM) was first proposed by He (see e.g. [49, 50])
and systematically elucidated in [51, 54, 126]. The method treats partial and ordinary differential equations without any need to postulate restrictive assumptions that
may change the physical structure of the solutions. It has been shown that the VIM
solves effectively, easily, and accurately a large class of nonlinear problems with
approximations converging rapidly to accurate solutions, see e.g. [127]. Examples

© Springer India 2016
M. Hermann and M. Saravi, *Nonlinear Ordinary Differential Equations*,
DOI 10.1007/978-81-322-2812-7_2

for such problems are the Fokker–Planck equation, the Lane–Emden equation, the Klein–Gordon equation, the Cauchy reaction–diffusion equation, and biological population models.

To illustrate the basic idea of the VIM, we consider, the ODE

$$Ly + N(y) = f(x), \quad x \in I, \tag{2.1}$$

where L and N are linear and nonlinear differential operators, respectively, and $f(x)$ is an given inhomogeneous term defined for all $x \in I$. In the VIM, a *correction functional* of the Eq. (2.1) is defined in the following form

$$y_{n+1}(x) = y_n(x) + \int_0^x \lambda(\tau)\big(Ly_n(\tau) + N(\tilde{y}_n(\tau)) - f(\tau)\big)d\tau, \tag{2.2}$$

where $\lambda(\tau)$ is a general Lagrange multiplier, which can be identified using the variational theory [38]. Furthermore, $y_n(x)$ is the nth approximation of $y(x)$ and $\tilde{y}_n(x)$ is considered as a restricted variation, i.e., $\delta\tilde{y}_n(x) = 0$.

By imposing the variation and by considering the restricted variation, Eq. (2.2) is reduced to

$$\begin{aligned}
\delta y_{n+1}(x) &= \delta y_n(x) + \delta\left(\int_0^x \lambda(\tau)Ly_n(\tau)d\tau\right) \\
&= \delta y_n(x) + \left[\lambda(\tau)\left(\int_0^\tau L\delta y_n(\xi)d\xi\right)\right]_{\tau=0}^{\tau=x} \\
&\quad - \int_0^x \lambda'(\tau)\left(\int_0^\tau L\delta y_n(\xi)d\xi\right)d\tau.
\end{aligned} \tag{2.3}$$

Obviously, in (2.3) we have used integration by parts, which is based on the following formula

$$\int \lambda(\tau)y_n'(\tau)d\tau = \lambda(\tau)y_n(\tau) - \int \lambda'(\tau)y_n(\tau)d\tau. \tag{2.4}$$

In the next sections, we will also use two other formulas for the integration by parts, namely

$$\int \lambda(\tau)y_n''(\tau)d\tau = \lambda(\tau)y_n'(\tau) - \lambda'(\tau)y_n(\tau) + \int \lambda''(\tau)y_n(\tau)d\tau, \tag{2.5}$$

and

$$\begin{aligned}
\int \lambda(\tau)y_n'''(\tau)d\tau &= \lambda(\tau)y_n''(\tau) - \lambda'(\tau)y_n'(\tau) + \lambda''(\tau)y_n(\tau) \\
&\quad - \int \lambda'''(\tau)y_n(\tau)d\tau.
\end{aligned} \tag{2.6}$$

Now, by applying the stationary conditions for (2.3), the optimal value of the Lagrange multiplier $\lambda(\tau)$ can be identified (see e.g. [50], formula (2.13) and the next section). Once $\lambda(\tau)$ is obtained, the solution of the Eq. (2.1) can be readily determined by calculating the successive approximations $y_n(x)$, $n = 0, 1, \ldots$, using the formula (see Eq. (2.2))

$$y_{n+1}(x) = y_n(x) + \int_0^x \lambda(\tau)\big(Ly_n(\tau) + N(y_n(\tau)) - f(\tau)\big)d\tau, \qquad (2.7)$$

where $y_0(x)$ is a starting function, which has to be prescribed by the user.

In the paper [50] it is shown, that the approximate solution $y_n(x)$ of the exact solution $y(x)$ can be achieved using any selected function $y_0(x)$. Consequently, the approximate solution is given as the limit $y(x) = \lim_{n\to\infty} y_n(x)$. In other words, the correction functional (2.2) will give a sequence of approximations and the exact solution is obtained at the limit of the successive approximations. In general, it is difficult to calculate this limit. Consequently, an accurate solution can be obtained by considering a large value for n. This value depends on the interval I where a good approximation of the solution is desired.

Let us consider, the following IVP

$$y'(x) + y(x)^2 = 0, \quad y(0) = 1. \qquad (2.8)$$

The corresponding exact solution is

$$y(x) = (1+x)^{-1} = 1 - x + x^2 - x^3 + x^4 - x^5 + x^6 + \cdots . \qquad (2.9)$$

Here, we have

$$Ly = y', \quad N(y) = y^2, \quad f(x) \equiv 0.$$

To determine the Lagrange multiplier, we insert these expressions into (2.2) and obtain

$$y_{n+1}(x) = y_n(x) + \int_0^x \lambda(\tau)\left(\frac{dy_n(\tau)}{d\tau} + \tilde{y}_n(\tau)^2\right)d\tau. \qquad (2.10)$$

Making the above correction functional stationary w.r.t. y_n, noticing that $\delta\tilde{y}_n(x) = 0$ and $\delta y_n(0) = 0$, it follows with (2.3)

$$\delta y_{n+1}(x) = \delta y_n(x) + \delta\left(\int_0^x \lambda(\tau)\frac{dy_n(\tau)}{d\tau}d\tau\right)$$

$$= \delta y_n(x) + \left[\lambda(\tau)\int_0^\tau \frac{d\,\delta y_n(\xi)}{d\xi}d\xi\right]_{\tau=0}^{\tau=x}$$

$$- \int_0^x \lambda'(\tau)\left(\int_0^\tau \frac{d\,\delta y_n(\xi)}{d\xi}d\xi\right)d\tau$$

$$= \delta y_n(x) + \lambda(\tau)\delta y_n(\tau)|_{\tau=x} - \int_0^x \lambda'(\tau)\delta y_n(\tau)d\tau$$

$$\doteq 0.$$

Thus, we obtain the equations

$$1 + \lambda(x) = 0 \text{ and } \lambda'(x) = 0. \tag{2.11}$$

Now, we substitute the solution $\lambda = -1$ of (2.11) into (2.10). It results the successive iteration formula

$$y_{n+1}(x) = y_n(x) - \int_0^x \left(\frac{dy_n(\tau)}{d\tau} + y_n(\tau)^2\right)d\tau. \tag{2.12}$$

We have to choose a starting function $y_0(x)$, which satisfies the given initial condition $y(0) = 1$. Starting with $y_0(x) \equiv 1$, we compute the following successive approximations

$$y_0(x) = 1,$$

$$y_1(x) = 1 - x,$$

$$y_2(x) = 1 - x + x^2 - \frac{1}{3}x^3,$$

$$y_3(x) = 1 - x + x^2 - x^3 + \frac{2}{3}x^4 - \frac{1}{3}x^5 + \frac{1}{9}x^6 - \frac{1}{63}x^7,$$

$$y_4(x) = 1 - x + x^2 - x^3 + x^4 - \frac{13}{15}x^5 + \cdots - \frac{1}{59535}x^{15},$$

$$y_5(x) = 1 - x + x^2 - x^3 + x^4 - x^5 + \frac{43}{45}x^6 - \cdots - \frac{1}{109876902975}x^{31}.$$

In Fig. 2.1 the first iterates $y_0(x), \ldots, y_4(x)$ are plotted.

Comparing the iterates with the Taylor series of the exact solution (see (2.9)), we see that in $y_5(x)$ the first six terms are correct. The value of the exact solution at $x = 1$ is $y(1) = 1/2 = 0.5$. In Table 2.1, the corresponding value is given for the iterates $y_i(x), i = 0, \ldots, 10$.

In the above example, the linear operator is $L = \dfrac{d}{dx}$. More generally, let us assume that $L = \dfrac{d^m}{dx^m}, m \geq 1$.

In [86], the corresponding optimal values of the Lagrange multipliers are given. It holds

Fig. 2.1 The first successive iterates $y_i(x)$ for the IVP (2.8). The *solid line* represents the exact solution $y(x)$

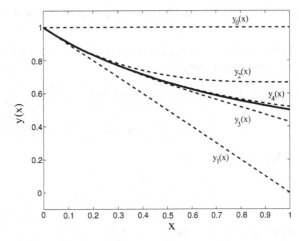

Table 2.1 The successive iterates at the right boundary for the IVP (2.8)

i	$y_i(1)$
0	1
1	0.0
2	0.66666667
3	0.42857143
4	0.51954733
5	0.49529971
6	0.50094000
7	0.49983557
8	0.50002547
9	0.49999645
10	0.50000045

$$\begin{aligned}
\lambda &= -1, && \text{for } m = 1, \\
\lambda &= \tau - x, && \text{for } m = 2, \\
\lambda &= \frac{(-1)^m}{(m-1)!}(\tau - x)^{m-1}, && \text{for } m \geq 1.
\end{aligned} \qquad (2.13)$$

Substituting (2.13) into the correction functional (2.2), we get the following iteration formula

$$y_{n+1}(x) = y_n(x) + \int_0^x \frac{(-1)^m}{(m-1)!}(\tau - x)^{m-1} \left(Ly_n(\tau) + N(y_n(\tau)) - f(\tau)\right)d\tau, \qquad (2.14)$$

where $y_0(x)$ must be given by the user.

2.3   Application of the Variational Iteration Method

In this section, we will consider some nonlinear ODEs and show how the VIM can be used to approximate the exact solution of these problems.

Example 2.1 Solve the following IVP for the Riccati equation

$$y'(x) + \sin(x)y(x) = \cos(x) + y^2, \quad y(0) = 0.$$

Solution. In (2.1), we set

$$Ly \equiv \frac{dy}{dx}, \quad N(y) \equiv \sin(x)y - y^2, \quad f(x) \equiv \cos(x).$$

Thus, the correction functional (2.2) is

$$y_{n+1}(x) = y_n(x) + \int_0^x \lambda(\tau)\left(\frac{dy_n(\tau)}{d\tau} + \sin(\tau)\tilde{y}_n(\tau) - \tilde{y}_n(\tau)^2 - \cos(\tau)\right)d\tau.$$

Since L is the first derivative, i.e., $m = 1$, a look at formula (2.13) shows that $\lambda = -1$ is the optimal value of the Langrange multiplier. The resulting successive iteration formula is

$$y_{n+1}(x) = y_n(x) - \int_0^x \left(y_n'(\tau) + \sin(\tau)y_n(\tau) - y_n(\tau)^2 - \cos(\tau)\right)d\tau. \qquad (2.15)$$

Let us choose $y_0(x) \equiv 0$ as starting function. Notice that $y_0(x)$ satisfies the given initial condition. Now, with (2.15) we obtain the following successive approximations

$$y_1(x) = \sin(x),$$
$$y_2(x) = \sin(x),$$

$$\vdots$$

$$y_n(x) = \sin(x).$$

Obviously, it holds $\lim_{n\to\infty} y_n(x) = \sin(x)$. The exact solution is $y(x) = \sin(x)$. $\square$

Example 2.2 Determine with the VIM a solution of the following IVP for the second order ODE

$$y''(x) + \omega^2 y(x) = g(y(x)), \quad y(0) = a, \quad y'(0) = 0. \qquad (2.16)$$

This problem is the prototype of nonlinear oscillator equations (see, e.g., [30, 99]). The real number ω is the angular frequency of the oscillator and must be determined in advance. Moreover, g is a known discontinuous function.

Solution. To apply the VIM, we set (see formula (2.1))

$$Ly \equiv y'' + \omega^2 y, \quad N(y) \equiv -g(y), \quad f(x) \equiv 0.$$

The corresponding correction functional is

$$y_{n+1}(x) = y_n(x) + \int_0^x \lambda(\tau, x)\left(\frac{d^2 y_n(\tau)}{d\tau^2} + \omega^2 y_n(\tau) - g(\tilde{y}_n(\tau))\right)d\tau.$$

Before we identify an optimal λ, we apply the formula (2.5) for the following integration by parts

$$\int_0^x \lambda(\tau, x)y_n''(\tau)d\tau = \lambda(\tau, x)y_n'(\tau)\,|_{\tau=0}^{\tau=x} - \frac{\partial\lambda(\tau, x)}{\partial\tau}y_n(\tau)\bigg|_{\tau=0}^{\tau=x}$$
$$+ \int_0^x \frac{\partial^2\lambda(\tau, x)}{\partial\tau^2}y_n(\tau)d\tau.$$

Using this relation in the correction functional, imposing the variation, and making the correction functional stationary, we obtain

$$\delta y_{n+1}(x) = \delta y_n(x) + \lambda(\tau, x)\delta y_n'(\tau)\,|_{\tau=x} - \frac{\partial\lambda(\tau, x)}{\partial\tau}\delta y_n(\tau)\bigg|_{\tau=x}$$
$$+ \int_0^x \left(\frac{\partial^2\lambda(\tau, x)}{\partial\tau^2} + \omega^2\lambda(\tau, x)\right)\delta y_n(\tau)d\tau$$
$$\doteq 0.$$

Thus, the stationary conditions are

$$\delta y_n: \quad \frac{\partial^2\lambda(\tau, x)}{\partial\tau^2} + \omega^2\lambda(\tau, x) = 0,$$
$$\delta y_n': \quad \lambda(\tau, x)|_{\tau=x} = 0, \tag{2.17}$$
$$\delta y_n: \quad 1 - \frac{\partial\lambda(\tau, x)}{\partial\tau}\bigg|_{\tau=x} = 0.$$

The solution of the Eqs. in (2.17) is

$$\lambda(\tau, x) = \frac{1}{\omega}\sin(\omega(x - \tau)), \tag{2.18}$$

which leads to the following iteration formula

$$y_{n+1}(x) = y_n(x)$$
$$+ \frac{1}{\omega} \int_0^x \sin(\omega(x - \tau)) \left(\frac{d^2 y_n(\tau)}{d\tau^2} + \omega^2 y_n(\tau) - g(y_n(\tau)) \right) d\tau. \tag{2.19}$$

□

Example 2.3 Let us consider the following IVP of the Emden-Lane-Fowler equation (see e.g. [48])

$$y'' + \frac{2}{x} y' + x^k y^\mu = 0, \quad y(0) = 1, \quad y'(0) = 0.$$

This ODE is used to model the thermal behavior of a spherical cloud of gas acting under the mutual attraction of its molecules.

Solve this equation for $k = 0$ and $\mu = 5$, which has a closed form solution.

Solution. For the given parameters, the IVP to be solved is

$$y'' + \frac{2}{x} y' + y^5 = 0, \quad y(0) = 1, \quad y'(0) = 0.$$

Obviously, there is a singularity at $x = 0$. To overcome this singularity, we set $y \equiv z/x$. Then, we get

$$z'' + x^{-4} z^5 = 0, \quad z(0) = 0, \quad z'(0) = 1.$$

We set

$$Lz \equiv \frac{d^2 z}{dx^2}, \quad N(z) \equiv x^{-4} z^5, \quad f(x) \equiv 0.$$

Thus, the correction functional (2.2) is

$$z_{n+1}(x) = z_n(x) + \int_0^x \lambda(\tau) \left(\frac{d^2 z_n(\tau)}{dx^2} + \tau^{-4} \tilde{z}_n^5(\tau) \right) d\tau.$$

Looking at formula (2.13), we obtain for $m = 2$ the Lagrange multiplier $\lambda = \tau - x$. Therefore, the corresponding iteration formula (2.14) is

$$z_{n+1}(x) = z_n(x) + \int_0^x (\tau - x) \left(\frac{d^2 z_n(\tau)}{dx^2} + \tau^{-4} z_n^5(\tau) \right) d\tau.$$

Starting with $z_0(x) = x$, we obtain the following successive approximations

$$z_0(x) = x,$$

$$z_1(x) = x - \frac{x^3}{6},$$

$$z_2(x) = x - \frac{x^3}{6} + \frac{x^5}{24},$$

$$z_3(x) = x - \frac{x^3}{6} + \frac{x^5}{24} - \frac{x^7}{432}.$$

It is not difficult to show that

$$\lim_{n \to \infty} z_n(x) = z(x) = x\left(1 - \frac{x^2}{6} + \frac{x^4}{24} - \frac{5x^6}{432} + \cdots\right) = x\left(1 + \frac{x^2}{3}\right)^{-1/2}.$$

Thus

$$y(x) = \frac{z(x)}{x} = \left(1 + \frac{x^2}{3}\right)^{-1/2}$$

is the exact solution of the given IVP.   □

Example 2.4 One of the problems that has been studied by several authors is Bratu's BVP (see, e.g., [18, 71, 80, 87, 100, 108]), which is given in one-dimensional planar coordinates by

$$y'' = -\alpha e^y, \quad y(0) = 0, \quad y(1) = 0, \tag{2.20}$$

where $\alpha > 0$ is a real parameter. This BVP plays an important role in the theory of the electric charge around a hot wire and in certain problems of solid mechanics.

The exact solution of (2.20) is

$$y(x) = -2\ln\left(\frac{\cosh(0.5(x - 0.5)\theta)}{\cosh(0.25\theta)}\right),$$

where θ satisfies

$$\theta = \sqrt{2\alpha}\cosh(0.25\theta).$$

Bratu's problem has zero, one or two solutions when $\alpha > \alpha_c$, $\alpha = \alpha_c$, and $\alpha < \alpha_c$, respectively, where the critical value α_c satisfies

$$1 = 0.25\sqrt{2\alpha_c}\sinh(0.25\theta).$$

In Chap. 5, the value α_c is determined as

$$\alpha_c = 3.51383071912.$$

Use the VIM to solve the BVP (2.20).

Solution. Let us expand e^y and use three terms of this expansion. We obtain

$$y'' + \alpha e^y = y'' + \alpha \sum_{i=0}^{\infty} \frac{y^i}{i!} \approx y'' + \alpha \left(1 + y + \frac{y^2}{2} \right).$$

Setting

$$Ly \equiv \frac{d^2 y}{dx^2}, \quad N(y) \equiv \alpha \left(1 + y + \frac{y^2}{2} \right), \quad f(x) \equiv 0,$$

the corresponding correction functional (2.2) is

$$y_{n+1}(x) = y_n(x) + \int_0^x \lambda(\tau) \left(\frac{d^2 y_n(\tau)}{d\tau^2} + \alpha \left(1 + \tilde{y}_n(\tau) + \frac{\tilde{y}_n(\tau)^2}{2} \right) \right) d\tau.$$

Looking at formula (2.13), we obtain for $m = 2$ the Lagrange multiplier $\lambda = \tau - x$. Therefore, the corresponding iteration formula (2.14) is

$$y_{n+1}(x) = y_n(x) + \int_0^x (\tau - x) \left(\frac{d^2 y_n(\tau)}{d\tau^2} + \alpha \left(1 + y_n(\tau) + \frac{y_n(\tau)^2}{2} \right) \right) d\tau. \tag{2.21}$$

Let us start with $y_0(x) = kx$, where k is a real number. The next iterate is

$$y_1(x) = kx + \alpha \int_0^x (\tau - x) \left(1 + k\tau + \frac{k^2 \tau^2}{2} \right) d\tau.$$

Integrating by parts leads to

$$y_1(x) = kx - \frac{\alpha x^2}{2!} - \frac{\alpha k x^3}{3!} - \frac{\lambda k^2 x^4}{4!}.$$

Substituting $y_1(x)$ into the right-hand side of (2.21), we obtain the next iterate

$$y_2(x) = kx - \frac{\alpha x^2}{2!} - \frac{\alpha k x^3}{3!} - \frac{\alpha k^2 x^4}{4!}$$
$$+ \int_0^x (\tau - x) \left(-\frac{\alpha \tau^2}{2} - \frac{2\alpha k \tau^3}{3} + \frac{\alpha}{24}(3\alpha - 5k^2)\tau^4 + \frac{\alpha k}{24}(2\alpha - k^2)\tau^5 \right.$$
$$\left. + \frac{5\alpha^2 k^2 \tau^6}{144} + \frac{\alpha^2 k^3 \tau^7}{144} + \frac{\alpha^2 k^4 \tau^8}{1152} \right) d\tau.$$

Again, integration by parts yields

$$y_2(x) = kx - \frac{\alpha x^2}{2!} - \frac{\alpha k x^3}{3!} - \frac{\alpha(k^2 - \alpha)x^4}{4!} + \frac{4\alpha^2 k x^5}{5!} + \frac{\alpha^2(5k^2 - 3\alpha)x^6}{6!}$$
$$+ \frac{5\alpha^2 k(k^2 - 2\alpha)x^7}{7!} - \frac{25\alpha^3 k^2 x^8}{8!} - \frac{35\alpha^3 k^3 x^9}{9!} - \frac{35\alpha^3 k^4 x^{10}}{10!}.$$

(2.22)

The function $y_2(x)$ must satisfy the initial conditions (see formula (2.20)). For a given α, the equation $y_2(1) = 0$ is a fourth degree polynomial in k. When an appropriate k is chosen from the corresponding four roots, the function $y_2(x)$ can be accepted as an approximation of the exact solution $y(x)$ for $x \in (0, 1)$.

Let us consider Bratu's problem with $\alpha = 1$. The polynomial in k is

$$35k^4 - 3250k^3 + 128250k^2 - 3137760k + 1678320 = 0.$$

Solving this algebraic equation by a numerical method, the following approximated roots are obtained:

$$k_1 = 0.546936690480377, \quad k_2 = 55.687874088793869,$$
$$k_{3,4} = 18.311166038934306 \pm 35.200557613929831 \cdot i.$$

When we substitute $k = k_1$ and $\alpha = 1$ into (2.22), the next iterate is determined. In Table 2.2, $y_2(x)$ is compared with $y(x)$ for $x = 0.1, 0.2, \ldots, 0.9$.

Next, let us consider Bratu's problem with $\alpha = 2$. The polynomial in k is

$$7k^4 - 290k^3 + 5490k^2 - 71136k + 78624 = 0.$$

Solving this algebraic equation by a numerical method, the following approximated roots are obtained:

Table 2.2 Numerical results for Bratu's problem with $\alpha = 1$; $\theta = 1.51716459905075436852$ 18444212962

x	$y_2(x)$	$y(x)$	Relative error
0.1	0.049605613312791	0.049846791245413	0.004838384309118
0.2	0.088710501172383	0.089189934628823	0.005375421099202
0.3	0.116898896821669	0.117609095767941	0.006038639627618
0.4	0.133864761192820	0.134790253884190	0.006866169212538
0.5	0.139428101877514	0.140539214400472	0.007906067553444
0.6	0.133548005691170	0.134790253884190	0.009216157379500
0.7	0.116331618650518	0.117609095767941	0.010862060532661
0.8	0.088038206883444	0.089189934628823	0.012913203156517
0.9	0.049077324553953	0.049846791245413	0.015436634379762

Table 2.3 Numerical results for Bratu's problem with $\alpha = 2$; $\theta = 2.357551053877402042593$ 9799885899

x	$y_2(x)$	$y(x)$	Relative error
0.1	0.110752223723751	0.114410743267745	0.031977062988159
0.2	0.199192293844683	0.206419116487609	0.035010432976831
0.3	0.263300560244860	0.273879311825552	0.038625595742077
0.4	0.301549640325367	0.315089364225670	0.042971059761337
0.5	0.313080790744633	0.328952421341114	0.048249015866102
0.6	0.297850396206136	0.315089364225670	0.054711361209852
0.7	0.256726478126018	0.273879311825552	0.062629168976661
0.8	0.191509696354622	0.206419116487609	0.072228872919732
0.9	0.104847118151250	0.114410743267745	0.083590271711759

$$k_1 = 1.211500000137995, \quad k_2 = 25.631365803713045,$$
$$k_{3,4} = 7.292852812360195 \pm 17.564893217135829 \cdot i.$$

As before, when we substitute $k = k_1$ and $\alpha = 2$ into (2.22), the next iterate is determined. In Table 2.3, $y_2(x)$ is compared with $y(x)$ for $x = 0.1, 0.2, \ldots, 0.9$.

The results in the Tables 2.2 and 2.3 show that the VIM is efficient and quite reliable. Only two iterations have lead to acceptable results. There is no doubt, if more terms of the expansion and/or more iterates are used, the VIM will generate far better results. □

2.4 The Adomian Decomposition Method

The Adomian decomposition method (ADM) is a semi-analytical technique for solving ODEs and PDEs. The method was developed by the Armenian-American mathematician George Adomian [7, 8, 9]. The ADM is based on a decomposition of the solution of nonlinear operator equations in appropriate function spaces into a series of functions. The method, which accurately computes the series solution, is of great interest to the applied sciences. The method provides the solution in a rapidly convergent series with components that are computed elegantly. The convergence of this method is studied in [1, 2, 73].

Let the general form of an ODE be

$$F(y) = f,$$

where F is the nonlinear differential operator with linear and nonlinear terms. In the ADM, the linear term is decomposed as $L + R$, where L is an easily invertible operator and R is the remainder of the linear term. For convenience L is taken as the highest-order derivative. Thus the ODE may be written as

$$Ly + Ry + N(y) = f(x), \tag{2.23}$$

where $N(y)$ corresponds to the nonlinear terms.

Let L be a first-order differential operator defined by $L \equiv \dfrac{d}{dx}$. If L is invertible, then the inverse operator L^{-1} is given by

$$L^{-1}(\cdot) = \int_0^x (\cdot) d\tau.$$

Thus,

$$L^{-1}Ly = y(x) - y(0). \tag{2.24}$$

Similarly, if $L^2 \equiv \dfrac{d^2}{dx^2}$, then the inverse operator L^{-1} is regarded as a double integration operator given by

$$L^{-1}(\cdot) = \int_0^x \int_0^\tau (\cdot) dt\, d\tau.$$

It follows

$$L^{-1}Ly = y(x) - xy'(0). \tag{2.25}$$

We can use the same operations to find relations for higher-order differential operators. For example, if $L^3 \equiv \dfrac{d^3}{dx^3}$, then it is not difficult to show that

$$L^{-1}Ly = y(x) - y(0) - xy'(0) - \frac{1}{2!}x^2 y''(0). \tag{2.26}$$

The basic idea of the ADM is to apply the operator L^{-1} formally to the expression

$$Ly(x) = f(x) - Ry(x) - N(y(x)).$$

This yields

$$y(x) = \Psi_0(x) + g(x) - L^{-1}Ry(x) - L^{-1}N(y(x)), \tag{2.27}$$

where the function $g(x)$ represents the terms, which result from the integration of $f(x)$, and

$$\Psi_0(x) \equiv \begin{cases} y(0), & \text{for } L = \dfrac{d}{dx}, \\[2mm] y(0) + xy'(0), & \text{for } L^2 = \dfrac{d^2}{dx^2}, \\[2mm] y(0) + xy'(0) + \dfrac{1}{2!}x^2 y''(0), & \text{for } L^3 = \dfrac{d^3}{dx^3}, \\[2mm] y(0) + xy'(0) + \dfrac{1}{2!}x^2 y''(0) + \dfrac{1}{3!}x^3 y'''(0), & \text{for } L^4 = \dfrac{d^4}{dx^4}. \end{cases}$$

Now, we write

$$y(x) = \sum_{n=0}^{\infty} y_n(x) \quad \text{and} \quad N(y(x)) = \sum_{n=0}^{\infty} A_n(x),$$

where

$$A_n(x) \equiv A_n(y_0(x), y_1(x), \ldots, y_{n-1}(x))$$

are known as the *Adomian polynomials*. Substituting these two infinite series into (2.27), we obtain

$$\sum_{n=0}^{\infty} y_n(x) = \Psi_0(x) + g(x) - L^{-1}R\sum_{n=0}^{\infty} y_n(x) - L^{-1}\sum_{n=0}^{\infty} A_n(x). \tag{2.28}$$

Identifying the zeroth component $y_0(x)$ by $\Psi_0(x) + g(x)$, the remaining components $y_k(x)$, $k \geq 1$, can be determined by using the recurrence relation

$$\begin{aligned} y_0(x) &= \Psi_0(x) + g(x), \\ y_k(x) &= -L^{-1}Ry_{k-1}(x) - L^{-1}A_{k-1}(x), \quad k = 1, 2, \ldots \end{aligned} \tag{2.29}$$

Obviously, when some of the components $y_k(x)$ are determined, the solution $y(x)$ can be approximated in form of a series. Under appropriate assumptions, it holds

$$y(x) = \lim_{n\to\infty} \sum_{k=0}^{n} y_k(x).$$

The polynomials $A_k(x)$ are generated for each nonlinearity so that A_0 depends only on y_0, A_1 depends only on y_0 and y_1, A_2 depends on y_0, y_1, y_2, etc. [7]. An appropriate strategy to determine the Adomian polynomials is

$$\begin{aligned} A_0 &= N(y_0), \\ A_1 &= y_1 N'(y_0), \\ A_2 &= y_2 N'(y_0) + \frac{1}{2!}y_1^2 N''(y_0), \\ A_3 &= y_3 N'(y_0) + y_1 y_2 N''(y_0) + \frac{1}{3!}y_1^3 N^{(3)}(y_0), \\ A_4 &= y_4 N'(y_0) + \left(\frac{1}{2!}y_2^2 + y_1 y_3\right) N''(y_0) + \frac{1}{2!}y_1^2 y_2 N^{(3)}(y_0) \\ &\quad + \frac{1}{4!}y_1^4 N^{(4)}(y_0), \\ &\vdots \end{aligned}$$

where $N^{(k)}(y) \equiv \dfrac{d^k}{dy^k}N(y)$.

The general formula is

$$A_k = \frac{1}{k!} \frac{\partial^k}{\partial \lambda^k} \left[N \left(\sum_{n=0}^{\infty} y_n \lambda^n \right) \right]_{\lambda=0}, \quad k = 0, 1, 2, \ldots \quad (2.30)$$

A MATHEMATICA program that generates the polynomials A_k automatically can be found in [16]. Moreover, in [20] a simple algorithm for calculating Adomian polynomials is presented. According to this algorithm the following formulas for the Adomian polynomials result:

$$A_0 = N(y_0),$$
$$A_1 = y_1 N'(y_0),$$
$$A_2 = y_2 N'(y_0) + \frac{1}{2} y_1^2 N''(y_0),$$
$$A_3 = y_3 N'(y_0) + y_1 y_2 N''(y_0) + \frac{1}{6} y_1^3 N^{(3)}(y_0),$$
$$A_4 = y_4 N'(y_0) + \left(y_1 y_3 + \frac{1}{2} y_2^2 \right) N''(y_0) + \frac{1}{2} y_1^2 y_2 N^{(3)}(y_0) + \frac{1}{24} y_1^4 N^{(4)}(y_0).$$
$$(2.31)$$

Before we highlight a few examples and show how the ADM can be used to solve concrete ODEs, let us list the Adomian polynomials for some classes of nonlinearity.

1. $N(y) = \exp(y)$:

$$A_0 = \exp(y_0), \qquad A_1 = y_1 \exp(y_0),$$
$$A_2 = \left(y_2 + \frac{1}{2!} y_1^2 \right) \exp(y_0), \qquad A_3 = \left(y_3 + y_1 y_2 + \frac{1}{3!} y_1^3 \right) \exp(y_0); \qquad (2.32)$$

2. $N(y) = \ln(y), \quad y > 0$:

$$A_0 = \ln(y_0), \qquad A_1 = \frac{y_1}{y_0},$$
$$A_2 = \frac{y_2}{y_0} - \frac{1}{2} \frac{y_1^2}{y_0^2}, \qquad A_3 = \frac{y_3}{y_0} - \frac{y_1 y_2}{y_0^2} + \frac{1}{3} \frac{y_1^3}{y_0^2}; \qquad (2.33)$$

3. $N(y) = y^2$:

$$A_0 = y_0^2, \qquad A_1 = 2 y_0 y_1,$$
$$A_2 = 2 y_0 y_2 + y_1^2, \qquad A_3 = 2 y_0 y_3 + 2 y_1 y_2; \qquad (2.34)$$

4. $N(y) = y^3$:

$$A_0 = y_0^3, \qquad\qquad A_1 = 3y_0^2 y_1,$$
$$A_2 = 3y_0^2 y_2 + 3y_0 y_1^2, \quad A_3 = 3y_0^2 y_3 + 6y_0 y_1 y_2 + y_1^3; \tag{2.35}$$

5. $N(y) = yy'$:

$$A_0 = y_0 y_0', \qquad\qquad A_1 = y_0' y_1 + y_0 y_1',$$
$$A_2 = y_0' y_2 + y_1' y_1 + y_2' y_0, \quad A_3 = y_0' y_3 + y_1' y_2 + y_2' y_1 + y_3' y_0; \tag{2.36}$$

6. $N(y) = (y')^2$:

$$A_0 = (y_0')^2, \qquad\qquad A_1 = 2y_0' y_1',$$
$$A_2 = 2y_0' y_2' + (y_1')^2, \quad A_3 = 2y_0' y_3' + 2y_1' y_2'; \tag{2.37}$$

7. $N(y) = \cos(y)$:

$$A_0 = \cos(y_0), \qquad\qquad A_1 = -y_1 \sin(y_0),$$
$$A_2 = -y_2 \sin(y_0) - \frac{1}{2} y_1^2 \cos(y_0), \quad A_3 = -y_3 \sin(y_0) - y_1 y_2 \cos(y_0)$$
$$+ \frac{1}{6} y_1^3 \sin(y_0); \tag{2.38}$$

8. $N(y) = \sin(y)$:

$$A_0 = \sin(y_0), \qquad\qquad A_1 = y_1 \cos(y_0),$$
$$A_2 = y_2 \cos(y_0) - \frac{1}{2} y_1^2 \sin(y_0), \quad A_3 = y_3 \cos(y_0) - y_1 y_2 \sin(y_0) \tag{2.39}$$
$$- \frac{1}{6} y_1^3 \cos(y_0).$$

2.5 Application of the Adomian Decomposition Method

In this section, we consider some IVPs for first-order and second-order ODEs.

Example 2.5 Solve the IVP

$$y'(x) = 1 - x^2 y(x) + y(x)^3, \quad y(0) = 0.$$

Solution. This is Abel's equation and its exact solution is $y(x) = x$. We apply the ADM to solve it. First, let us look at formula (2.23). We have

$$Ly \equiv y', \quad Ry \equiv x^2 y, \quad N(y) \equiv -y^3, \quad f(x) \equiv 1.$$

Using (2.27), we obtain

$$y(x) = \Psi_0(x) + g(x) - L^{-1}Ry(x) - L^{-1}N(y(x))$$
$$= y(0) + x - \int_0^x \tau^2 y(\tau)d\tau + \int_0^x y(\tau)^3 d\tau.$$

Now, applying formula (2.28), we get

$$\sum_{n=0}^{\infty} y_n(x) = x - \int_0^x \left(\tau^2 \sum_{n=0}^{\infty} y_n(\tau) \right) d\tau - \int_0^x \left(\sum_{n=0}^{\infty} A_n(\tau) \right) d\tau.$$

The Adomian polynomials, which belong to the nonlinearity $N(y) = y^3$, are given in (2.35). Setting

$$y_0(x) = \Psi_0(x) + g(x) = y(0) + x = x,$$

the recurrence relation (2.29) yields

$$y_1(x) = -\int_0^x \tau^3 d\tau - \int_0^x A_0(\tau)d\tau = -\int_0^x \tau^3 d\tau + \int_0^x \tau^3 d\tau = 0,$$
$$y_2(x) = -\int_0^x \tau^2 y_1(\tau)d\tau + 3\int_0^x y_0(\tau)^2 y_1(\tau)d\tau = 0,$$
$$\vdots$$

Hence

$$y(x) = \sum_{k=0}^{\infty} y_k(x) = x + 0 + \cdots + 0 + \cdots = x.$$

$\square$

Example 2.6 Solve the IVP

$$y'(x) + x e^{y(x)} = 0, \quad y(0) = 0.$$

Solution. Here, we have

$$Ly \equiv y', \quad Ry \equiv 0, \quad N(y) \equiv x e^y, \quad f(x) \equiv 0.$$

Using (2.27), we obtain

$$y(x) = \Psi_0(x) + g(x) - L^{-1}Ry(x) - L^{-1}N(y(x))$$
$$= y(0) - \int_0^x \tau e^{y(\tau)} d\tau = -\int_0^x \tau e^{y(\tau)} d\tau.$$

Now, applying formula (2.28), we get

$$\sum_{n=0}^{\infty} y_n(x) = -\int_0^x \tau \left(\sum_{n=0}^{\infty} A_n(\tau) \right) d\tau.$$

The Adomian polynomials, which belong to the nonlinearity $\exp(y)$, are given in (2.32). Setting

$$y_0(x) = \Psi_0(x) + g(x) = y(0) = 0,$$

the recurrence relation (2.29) yields

$$y_1(x) = -\int_0^x \tau \cdot 1 \, d\tau = -\frac{x^2}{2},$$

$$y_2(x) = \int_0^x \tau \cdot \frac{\tau^2}{2} d\tau = \int_0^x \frac{\tau^3}{2} d\tau = \frac{x^4}{8},$$

$$y_3(x) = -\int_0^x \tau \left(\frac{\tau^4}{8} + \frac{1}{2} \left(-\frac{\tau^2}{2} \right)^2 \right) \cdot 1 \, d\tau = -\int_0^x \frac{\tau^5}{4} d\tau$$

$$= -\frac{x^6}{24},$$

$$\vdots$$

Thus,

$$y_n(x) = -\int_0^x \tau A_{n-1}(\tau) d\tau = \frac{1}{n} \left(-\frac{x^2}{2} \right)^n, \quad n = 1, 2, \ldots,$$

and it holds

$$y(x) = \sum_{n=0}^{\infty} y_n(x) = -\frac{1}{2}x^2 + \frac{1}{8}x^4 - \frac{1}{24}x^6 + \cdots = -\ln\left(1 + \frac{x^2}{2}\right).$$

$\square$

Example 2.7 Solve the IVP

$$y''(x) + 2y(x)y'(x) = 0, \quad y(0) = 0, \quad y'(0) = 1.$$

Solution. Here, we have

$$Ly \equiv y'', \quad Ry \equiv 0, \quad N(y) \equiv 2yy', \quad f(x) \equiv 0.$$

Using (2.27), we obtain

$$y(x) = \Psi_0(x) + g(x) - L^{-1}Ry(x) - L^{-1}N(y(x))$$
$$= y(0) + xy'(0) - 2\int_0^x \int_0^\tau y(t)y'(t)dtd\tau.$$

Now, applying formula (2.28), we get

$$\sum_{n=0}^\infty y_n(x) = x - 2\int_0^x \int_0^\tau \left(\sum_{n=0}^\infty A_n(t)\right)dtd\tau.$$

The Adomian polynomials, which belong to the nonlinearity yy', are given in (2.36). Setting

$$y_0(x) = \Psi_0(x) + g(x) = y(0) + xy'(0) = x,$$

the recurrence relation (2.29) yields

$$y_1(x) = -2\int_0^x \int_0^\tau \left(y_0(t)y_0'(t)\right)dtd\tau = -2\int_0^x \int_0^\tau t\,dtd\tau$$
$$= -2\int_0^x \frac{\tau^2}{2}d\tau = -\int_0^x \tau^2 d\tau = -\frac{x^3}{3},$$
$$y_2(x) = -2\int_0^x \int_0^\tau \left(y_0'(t)y_1(t) + y_0(t)y_1'(t)\right)dtd\tau$$
$$= 2\int_0^x \int_0^\tau \left(1 \cdot \frac{t^3}{3} + t \cdot t^2\right)dtd\tau = \frac{4 \cdot 2}{3}\int_0^x \int_0^\tau t^3 dtd\tau$$
$$= \frac{2}{3}\int_0^x \tau^4 d\tau = \frac{2}{15}x^5,$$
$$y_3(x) = -2\int_0^x \int_0^\tau \left(y_0'(t)y_2(t) + y_1'(t)y_1(t) + y_2'(t)y_0(t)\right)dtd\tau$$
$$= \int_0^x \int_0^\tau \left(1 \cdot \frac{2}{15}t^5 + t^2 \cdot \frac{t^3}{3} + \frac{2}{3}t^4 \cdot t\right)dtd\tau$$
$$= -\frac{2 \cdot 17}{15}\int_0^x \int_0^\tau t^5 dtd\tau = -\frac{17}{15 \cdot 3}\int_0^x \tau^6 d\tau = -\frac{17}{315}x^7.$$

Obviously, it holds

$$y(x) = \sum_{n=0}^\infty y_n(x) = x - \frac{x^3}{3} + \frac{2}{15}x^5 - \frac{17}{315}x^7 + \cdots$$

□

Example 2.8 Solve the IVP

$$y''(x) - y'(x)^2 + y(x)^2 = e^x, \quad y(0) = y'(0) = 1.$$

Solution. Here, we have

$$Ly \equiv y'', \quad Ry \equiv 0, \quad N(y) \equiv -(y')^2 + y^2, \quad f(x) \equiv e^x.$$

We write

$$N(y) = N_1(y) + N_2(y), \quad N_1(y) \equiv -(y')^2, \quad N_2(y) \equiv y^2.$$

Using (2.27), we obtain

$$y(x) = \Psi_0(x) + g(x) - L^{-1}N_1(y(x)) - L^{-1}N_2(y(x))$$
$$= y(0) + xy'(0) + \int_0^x \int_0^\tau e^t dt d\tau + \int_0^x \int_0^\tau y'(t)^2 dt d\tau$$
$$- \int_0^x \int_0^\tau y(t)^2 dt d\tau$$

Now, applying formula (2.28), we get

$$\sum_{n=0}^\infty y_n(x) = 1 + x + e^x - x - 1 + \int_0^x \int_0^\tau \left(\sum_{n=0}^\infty A_n(t) \right) dt d\tau$$
$$- \int_0^x \int_0^\tau \left(\sum_{n=0}^\infty B_n(t) \right) dt d\tau.$$

The Adomian polynomials A_n, which belong to the nonlinearity $(y')^2$, are given in (2.37), and for the nonlinearity y^2, the Adomian polynomials B_n are given in (2.34). Setting

$$y_0(x) = \Psi_0(x) + g(x) = e^x,$$

the recurrence relation (2.29) yields

$$y_1(x) = \int_0^x \int_0^\tau A_0(t) dt d\tau - \int_0^x \int_0^\tau B_0(t) dt d\tau$$
$$= \int_0^x \int_0^\tau (e^t)^2 dt d\tau - \int_0^x \int_0^\tau (e^t)^2 dt d\tau = 0.$$

This implies $y_n(x) \equiv 0$, $n = 1, 2, \ldots$, and we obtain the exact solution of the given problem:

$$y(x) = e^x + 0 + 0 + \cdots = e^x.$$

□

The convergence of the ADM can be accelerated if the so-called *noise terms phenomenon* occurs in the given problem (see, e.g., [9, 10]). The noise terms are the identical terms with opposite sign that appear within the components $y_0(x)$ and $y_1(x)$. They only exist in specific types of nonhomogeneous equations. If noise terms indeed exist in the $y_0(x)$ and $y_1(x)$ components, then, in general, the solution can be obtained after two successive iterations.

By canceling the noise terms in $y_0(x)$ and $y_1(x)$, the remaining non-canceled terms of $y_0(x)$ give the exact solution. It has been proved that a necessary condition for the existence of noise terms is that the exact solution is part of $y_0(x)$.

Example 2.9 Solve the IVP

$$y''(x) - y'(x)^2 + y(x)^2 = 1, \quad y(0) = 1, \quad y'(0) = 0.$$

Solution. As in the Example 2.8, we set

$$Ly \equiv y'', \quad Ry \equiv 0, \quad N_1(y) \equiv -(y')^2, \quad N_2(y) \equiv y^2, \quad f(x) \equiv 1.$$

Using (2.27), we obtain

$$y(x) = 1 + \frac{x^2}{2} + \int_0^x \int_0^\tau y'(t)^2 dt d\tau - \int_0^x \int_0^\tau y(t)^2 dt d\tau$$

Now, applying formula (2.28), we get

$$\sum_{n=0}^{\infty} y_n(x) = 1 + \frac{x^2}{2} + \int_0^x \int_0^\tau \left(\sum_{n=0}^{\infty} A_n(t) \right) dt d\tau$$
$$- \int_0^x \int_0^\tau \left(\sum_{n=0}^{\infty} B_n(t) \right) dt d\tau.$$

The Adomian polynomials A_n, which belong to the nonlinearity $(y')^2$, are given in (2.37), and for the nonlinearity y^2, the Adomian polynomials B_n are given in (2.34). Setting

$$y_0(x) = 1 + \frac{1}{2}x^2,$$

the recurrence relation (2.29) yields

$$y_1(x) = \int_0^x \int_0^\tau t^2 \, dt d\tau - \int_0^x \int_0^\tau \left(1 + \frac{t^2}{2}\right)^2 dt d\tau$$

$$= \int_0^x \left[\frac{t^3}{3}\right]_0^\tau d\tau - \int_0^x \left[\frac{1}{20}t^5 + \frac{1}{3}t^3 + t\right]_0^\tau d\tau$$

$$= \int_0^x \frac{\tau^3}{3} d\tau - \int_0^x \left(\frac{1}{20}\tau^5 + \frac{1}{3}\tau^3 + \tau\right) d\tau$$

$$= \frac{1}{12}x^4 - \frac{1}{120}x^6 - \frac{1}{12}x^4 - \frac{1}{2}x^2$$

$$= -\frac{1}{120}x^6 - \frac{1}{2}x^2.$$

Comparing $y_0(x)$ with $y_1(x)$, we see that there is the noise term $\frac{1}{2}x^2$. Therefore, we can conclude that the solution of the given IVP is $y(x) = 1$.   □

Several authors have proposed a variety of modifications of the AMD (see, e.g., [12]) by which the convergence of the iteration (2.29) can be accelerated. Wazwaz [124, 126] suggests the following reliable modification which is based on the assumption that the function $h(x) \equiv \Psi_0(x) + g(x)$ in formula (2.27) can be divided into two parts, i.e.,

$$h(x) \equiv \Psi_0(x) + g(x) = h_0(x) + h_1(x).$$

The idea is that only the part $h_0(x)$ is assigned to the component $y_0(x)$, whereas the remaining part $h_1(x)$ is combined with other terms given in (2.29). It results the modified recurrence relation

$$\begin{aligned}
y_0(x) &= h_0(x), \\
y_1(x) &= h_1(x) - L^{-1}R y_0(x) - L^{-1}A_0(x), \\
y_k(x) &= -L^{-1}R y_{k-1}(x) - L^{-1}A_{k-1}(x), \quad k = 2, 3, \ldots
\end{aligned} \qquad (2.40)$$

Example 2.10 Solve the IVP

$$y''(x) - y(x)^2 = 2 - x^4, \quad y(0) = y'(0) = 0.$$

Solution. Here, we have

$$Ly \equiv y'', \quad Ry \equiv 0, \quad N(y) \equiv -y^2, \quad f(x) \equiv 2 - x^4.$$

Using (2.27), we obtain

$$y(x) = \Psi_0(x) + g(x) - L^{-1}Ry(x) - L^{-1}N(y(x))$$
$$= x^2 - \frac{1}{30}x^6 + \int_0^x \int_0^\tau y(t)^2 dt d\tau.$$

Now, applying formula (2.28), we get

$$\sum_{n=0}^\infty y_n(x) = x^2 - \frac{1}{30}x^6 + \int_0^x \int_0^\tau \left(\sum_{n=0}^\infty A_n(t)\right) dt d\tau.$$

The Adomian polynomials, which belong to the nonlinearity y^2, are given in (2.34). Dividing

$$h(x) = x^2 - \frac{1}{30}x^6$$

into $h_0(x) \equiv x^2$ and $h_1(x) \equiv -\frac{1}{30}x^6$, and starting with $y_0(x) = x^2$, the recurrence relation (2.29) yields

$$y_1(x) = -\frac{1}{30}x^6 + \int_0^x \int_0^\tau (t^2)^2 dt d\tau$$
$$= -\frac{1}{30}x^6 + \int_0^x \frac{\tau^5}{5} d\tau$$
$$= -\frac{1}{30}x^6 + \frac{1}{30}x^6 = 0.$$

This implies

$$y_k(x) = 0, \quad k = 1, 2, \ldots$$

Thus, we can conclude that the exact solution of the given IVP is $y(x) = x^2$.

Let us compare the modified ADM with the standard method. The ADM is based on the recurrence relation (2.29). Here, we have to set

$$y_0(x) = x^2 - \frac{1}{30}x^6.$$

Now, the recurrence relation (2.29) yields

$$y_1(x) = \int_0^x \int_0^\tau \left(t^2 - \frac{1}{30}t^6\right)^2 dt d\tau = \int_0^x \left(\frac{1}{11700}\tau^{13} - \frac{1}{135}\tau^9 - \frac{1}{5}\tau^5\right) d\tau$$
$$= \frac{1}{30}x^6 - \frac{1}{1350}x^{10} + \frac{1}{163800}x^{14}.$$

$$y_2(x) = 2 \int_0^x \int_0^\tau \left(t^2 - \frac{1}{30} t^6 \right) \left(\frac{1}{163800} x^{14} - \frac{1}{1350} x^{10} + \frac{1}{30} x^6 \right) dt d\tau$$

$$= \int_0^x \left(-\frac{1}{51597000} \tau^{21} + \frac{227}{62653500} \tau^{17} - \frac{1}{3510} \tau^{13} + \frac{1}{135} \tau^9 \right) d\tau$$

$$= \frac{1}{1350} x^{10} - \frac{1}{49140} x^{14} + \frac{227}{1127763000} x^{18} - \frac{1}{1135134000} x^{22}.$$

Since

$$y(x) = y_0(x) + y_1(x) + y_2(x) + \cdots ,$$

we see that the terms in x^6 and x^{10} cancel each other. The cancelation of terms is continued when further components y_k, $k \geq 3$, are added.

This is an impressive example of how fast the modified ADM generates the exact solution $y(x) = x^2$, compared with the standard method.   □

Many problems in the mathematical physics can be formulated as ODEs of *Emden-Fowler type* (see, e.g., [26, 30]) defined in the form

$$y''(x) + \frac{\alpha}{x} y'(x) + \beta f(x) g(y) = 0, \quad \alpha \geq 0,$$

$$y(0) = a, \quad y'(0) = 0,$$
(2.41)

where f and g are given functions of x and y, respectively. The standard Emden-Lane-Fowler ODE results when we set $f(x) \equiv 1$ and $g(y) \equiv y^n$.

Obviously, a difficulty in the analysis of (2.41) is the singularity behavior that occurs at $x = 0$. Before the ADM can be applied, a slight change of the problem is necessary to overcome this difficulty. The strategy is to define the differential operator L in terms of the two derivatives, $y'' + (\alpha/x)y'$, which are contained in the ODE. First, the ODE (2.41) is rewritten as

$$Ly(x) = -\beta f(x) g(y),$$
(2.42)

with

$$L \equiv x^{-\alpha} \frac{d}{dx} \left(x^\alpha \frac{d}{dx} \right).$$

The corresponding inverse operator L^{-1} is

$$L^{-1}(\cdot) = \int_0^x \tau^{-\alpha} \int_0^\tau t^\alpha (\cdot) dt d\tau.$$

Applying L^{-1} to the first two terms of (2.41), we obtain

$$L^{-1}\left(y''(x) + \frac{\alpha}{x}y'(x)\right)$$

$$= \int_0^x \tau^{-\alpha} \int_0^\tau t^\alpha \left(y''(t) + \frac{\alpha}{t}y'(t)\right) dt d\tau$$

$$= \int_0^x \left[\tau^\alpha y'(\tau) - \int_0^\tau \alpha t^{\alpha-1} y'(t) dt + \int_0^\tau \alpha t^{\alpha-1} y'(t) dt\right] d\tau$$

$$= \int_0^x y'(\tau) d\tau = y(x) - y(0) = y(x) - a.$$

Now, operating with L^{-1} on (2.41), we find

$$y(x) = a - \beta L^{-1}(f(x)g(y)). \tag{2.43}$$

It is interesting, that only the first initial condition is sufficient to represent the solution $y(x)$ in this form. The second initial condition can be used to show that the obtained solution satisfies this condition.

Let us come back to the Adomian decomposition method. As before, the solution $y(x)$ is represented by an infinite series of components

$$y(x) = \sum_{n=0}^\infty y_n(x). \tag{2.44}$$

In addition, the given nonlinear function $g(y)$ is represented by an infinite series of Adomian polynomials (as we have done it for $N(y)$)

$$g(y(x)) = \sum_{n=0}^\infty A_n(x), \tag{2.45}$$

where

$$A_n(x) \equiv A_n(y_0(x), y_1(x), \ldots, y_{n-1}(x)).$$

Substituting (2.44) and (2.45) into (2.43) gives

$$\sum_{n=0}^\infty y_n(x) = a - \beta L^{-1}\left(f(x) \sum_{n=0}^\infty A_n(x)\right). \tag{2.46}$$

Now, the components $y_n(x)$ are determined recursively. The corresponding recurrence relation is

$$y_0(x) = a,$$
$$y_k(x) = -\beta L^{-1}(f(x)A_{k-1}(x)), \quad k = 1, 2, \ldots,$$

or equivalently

$$y_0(x) = a,$$

$$y_k(x) = -\beta \left(\int_0^x \tau^{-\alpha} \int_0^\tau t^\alpha (f(t) A_{k-1}(x)) dt d\tau \right), \quad k = 1, 2, \ldots \tag{2.47}$$

Example 2.11 Solve the IVP

$$y''(x) + \frac{2}{x} y'(x) + e^{y(x)} = 0, \quad y(0) = y'(0) = 0.$$

Solution. This problem is a special case of (2.41), where $\alpha = 2$, $\beta = 1$, $f(x) \equiv 1$, $g(y) = \exp(y)$ and $a = 0$. A particular solution of the ODE is

$$y(x) = \ln \left(\frac{2}{x^2} \right).$$

Obviously, this solution does not satisfy the initial conditions. We will see that the ADM can be used to determine a solution which satisfies the ODE as well as the initial conditions.

For the nonlinearity $g(y) = \exp(y)$, the Adomian polynomials are given in Eq. (2.32). Using the recurrence relation (2.47), we obtain

$$y_0(x) = 0,$$

$$y_1(x) = - \int_0^x \tau^{-2} \int_0^\tau t^2 \cdot 1 \, dt d\tau = - \int_0^x \tau^{-2} \frac{\tau^3}{3} d\tau = - \int_0^x \frac{\tau}{3} d\tau$$

$$= -\frac{1}{6} x^2,$$

$$y_2(x) = - \int_0^x \tau^{-2} \int_0^\tau t^2 \cdot \left(\frac{-t^2}{6} \right) dt d\tau = \int_0^x \tau^{-2} \int_0^\tau \frac{t^4}{6} dt d\tau$$

$$= \int_0^x \tau^{-2} \frac{\tau^5}{30} d\tau = \int_0^x \frac{\tau^3}{30} d\tau$$

$$= \frac{1}{120} x^4,$$

$$y_3(x) = - \int_0^x \tau^{-2} \int_0^\tau t^2 \cdot \left(\frac{t^4}{120} + \frac{t^4}{72} \right) dt d\tau = - \int_0^x \tau^{-2} \int_0^\tau \frac{t^6}{45} dt d\tau$$

$$= - \int_0^x \tau^{-2} \frac{\tau^7}{315} d\tau = - \int_0^\tau \frac{\tau^5}{315} d\tau$$

$$= -\frac{1}{1890} x^6,$$

$$y_4(x) = -\int_0^x \tau^{-2} \int_0^\tau t^2 \left(-\frac{1}{1890}t^6 - \frac{1}{6}t^2 \cdot \frac{1}{120}t^4 - \frac{1}{6} \cdot \frac{1}{216}t^6\right) dt d\tau$$

$$= \int_0^x \tau^{-2} \left(\frac{61}{204120}\tau^9\right) d\tau = \int_0^x \frac{61}{204120}\tau^7 d\tau$$

$$= \frac{61}{1632960}x^8.$$

Thus, we have

$$y(x) = -\frac{1}{6}x^2 + \frac{1}{120}x^4 - \frac{1}{1890}x^6 + \frac{61}{1632960}x^8 + \cdots$$

In [125] further variants of the general Emden-Fowler equation are discussed, and solved by the ADM. □

2.6   Exercises

Exercise 2.1 Solve following ODEs by VIM or ADM:

(1) $y'(x) - y(x) = -y(x)^2, \quad y(0) = 1,$

(2) $y'(x) = 1 + \frac{y(x)}{x} + \left(\frac{y(x)}{x}\right)^2, \quad y(0) = 0,$

(3) $y''(x) + \frac{5}{x}y'(x) + \exp(y(x)) + 2\exp\left(\frac{y(x)}{2}\right) = 0, \quad y(0) = y'(0) = 0,$

(4) $y''(x) + 2y(x)y'(x) - y(x) = \sinh(2x), \quad y(0) = 0, \ y'(0) = 1,$

(5) $y''(x) + \cos(y(x)) = 0, \quad y(0) = 0, \ y'(0) = \frac{\pi}{2},$

(6) $y''(x) + \frac{8}{x}y'(x) + 2y(x) = -4y(x)\ln(y(x)), \quad y(0) = 1, \ y'(0) = 0,$

(7) $y'''(x) + y''(x)^2 + y'(x)^2 = 1 - \cos(x), \quad y(0) = y''(0) = 0, \ y'(0) = 1.$

Exercise 2.2 Given the following IVP for the Emden-Fowler ODE

$$y''(x) + \frac{2}{x}y'(x) + \alpha x^m y(x)^\mu = 0, \quad y(0) = 1, \ y'(0) = 0.$$

Approximate the solution of this IVP by the VIM and the ADM.

Exercise 2.3 Given the following IVP for the Emden-Fowler ODE

$$y''(x) + \frac{2}{x}y'(x) + \alpha x^m e^{y(x)} = 0, \quad y(0) = y'(0) = 0.$$

Approximate the solution of this IVP by the VIM and the ADM.

Chapter 3
Further Analytical Approximation Methods and Some Applications

3.1 Perturbation Method

In this chapter four additional analytical approximation methods are presented, namely the perturbation method, the energy balance method, the Hamiltonian approach and the homotopy analysis method. After a short description of their theoretical background, we show exemplary how these methods can be used to solve problems from the applied physics and engineering.

3.1.1 Theoretical Background

Let us start with the perturbation method (PM). This method is closely related to techniques used in numerical analysis (see Chap. 4). The corresponding perturbation theory comprises mathematical methods for finding an approximate solution to a problem which cannot be solved analytically. The idea is to start with a simplified problem for which a mathematical solution is known or can be determined by mathematical standard techniques. Then, an additional *perturbation term*, which depends on a small parameter ε is added to the simplified problem. The parameter ε can appear naturally in the original problem or is introduced artificially. If the perturbation is not too large, the various quantities associated with the perturbed system can be expressed as *corrections* to those of the simplified system. These corrections, being small compared to the size of the quantities themselves, are now calculated using approximate methods such as asymptotic series. The original problem can therefore be studied based on the knowledge of the simpler one.

To be more precise, let us assume that an ODE, which is subject to initial or boundary conditions, is given. The essential precondition of the perturbation method is that the solution of the ODE can be expanded in ε as an infinite sum

© Springer India 2016

M. Hermann and M. Saravi, *Nonlinear Ordinary Differential Equations*,
DOI 10.1007/978-81-322-2812-7_3

$$y(x, \varepsilon) = \sum_{k=0}^{n} \varepsilon^k y_k(x) + O(\varepsilon^{n+1}). \tag{3.1}$$

where $y_0, \ldots, y_n$ are independent of ε, and $y_0(x)$ is the solution of the problem for $\varepsilon = 0$, for which we assume that it can be solved very easily.

Then, the perturbation method is based on the following four steps:

- STEP 1: Substitute the expansion (3.1) into the ODE and the corresponding initial (boundary) conditions.
- STEP 2: Equate the successive terms of $\varepsilon, \varepsilon, \ldots, \varepsilon^n$ to zero.
- STEP 3: Solve the sequence of equations that arise in Step 2.
- STEP 4: Substitute $y_0(x), y_1(x), \ldots, y_n(x)$ into (3.1).

To understand conceptually the perturbation technique, we will consider the following IVP:

$$y'(x) = y(x) + \varepsilon y(x)^2, \quad y(0) = 1, \tag{3.2}$$

where ε is a small (positive) real parameter. This is a Bernoulli equation, and one can show that its exact solution is

$$y(x) = \frac{e^x}{1 - \varepsilon \left(e^x - 1\right)}.$$

If we use the formula of the geometric series

$$(1 - z)^{-1} = 1 + z + z^2 + \cdots, \quad |z| < 1,$$

the solution can be written in the form

$$\begin{aligned} y(x) &= \frac{e^x}{1 - \varepsilon(e^x - 1)} = e^x \left(1 - \varepsilon(e^x - 1)\right)^{-1} \\ &= e^x \left(1 + \varepsilon \left(e^x - 1\right) + \varepsilon^2 \left(e^x - 1\right)^2 + O(\varepsilon^3)\right) \\ &= e^x + \varepsilon \left(e^{2x} - e^x\right) + \varepsilon^2 \left(e^{3x} - 2e^{2x} + e^x\right) + O(\varepsilon^3), \end{aligned} \tag{3.3}$$

where it has to be assumed that $|\varepsilon(e^x - 1)| < 1$.

Now, we will show that the expansion (3.3) can be obtained by the perturbation method without knowledge of the exact solution. Let us proceed as described above.

- STEP 1: We substitute the ansatz

$$y(x; \varepsilon) = y_0(x) + \varepsilon y_1(x) + \varepsilon^2 y_2(x) + O(\varepsilon^3)$$

into (3.2), and obtain

$$y_0'(x) + \varepsilon y_1'(x) + \varepsilon^2 y_2'(x) + O(\varepsilon^3)$$
$$= y_0(x) + \varepsilon y_1(x) + \varepsilon^2 y_2(x) + O(\varepsilon^3)$$
$$+ \varepsilon \left(y_0(x) + \varepsilon y_1(x) + \varepsilon^2 y_2(x) + O(\varepsilon^3) \right)^2,$$
$$y_0(0) + \varepsilon y_1(0) + \varepsilon^2 y_2(0) + O(\varepsilon^3) = 1.$$

- STEP 2: We equate the successive terms of ε^0, ε^1, and ε^2 to zero. This yields

$$\varepsilon^0 : \quad y_0'(x) = y_0(x), \quad y_0(0) = 1,$$
$$\varepsilon^1 : \quad y_1'(x) = y_1(x) + y_0(x)^2, \quad y_1(0) = 0,$$
$$\varepsilon^2 : \quad y_2'(x) = y_2(x) + 2y_0(x)y_1(x), \quad y_2(0) = 0.$$

- STEP 3: We solve the three IVPs determined in Step 2, and get

$$y_0(x) = e^x, \quad y_1(x) = e^{2x} - e^x, \quad y_2(x) = e^{3x} - 2e^{2x} + e^x.$$

- STEP 4: We substitute the above three functions $y_0(x)$, $y_1(x)$, and $y_2(x)$ into the ansatz for $y(x; \varepsilon)$ given in Step 1, and obtain

$$y(x; \varepsilon) = e^x + \varepsilon \left(e^{2x} - e^x \right) + \varepsilon^2 \left(e^{3x} - 2e^{2x} + e^x \right) + O(\varepsilon^3).$$

It can be seen that the solution generated by the perturbation method is identical to (3.3).

3.1.2 Application of the Perturbation Method

Let us now consider two examples by which the application of the PM to nonlinear problems is demonstrated.

Example 3.1 In dynamics, the van der Pol oscillator is a non-conservative oscillator with nonlinear damping. The mathematical model of this dynamical system is an IVP for a second-order ODE given by

$$y''(x) - \varepsilon \left(1 - y(x)^2 \right) y'(x) + y(x) = 0, \quad y(0) = A, \quad y'(0) = 0, \qquad (3.4)$$

where y is the position coordinate, which is a function of the time x, and ε is a real parameter indicating the nonlinearity and the strength of the damping. Moreover, y' is the velocity and y'' is the acceleration.

Approximate the solution of the IVP (3.4) with the PM.

Solution. When $\varepsilon = 0$, i.e., there is no damping function, the ODE becomes $y''(x) + y(x) = 0$. This is the form of a simple harmonic oscillator, which indicates there is always conservation of energy. The general solution of this equation is $y(x) = a\cos(x+b)$, where a and b are constants. Imposing the initial conditions $y(0) = A$ and $y'(0) = 0$, we get $y(x) = A\cos(x)$.

To determine an improved approximation of the solution of the given problem (3.4), we use the perturbation expansion (3.1). Substituting this expansion into (3.4), we obtain

$$[y_0''(x) + y_0(x)] + \varepsilon[y_1''(x) + y_1(x)] + \varepsilon^2[y_2''(x) + y_2(x)] + \cdots$$
$$= \varepsilon\left(1 - y_0(x)^2\right)y_0'(x) + \varepsilon^2\left(\begin{array}{c}\left(1 - y_0(x)^2\right)y_1'(x) \\ -2y_0(x)y_1(x)y_0'(x)\end{array}\right) + \cdots$$

Comparing both sides w.r.t. powers of ε, we get

- coefficient of ε^0

$$y_0''(x) + y_0(x) = 0, \quad y_0(0) = A, \quad y_0'(0) = 0; \tag{3.5}$$

- coefficient of ε^1

$$y_1''(x) + y_1(x) = \left(1 - y_0(x)^2\right)y_0'(x), \quad y_1(0) = y_1'(0) = 0; \tag{3.6}$$

- coefficient of ε^2

$$y_2''(x) + y_2(x) = \left(1 - y_0(x)^2\right)y_1'(x) - 2y_0(x)y_1(x)y_0'(x),$$
$$y_2(0) = y_2'(0) = 0. \tag{3.7}$$

The solution of the IVP (3.5) is $y_0(x) = A\cos(x)$. Substituting $y_0(x)$ into (3.6) gives

$$y_1''(x) + y_1(x) = -\left(1 - A^2\cos(x)^2\right)A\sin(x),$$
$$y_1(0) = y_1'(0) = 0. \tag{3.8}$$

Using the trigonometric formula

$$\cos(x)^2\sin(x) = \frac{\sin(x) + \sin(3x)}{4},$$

the IVP (3.8) can be written as

$$y_1''(x) + y_1(x) = \frac{A^3 - 4A}{4}\sin(x) + \frac{1}{4}A^3\sin(3x),$$
$$y_1(0) = y_1'(0) = 0. \tag{3.9}$$

It can be easily seen that the solution of (3.9) is

$$y_1(x) = -\frac{1}{32}A^3 \sin(3x) - \frac{A^3 - 4A}{8} x \cos(x).$$

Now using $y_0(x)$ and $y_1(x)$, the next function $y_2(x)$ can be determined from (3.7) in a similar manner. At this point, we have obtained the following approximation of the exact solution $y(x)$ of the given IVP (3.4)

$$y(x) \approx y_0(x) + \varepsilon y_1(x)$$
$$= A \cos(x) + \varepsilon \left(-\frac{1}{32}A^3 \sin(3x) - \frac{A^3 - 4A}{8} x \cos(x) \right). \qquad (3.10)$$

The approximation (3.10) will be acceptable if ε is very small. Otherwise, more functions $y_i(x), i > 1$, must be determined. But this task is extremely work-intensive, and numerical methods should be used instead. $\qquad \square$

Example 3.2 In Applied Mechanics, *bending* characterizes the behavior of a slender structural element subjected to an external load applied perpendicularly to a longitudinal axis of the element (see [23]). The structural element is assumed to be such that at least one of its dimensions is a small fraction of the other two. When the length is considerably longer than the width and the thickness, the element is called a *beam* (see Fig. 3.1). The classical beam bending theory is an important consideration in nearly all structural designs and analyses. The following nonlinear ODE describes the bending of a beam under a statical load (see e.g. [25])

$$y''(x) = \frac{M(x)}{EI}\left(1 + y'(x)^2\right)^{3/2}, \qquad (3.11)$$

where $y(x)$ is the beam deflection (vertical displacement), $y'(x)$ the corresponding slope and $M(x)$ the bending moment. The constants E and I are the modulus

Fig. 3.1 Schematic representation of a beam

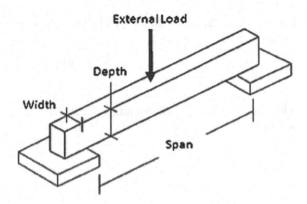

Fig. 3.2 Beam with two
fixed ends subjected to a
concentrated load

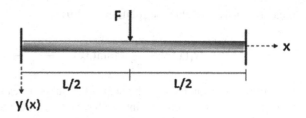

Fig. 3.2 Beam with two fixed ends subjected to a concentrated load

of elasticity and the moment of inertia of the cross section about its neutral axis, respectively.

In this example, let us consider a beam of the length L with two fixed ends subjected to a load F, which is concentrated at the middle of the span (see Fig. 3.2). The governing equations for this problem are given by the following IVP

$$y''(x) = \frac{F(4x - L)}{8E\,I}\left(1 + y'(x)^2\right)^{3/2}, \quad y(0) = y'(0) = 0. \qquad (3.12)$$

Approximate the solution of the IVP (3.12) with the PM.

Solution. This IVP does not depend on a parameter. The idea of the PM is to introduce an artificial parameter ε into the problem by multiplying the nonlinear term with ε. Thus, we write

$$y''(x) = \frac{F(4x - L)}{8E\,I}\left(1 + \varepsilon\left(y'(x)\right)^2\right)^{3/2}, \quad y(0) = y'(0) = 0.$$

Obviously, for $\varepsilon = 0$ we have a *linear* problem that can be solved very easily, whereas for $\varepsilon = 1$ we have the given *nonlinear* problem. For simplicity, we write the ODE in the following form:

$$(y''(x))^2 = \left(\frac{F(4x - L)}{8E\,I}\right)^2\left(1 + \varepsilon\left(y'(x)\right)^2\right)^3, \quad y(0) = y'(0) = 0. \qquad (3.13)$$

Let us assume that the solution of (3.13) can be represented by the power series (3.1), i.e.,

$$y(x) = \sum_{k=0}^{n} \varepsilon^k y_k(x) + R_{n+1},$$

where $R_{n+1} = O(\varepsilon^{n+1})$. Substituting the above expansion into the ODE in (3.13), we get

$$\left(\sum_{k=0}^{n} \varepsilon^k y_k''(x) + R_{n+1}'' \right)^2$$

$$= \left(\frac{F(4x-L)}{8EI} \right)^2 \left(1 + \varepsilon \left(\sum_{k=0}^{n} \varepsilon^k y_k'(x) + R_{n+1}' \right)^2 \right)^3,$$

where $R_{n+1}' = O(\varepsilon^{n+1})$ and $R_{n+1}'' = O(\varepsilon^{n+1})$.

Comparing both sides of this equation w.r.t. powers of ε, we obtain

- coefficient of ε^0

$$\left(y_0''(x) \right)^2 = \frac{F^2(4x-L)^2}{64(EI)^2}, \quad y_0(0) = y_0'(0) = 0; \tag{3.14}$$

- coefficient of ε^1

$$2y_0''(x)y_1''(x) = \frac{3F^2(4x-L)^2}{64(EI)^2} \left(y_0'(x) \right)^2, \quad y_1(0) = y_1'(0) = 0; \tag{3.15}$$

- coefficient of ε^2

$$2y_0''(x)y_2''(x) + \left(y_1''(x) \right)^2 = \frac{F^2(4x-L)^2}{64(EI)^2} \left(6y_0'(x)y_1'(x) + 3 \left(y_0'(x) \right)^4 \right), \tag{3.16}$$

$$y_2(0) = y_2'(0) = 0.$$

Now, we solve successively the (linear) IVPs (3.14)–(3.16). The results are

$$y_0(x) = \frac{F}{12EI} \left(x^3 - \frac{3}{4}Lx^2 \right),$$

$$y_1(x) = \frac{3F^3}{1024(EI)^3} \left(\frac{8}{21}x^7 - \frac{2}{3}Lx^6 + \frac{2}{5}L^2x^5 - \frac{1}{12}L^3x^4 \right),$$

$$y_2(x) = \frac{15F^5}{262144(EI)^5} \left(\frac{32}{55}x^{11} - \frac{8}{5}Lx^{10} + \frac{16}{9}L^2x^9 - L^3x^8 \right.$$

$$\left. + \frac{2}{7}L^4x^7 - \frac{1}{30}L^5x^6 \right).$$

HINT: Since we have used in (3.13) the squared ODE (3.12), the IVP (3.14) has two solutions, which differ only in the sign. Here, we use the solution that fits the elastomechanical problem. □

According to the PM, we may write

$$y(x) \approx y_0(x) + \varepsilon y_1(x) + \varepsilon^2 y_2(x). \tag{3.17}$$

If we set $\varepsilon = 1$, and insert the expressions for $y_0(x)$, $y_1(x)$, and $y_2(x)$ into (3.17), we obtain the following approximation for the given nonlinear IVP:

$$
\begin{aligned}
y(x) \approx & \frac{F}{12E\,I}\left(x^3 - \frac{3}{4}Lx^2\right) \\
& + \frac{3F^3}{1024(E\,I)^3}\left(\frac{8}{21}x^7 - \frac{2}{3}Lx^6 + \frac{2}{5}L^2x^5 - \frac{1}{12}L^3x^4\right) \\
& + \frac{15F^5}{262144(E\,I)^5}\left(\frac{32}{55}x^{11} - \frac{8}{5}Lx^{10} + \frac{16}{9}L^2x^9 - L^3x^8 \right. \\
& \left. \qquad\qquad + \frac{2}{7}L^4x^7 - \frac{1}{30}L^5x^6\right).
\end{aligned}
\tag{3.18}
$$

This example shows that an additional parameter ε can often be introduced into the given problem such that for $\varepsilon = 0$ we have a *linear* problem which can be solved quite simply, and for $\varepsilon \neq 0$ the solution of the *nonlinear* problem can be represented in form of a power series in ε. Then, setting $\varepsilon = 1$, the truncated power series with just a few terms is a sufficiently good approximation of the given nonlinear problem. In Table 3.1, for different parameter sets we present the errors err_1, err_2, and err_3 of the one-term approximation $y_0(x)$, two-term approximation $y_0(x) + y_1(x)$, and three-term approximation $y_0(x) + y_1(x) + y_2(x)$, respectively.

In the mechanics of materials the deflection of a beam with two fixed ends subjected to a concentrated load at the middle of the span is often computed by the one-term approximation

$$
y(x) = \frac{F}{12E\,I}\left(x^3 - \frac{3}{4}Lx^2\right).
$$

With our studies we could show that this formula is justified in most cases.   □

The perturbation method has been developed for the solution of ODEs with a nonperiodic solution (static systems). As we could see, the method works very well in these cases. However, for problems with a periodic solution (dynamical systems), the perturbation method is divergent for large values of x. The divergence is caused by terms $x\cos(x)$ and/or $x\sin(x)$, which will appear in the expansion (3.1). To

Table 3.1 Comparison of the accuracy of the n-term approximations, $n = 1, 2, 3$

	L	$E\,I$	F	L	$E\,I$	F	L	$E\,I$	F
	1	500	100	3	500	100	5	500	100
err_1	7.3 e-7			1.6 e-3			7.1 e-2		
err_2	2.2 e-10			4.0 e-5			1.4 e-2		
err_3	8.4 e-14			3.1 e-3			2.5 e-1		

overcome this problem, the independent variable x is also expanded into a power series in ε as follows:

$$x = X(\varepsilon, \chi) = \chi + \varepsilon X_1(\chi) + \varepsilon^2 X_2(\chi) + \cdots \tag{3.19}$$

Then, the ansatz (3.19) is substituted into (3.1). By an appropriate selection of $X_1(\chi), X_2(\chi), \ldots$, the disturbing terms $x \cos(x)$, $x \sin(x)$ can be eliminated. In the following example, the modified approach is demonstrated.

Example 3.3 Determine an approximate solution of the following IVP by the perturbation method

$$y''(x) + y(x) + \varepsilon y(x)^3 = 0, \quad y(0) = 1, \quad y'(0) = 0. \tag{3.20}$$

This ODE is a special case of the so-called Duffing equation. It models the free vibration of a mass, which is connected to two nonlinear springs on a frictionless contact surface (see Example 3.5).

Solution. If we expand the solution $y(x)$ of the IVP (3.20) in the form

$$y(x; \varepsilon) = y_0(x) + \varepsilon y_1(x) + O(\varepsilon^2),$$

the following IVPs have to be solved:

$$\begin{aligned} y_0''(x) &= -y_0(x), & y_0(0) &= 1, \quad y_0'(x) = 0, \\ y_1''(x) &= -y_1(x) - y_0(x)^3, & y_1(0) &= 0, \quad y_1'(0) = 0. \end{aligned}$$

It can be easily seen that the corresponding solutions are

$$y_0(x) = \cos(x),$$
$$y_1(x) = \frac{1}{32}\big(\cos(3x) - \cos(x) - 12x \sin(x)\big).$$

Thus,

$$\begin{aligned} y(x; \varepsilon) &= \cos(x) + \frac{\varepsilon}{32}\big(\cos(3x) - \cos(x) - 12x \sin(x)\big) + O(\varepsilon^2) \\ &= \left(1 - \frac{x^2}{2!} + \frac{x^4}{4!} - \cdots\right) + \frac{\varepsilon}{32}\left[\left(1 - \frac{9x^2}{2!} + \frac{81x^4}{4!} - \cdots\right)\right. \\ &\quad \left. - \left(1 - \frac{x^2}{2!} + \frac{x^4}{4!} - \cdots\right) - 12x\left(x - \frac{x^3}{3!} + \frac{x^5}{5!} + \cdots\right)\right] + O(\varepsilon^2). \end{aligned}$$

$$\tag{3.21}$$

Because of the term $-12x \sin(x)$, this expansion is not uniform for $x \geq 0$. Therefore, we substitute the expansion (3.19) into (3.21), and obtain

$$y(x; \varepsilon) = \left(1 - \frac{(\chi + \varepsilon X_1 + O(\varepsilon^2))^2}{2!} + \frac{(\chi + \varepsilon X_1 + O(\varepsilon^2))^4}{4!} - \cdots\right)$$

$$+ \frac{\varepsilon}{32}\left[\left(1 - \frac{9(\chi + \varepsilon X_1 + O(\varepsilon^2))^2}{2!}\right.\right.$$

$$\left. + \frac{81(\chi + \varepsilon X_1 + O(\varepsilon^2))^4}{4!} - \cdots\right)$$

$$- \left(1 - \frac{(\chi + \varepsilon X_1 + O(\varepsilon^2))^2}{2!} + \frac{(\chi + \varepsilon X_1 + O(\varepsilon^2))^4}{4!} - \cdots\right)$$

$$- 12\left(\chi + \varepsilon X_1 + O(\varepsilon^2)\right)\left(\frac{\chi + \varepsilon X_1 + O(\varepsilon^2)}{1!}\right.$$

$$\left.\left. - \frac{(\chi + \varepsilon X_1 + O(\varepsilon^2))^3}{3!} + \frac{(\chi + \varepsilon X_1 + O(\varepsilon^2))^5}{5!} + \cdots\right)\right]$$

$$+ O(\varepsilon^2).$$

It follows

$$y(x; \varepsilon) = \left(1 - \frac{\chi^2}{2!} + \frac{\chi^4}{4!} - \cdots\right)$$

$$- \varepsilon X_1\left(\chi - \frac{\chi^3}{3!} + \frac{\chi^5}{5!} - \frac{\chi^7}{7!} + \cdots\right)$$

$$+ \frac{\varepsilon}{32}\left[\left(1 - \frac{9\chi^2}{2!} + \frac{81\chi^4}{4!} - \cdots\right) - \left(1 - \frac{\chi^2}{2!} + \frac{\chi^4}{4!} - \cdots\right)\right.$$

$$\left. - 12\chi\left(\frac{\chi}{1!} - \frac{\chi^3}{3!} + \frac{\chi^5}{5!} - \cdots\right)\right] + O(\varepsilon^2).$$

$$= \cos(\chi) - \varepsilon X_1 \sin(\chi) + \frac{\varepsilon}{32}\left[\cos(3\chi) - \cos(\chi) - 12\chi \sin(\chi)\right] + O(\varepsilon^2)$$

$$= \cos(\chi) + \frac{\varepsilon}{32}\left[\cos(3\chi) - \cos(\chi) - (12\chi + 32X_1)\sin(\chi)\right] + O(\varepsilon^2).$$

Since X_1 can be freely chosen, we eliminate the term $\chi \sin(\chi)$ by setting

$$X_1 = X_1(\chi) = -\frac{3}{8}\chi.$$

Thus, the approximate solution can be written in the form

$$y(x; \varepsilon) = Y(\chi; \varepsilon) = \cos(\chi) + \frac{\varepsilon}{32}\left[\cos(3\chi) - \cos(\chi)\right] + O(\varepsilon^2), \qquad (3.22)$$

where $x = \left(1 - \frac{3}{8}\varepsilon\right)\chi + O(\varepsilon^2).$ □

In the next section, we consider the Duffing equation once more and obtain a better approximation.

3.2 Energy Balance Method

3.2.1 Theoretical Background

Nonlinear oscillator models have been widely used in many areas of physics and engineering and are of significant importance in mechanical and structural dynamics for the comprehensive understanding and accurate prediction of motion. One important method that can be used to study such models is the energy balance method (EBM). This method converges very rapidly to the exact solution and can be easily extended to nonlinear oscillations. Briefly speaking, the EBM is characterized by an extended scope of applicability, simplicity, flexibility in applications, and it avoids any complicated numerical and analytical integration compared with the methods presented in the previous chapter.

In order to introduce the EBM, let us consider the motion of a general oscillator, which is modeled by the following IVP:

$$y''(x) + f(y(x)) = 0, \quad y(0) = A, \quad y'(0) = 0, \tag{3.23}$$

where A is the initial amplitude. The characteristic of this problem is the periodicity of the solution. Its variational can be written as

$$J(y) \equiv \int_0^{T/4} \left\{ -\frac{1}{2} \left(y'(x) \right)^2 + F(y) \right\} dx, \tag{3.24}$$

where $T = 2\pi/\omega$ is the period of the nonlinear oscillation, $\omega \geq 0$ is the angular frequency, and $dF/dy = f(y)$.

The Hamiltonian of (3.24) can be written in the form

$$H(y) = \frac{1}{2} y'(x)^2 + F(y), \tag{3.25}$$

where $E(y) \equiv (y'(x))^2/2$ is the kinetic energy, and $P(y) \equiv F(y)$ is the potential energy. Since the system is conservative throughout the oscillation, the total energy remains unchanged during the motion, i.e., the Hamiltonian of the oscillator must be a constant value,

$$H(y) = E(y) + P(y) \doteq F(A). \tag{3.26}$$

Thus, the associated residual satisfies

$$R(x) \equiv \frac{1}{2} \left(\frac{dy}{dx} \right)^2 + F(y) - F(A) = 0. \tag{3.27}$$

For a first-order approximation, we use the following trial function

$$y^{(1)}(x) = A\cos(\omega x),\tag{3.28}$$

where ω is an approximation of the frequency of the system. The choice of the trail function depends on the given initial or boundary conditions. Sometimes this choice may be $y^{(1)}(x) = A\sin(\omega x)$, or a combination of $\cos(\omega x)$ and $\sin(\omega x)$. Substituting (3.28) into $R(x)$ yields

$$R(x) = \frac{1}{2}A^2\omega^2\sin(\omega x)^2 + F(A\cos(\omega x)) - F(A).\tag{3.29}$$

If, by chance, the exact solution had been chosen as the trial function, then it would be possible to make R zero for all values of x by appropriate choice of ω. Since (3.28) is only an approximation of the exact solution, R cannot be made zero for all x. The idea is to force the residual to zero, in an average sense. Various formulations have been proposed in the literature, for example, the least square method, the Ritz–Galerkin method and the collocation method (see e.g. [14, 35, 94, 135]).

Let us start with collocation at $\omega x = \pi/4$. The first step is to substitute $x = \pi/(4\omega)$ into $R(x)$. We obtain

$$R(x) = \frac{1}{2}A^2\omega^2\sin\left(\frac{\pi}{4}\right)^2 + F\left(A\cos\left(\frac{\pi}{4}\right)\right) - F(A)$$

$$= \frac{1}{2}A^2\omega^2\left(\frac{1}{\sqrt{2}}\right)^2 + F\left(\frac{A}{\sqrt{2}}\right) - F(A).$$

Setting $R(x) \doteq 0$, the following quadratic equation for $\omega_C^{(1)}$ results

$$\frac{A^2}{4}\left(\omega_C^{(1)}\right)^2 = F(A) - F\left(\frac{A}{\sqrt{2}}\right).$$

The corresponding positive solution is

$$\omega_C^{(1)} = \frac{2}{A}\sqrt{F(A) - F\left(\frac{A}{\sqrt{2}}\right)}.\tag{3.30}$$

For the period $T = 2\pi/\omega$ we obtain

$$T_C^{(1)} = \frac{A\pi}{\sqrt{F(A) - F\left(\frac{A}{\sqrt{2}}\right)}}.\tag{3.31}$$

Substituting $\omega = \omega_C^{(1)}$ into (3.28), the following approximation for the exact solution $y(x)$ results

$$y_C^{(1)}(x) = A \cos\left(\frac{2x}{A}\sqrt{F(A) - F\left(\frac{A}{\sqrt{2}}\right)}\right). \tag{3.32}$$

Remark 3.4

- In the above formulas, we use the subscript "C" to indicate that the collocation method has been applied. The superscript "(1)" indicates that the approximation is first-order. Later we will use the subscript "R" when the Ritz–Galerkin method is applied.
- The advantage of the EBM is that it does not require a linearization or a small perturbation parameter. However, the disadvantage of this method is that it is restricted to dynamical systems (with periodic solutions). We also cannot use this method for dynamical systems where the independent variable x is multiplied by the dependent variable.   □

It is also possible to use a more general ansatz for the trial function than (3.28). In [35], the following trial function is proposed:

$$y^{(n)}(x) = A_1 w_1(x) + A_2 w_2(x) + \cdots + A_n w_n(x), \tag{3.33}$$

where the functions $w_k(x)$ are given by

$$w_k(x) \equiv \cos((2k - 1)\omega x), \quad k = 1, \ldots, n.$$

Obviously, the first initial condition implies

$$A = A_1 + A_2 + \cdots + A_n.$$

Therefore, one of these parameters can be chosen as a dependent parameter, i.e.,

$$A_n = A - A_1 - A_3 - \cdots - A_{n-1}.$$

If the collocation method is applied again to determine ω and the remaining $A_1, \ldots, A_{n-1}$, the user must specify n collocation points such that the corresponding computations are relatively simple. In Example 3.6 this strategy is demonstrated.

Another way to find an approximation of the exact solution of the IVP (3.23) by the EBM is based on the ansatz (3.33) and the Ritz–Galerkin method. Here, ω and the parameter $A_2, \ldots, A_n$ are determined as an appropriate solution of the following system of n nonlinear algebraic equations

$$\int_0^{T/4} R(x) w_k(x) dx = 0, \quad k = 1, 2, \ldots, \tag{3.34}$$

where as before, $w_k(x) = \cos((2k-1)\omega x), k = 1, \ldots, n$. In Example 3.7 we show an application of this strategy.

3.2.2   Application of the Energy Balance Method

To illustrate the applicability and accuracy of the EBM, we apply it to a nonlinear dynamical system.

Example 3.5 Consider the free vibration of a mass connected to two nonlinear springs on a frictionless contact surface (see Fig. 3.3).

The problem can be modeled by a special form of the general Duffing equation

$$y''(x) + \delta y'(x) + \alpha y(x) + \beta y(x)^3 = \gamma \cos(\omega x). \tag{3.35}$$

Here, we set $\delta = \gamma = 0$, and consider the following IVP

$$y''(x) + \alpha y(x) + \beta y(x)^3 = 0, \quad y(0) = A, \quad y'(0) = 0. \tag{3.36}$$

In this problem, $y(x)$ is the displacement at time x, $y''(x)$ is the corresponding acceleration, β is a small positive parameter, which controls the amount of the non-linearity in the restoring force, and

$$\alpha \equiv \frac{K_1 + K_2}{m}, \tag{3.37}$$

where the constants K_1 and K_2 are the stiffnesses of the nonlinear springs, and m is the mass of the system.

Determine a first-order approximation of the exact solution of the IVP (3.36) by the EBM using the collocation technique.

Solution. For this dynamical system, we have

$$f(y(x)) = \alpha y(x) + \beta y(x)^3, \quad F(y(x)) = \frac{1}{2}\alpha y(x)^2 + \frac{1}{4}\beta y(x)^4.$$

Fig. 3.3 A mass connected to two nonlinear springs

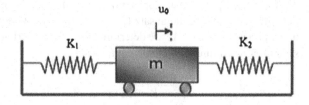

Its variational is

$$J(y) = \int_0^{T/4} \left(-\frac{1}{2}y'(x)^2 + \frac{1}{2}\alpha y(x)^2 + \frac{1}{4}\beta y(x)^4 \right) dx.$$

The corresponding Hamiltonian must satisfy

$$H(y) = \frac{1}{2}y'(x)^2 + \frac{1}{2}\alpha y(x)^2 + \frac{1}{4}\beta y(x)^4 \doteq \frac{1}{2}\alpha A^2 + \frac{1}{4}\beta A^4.$$

Thus, the residuum $R(x)$ is

$$\begin{aligned}
R(x) =& \frac{1}{2}y'(x)^2 + \frac{1}{2}\alpha y(x)^2 + \frac{1}{4}\beta y(x)^4 - \frac{1}{2}\alpha A^2 - \frac{1}{4}\beta A^4 \\
=& \frac{1}{2}y'(x)^2 + \frac{1}{2}y(x)^2 \left\{ \alpha + \frac{1}{2}\beta y(x)^2 \right\} - \frac{1}{2}A^2 \left\{ \alpha + \frac{1}{2}\beta A^2 \right\}.
\end{aligned} \tag{3.38}$$

If the exact solution is substituted into the residuum, it holds $R(x) = 0$.

To determine a first-order approximation of the exact solution, we use the ansatz (3.28) as a trial function. Substituting this ansatz into (3.38), we obtain

$$\begin{aligned}
R(x) =& \frac{A^2\omega^2}{2}\sin(\omega x)^2 + \frac{A^2}{2}\cos(\omega x)^2 \left\{ \alpha + \frac{A^2}{2}\beta\cos(\omega x)^2 \right\} \\
& - \frac{A^2}{2} \left\{ \alpha + \frac{A^2}{2}\beta \right\}.
\end{aligned}$$

As in the previous section, we apply collocation at $\omega x = \pi/4$. Using formula (3.30), we obtain

$$\omega_C^{(1)} = \sqrt{\alpha + \frac{3}{4}\beta A^2}. \tag{3.39}$$

The corresponding period is given by formula (3.31). It is

$$T_C^{(1)} = \frac{2\pi}{\sqrt{\alpha + \frac{3}{4}\beta A^2}}. \tag{3.40}$$

Thus, we get the following first-order approximation of the exact solution of the IVP (3.36)

$$y_C^{(1)}(x) = A\cos\left(\sqrt{\alpha + \frac{3}{4}\beta A^2} \, x \right). \tag{3.41}$$

In Table 3.2, for different parameter sets (A, α, β) the corresponding frequencies $\omega_C^{(1)}$ are given. To study the accuracy of the approximations $y_C^{(1)}(x)$, we have determined

Table 3.2 Values of $\omega_C^{(1)}$ determined for different parameter sets

No.	A	α	β	$\omega_C^{(1)}$
(1)	0.1	1	0.1	1.0003749297
(2)	2	1	0.1	1.1401754251
(3)	2	3	0.2	1.8973665961
(4)	4	3	0.2	2.3237900078

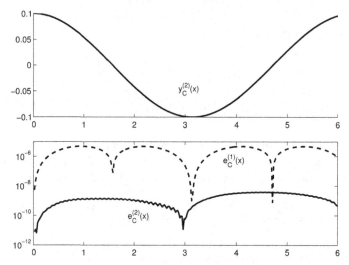

Fig. 3.4 Parameter set (1): second-order approximation $y_C^{(2)}$, and the the error functions $e_C^{(1)}, e_C^{(2)}$

numerical approximations $y_{ODE}^{(1)}$ of the IVP (3.36) by the IVP-solver ODE45, which is part of the MATLAB. Let us denote the difference between these solutions by

$$e_C^{(1)}(x) \equiv \left| y_{ODE}^{(1)} - y_C^{(1)}(x) \right|.$$

In the Figs. 3.4 and 3.5, the first-order approximation $y_C^{(1)}(x)$ and the corresponding $e_C^{(1)}(x)$ can be seen for the parameter sets (1) and (4), respectively.

The plots show that the first-order approximation $y_C^{(1)}(x)$ is already quite accurate. □

Example 3.6 Determine a second-order approximation of the exact solution of the IVP (3.36) by the EBM using the collocation technique.

Solution. Looking at formula (3.33), we set $n = 2$ and use the functions $w_1(x) = \cos(\omega x)$ and $w_2(x) = \cos(3\omega x)$. Thus, the appropriate ansatz is

$$y^{(2)}(x) = A_1 \cos(\omega x) + (A - A_1) \cos(3\omega x). \tag{3.42}$$

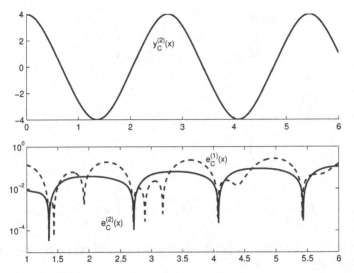

Fig. 3.5 Parameter set (4): second-order approximation $y_C^{(2)}$, and the the error functions $e_C^{(1)}$, $e_C^{(2)}$

It follows

$$\frac{d}{dx}y^{(2)}(x) = -A_1\omega\sin(\omega x) - 3(A - A_1)\omega\sin(3\omega x). \qquad (3.43)$$

Let the collocation points be $\omega x = \pi/4$ and $\omega x = \pi/2$.

For $\omega x = \pi/4$ we obtain

$$y^{(2)}\left(\frac{\pi}{4\omega}\right) = A_1\frac{1}{\sqrt{2}} - (A - A_1)\frac{1}{\sqrt{2}} = \frac{1}{\sqrt{2}}(2A_1 - A),$$

$$\left(y^{(2)}\left(\frac{\pi}{4\omega}\right)\right)^2 = \frac{1}{2}(2A_1 - A)^2,$$

and

$$\frac{d}{dx}y^{(2)}\left(\frac{\pi}{4\omega}\right) = -\frac{1}{\sqrt{2}}A_1\omega - \frac{3}{\sqrt{2}}\omega(A - A - 1) = \frac{2\omega}{\sqrt{2}}\left(A_1 - \frac{3}{2}A\right),$$

$$\left(\frac{d}{dx}y^{(2)}\left(\frac{\pi}{4\omega}\right)\right)^2 = 2\omega^2\left(A_1 - \frac{3}{2}A\right)^2.$$

Thus,

$$\begin{aligned}
R\left(\frac{\pi}{4\omega}\right) = &\omega^2\left(A_1 - \frac{3}{2}A\right)^2 + \frac{1}{4}(2A_1 - A)^2\left\{\alpha + \frac{1}{4}\beta(2A_1 - A)^2\right\} \\
&- \frac{1}{2}A^2\left\{\alpha + \frac{1}{2}\beta A^2\right\}.
\end{aligned} \qquad (3.44)$$

For $\omega x = \pi/2$ we obtain

$$y^{(2)} \left(\frac{\pi}{2\omega}\right) = 0, \quad \left(y^{(2)} \left(\frac{\pi}{2\omega}\right)\right)^2 = 0,$$

and

$$\frac{d}{dx} y^{(2)} \left(\frac{\pi}{2\omega}\right) = -A_1\omega + 3(A - A_1)\omega = \omega(3A - 4A_1),$$

$$\left(\frac{d}{dx} y^{(2)} \left(\frac{\pi}{2\omega}\right)\right)^2 = \omega^2(3A - 4A_1)^2.$$

Thus,

$$R\left(\frac{\pi}{2\omega}\right) = \frac{1}{2}\omega^2(3A - 4A_1)^2 - \frac{1}{2}A^2\left(\alpha + \frac{1}{2}\beta A^2\right). \tag{3.45}$$

From (3.44) and (3.45), we get the following two nonlinear algebraic equations for the two unknowns A_1 and ω:

$$0 = 4\omega^2 \left(A_1 - \frac{3}{2}A\right)^2 + (2A_1 - A)^2\left\{\alpha + \frac{1}{4}\beta(2A_1 - A)^2\right\}$$

$$- 2A^2\left(\alpha + \frac{1}{2}\beta A^2\right), \tag{3.46}$$

$$0 = \omega^2(3A - 4A_1)^2 - A^2\left\{\alpha + \frac{1}{2}\beta A^2\right\}.$$

The determination of the solutions of this system in closed form leads to very large and complicated expressions. A better strategy is to prescribe fixed values for A, α, and β, and to solve numerically the algebraic equations (3.46). In Table 3.3, for the parameter sets (1) and (4) presented in Table 3.2, the numerically determined appropriate solutions of the nonlinear algebraic equations (3.46) are given.

Note, there are 12 solutions of the system (3.46). Since 4 solutions are complex-valued, and for 4 solutions the ω-component is negative, there still remain 4 real-valued solutions with a positive ω-component. We have chosen the solution whose component $\omega_C^{(2)}$ is nearest to $\omega_C^{(1)}$. For the parameter sets (1) and (4), the corresponding solution $y_C^{(2)}$ as well as the error functions $e_C^{(1)}$ and $e_C^{(2)}$ are presented in the Figs. 3.4 and 3.5, respectively.

Table 3.3 Values of $\omega_C^{(2)}$ and $(A_1)_C^{(2)}$ determined for different parameter sets

No.	A	α	β	$\omega_C^{(2)}$	$(A_1)_C^{(2)}$
(1)	0.1	1	0.1	1.0003749102	0.0999968776
(4)	4	3	0.2	2.3090040391	3.9288684743

It is easy to see that in both parameter constellations the second-order approximation is better than the first-order approximation. However, the difference between these approximations in not significant. □

Example 3.7 Determine a first-order approximation of the exact solution of the IVP (3.36) by the EBM using the Ritz–Galerkin technique.

Solution. We start with the ansatz (3.28), i.e.,

$$y^{(1)} = A\cos(\omega x),$$

and obtain

$$R(x) = \frac{A^2}{2}\omega^2 \sin(\omega x)^2 + \frac{A^2}{2}\alpha\cos(\omega x)^2 + \frac{A^4}{4}\beta\cos(\omega x)^4 - \frac{A^2}{2}\alpha - \frac{A^4}{4}\beta.$$

The next step is to set the weighted integral of the residual to zero, i.e.,

$$\int_0^{T/4} R(x)\cos(\omega x)dx \doteq 0, \quad T = \frac{2\pi}{\omega}.$$

This yields

$$\frac{A^2}{6}\omega + \frac{A^2}{3\omega}\alpha + \frac{2A^4}{15\omega}\beta - \frac{A^2}{2\omega}\alpha - \frac{A^4}{4\omega}\beta = \omega^2 - \alpha - \frac{7}{10}A^2\beta = 0.$$

The positive solution of the equation

$$\omega^2 = \alpha + \frac{7}{10}A^2\beta \tag{3.47}$$

is

$$\omega_R^{(1)} = \sqrt{\alpha + \frac{7}{10}A^2\beta}. \tag{3.48}$$

Thus, the first-order approximate solution of the IVP (3.36) is

$$y_R^{(1)} = A\cos\left(\sqrt{\alpha + \frac{7}{10}A^2\beta}\, x\right). \tag{3.49}$$

For the parameter sets (1)–(4) given in Table 3.2, the approximated frequencies are presented in Table 3.4. □

Example 3.8 Determine a second-order approximation of the exact solution of the IVP (3.36) by the EBM using the Ritz–Galerkin technique.

Table 3.4 Values of $\omega_R^{(1)}$ determined for different parameter sets

No.	A	α	β	$\omega_R^{(1)}$
(1)	0.1	1	0.1	1.0003499388
(2)	2	1	0.1	1.1313708499
(3)	2	3	0.2	1.8867962264
(4)	4	3	0.2	2.2891046285

Solution. Substituting the ansatz

$$y^{(2)} = A_1 \cos(\omega x) + (A - A_1) \cos(3\omega x)$$

into the residual (3.38), we obtain

$$
\begin{aligned}
R(x) =& \frac{\omega^2}{2}\big(A_1 \sin(\omega x) + 3(A - A_1) \sin(3\omega x)\big)^2 \\
&+ \frac{\alpha}{2}\big(A_1 \cos(\omega x) + (A - A_1) \cos(3\omega x)\big)^2 \\
&+ \frac{\beta}{4}\big(A_1 \cos(\omega x) + (A - A_1) \cos(3\omega x)\big)^4 - \frac{1}{2}\alpha A^2 - \frac{1}{4}\beta A^4.
\end{aligned}
\tag{3.50}
$$

Now, from (3.34) we get the following two nonlinear algebraic equations for the two unknowns A_1 and ω:

$$
\begin{aligned}
\int_0^{T/4} R(x) \cos(\omega x)\,dx &= 0, \\
\int_0^{T/4} R(x) \cos(3\omega x)\,dx &= 0,
\end{aligned}
\tag{3.51}
$$

where $T = 2\pi/\omega$.

In the Table 3.5, for the parameter sets (1) and (4) presented in Table 3.2, the numerically determined appropriate solutions of the nonlinear algebraic equations (3.51) are given. Moreover, in the Figs. 3.6 and 3.7 the accuracy of the first-order approximations of the EBM (collocation and Ritz–Galerkin) and the second-order approximations (collocation and Ritz–Galerkin) are visualized for the same two parameter sets. □

Table 3.5 Values of $\omega_R^{(2)}$ and $(A_1)_R^{(2)}$ determined for different parameter sets

No.	A	α	β	$\omega_R^{(2)}$	$(A_1)_R^{(2)}$
(1)	0.1	1	0.1	1.0003749144	0.0999968773
(4)	4	3	0.2	2.3114595536	3.9249208620

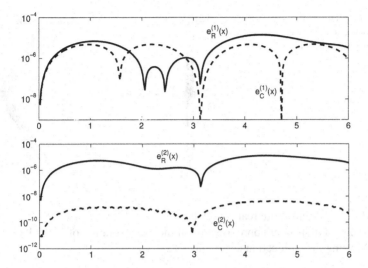

Fig. 3.6 Parameter set (1): Comparison of the accuracy of the approximations, which have been computed with the EBM (using collocation) and the EBM (using Ritz–Galerkin)

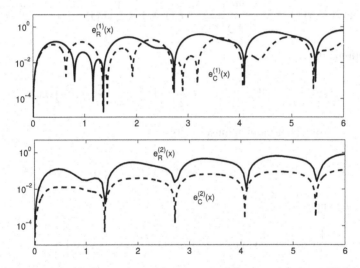

Fig. 3.7 Parameter set (4): Comparison of the accuracy of the approximations, which have been computed with the EBM (using collocation) and the EBM (using Ritz–Galerkin)

Example 3.9 Consider the free vibration of a mass attached to the center of a stretched elastic wire. The governing IVP corresponding to this dynamical system is given by (see [36, 118] and Fig. 3.8)

$$y''(x) + y(x) - \lambda \frac{y(x)}{\sqrt{1 + y(x)^2}} = 0, \quad y(0) = A, \quad y'(0) = 0, \quad (3.52)$$

Fig. 3.8 Oscillation of a
mass attached to the center
of a stretched elastic wire

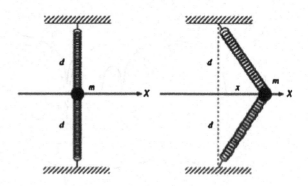

where $0 < \lambda \leq 1$ and A are real parameters.

Determine a first-order approximation of the exact solution of the IVP (3.52) by the EBM using the collocation technique.

Solution. For this dynamical system, we have

$$f(y(x)) = y(x) - \lambda \frac{y(x)}{\sqrt{1 + y(x)^2}}, \quad F(y(x)) = \frac{1}{2} y(x)^2 - \lambda \sqrt{1 + y(x)^2}.$$

Its variational is

$$J(y) = \int_0^{T/4} \left(-\frac{1}{2} y'(x)^2 + \frac{1}{2} y(x)^2 - \lambda \sqrt{1 + y(x)^2} \right) dx.$$

The corresponding Hamiltonian must satisfy

$$H(y) = \frac{1}{2} y'(x)^2 + \frac{1}{2} y(x)^2 - \lambda \sqrt{1 + y(x)^2} \doteq \frac{1}{2} A^2 - \lambda \sqrt{1 + A^2}.$$

Thus, the residuum $R(x)$ is

$$R(x) = \frac{1}{2} y'(x)^2 + \frac{1}{2} y(x)^2 - \lambda \sqrt{1 + y(x)^2} - \frac{1}{2} A^2 + \lambda \sqrt{1 + A^2}. \qquad (3.53)$$

To determine a first-order approximation of the exact solution, we use the ansatz (3.28) as a trial function. Substituting this ansatz into $R(x)$, we obtain

$$R(x) = \frac{1}{2} A^2 \omega^2 \sin(\omega x)^2 + \frac{1}{2} A^2 \cos(\omega x)^2$$
$$- \lambda \sqrt{1 + A^2 \cos(\omega x)^2} - \frac{1}{2} A^2 + \lambda \sqrt{1 + A^2}.$$

Using the collocation point $\omega x = \pi/4$, we get $\sin(\pi/4) = \cos(\pi/4) = 1/\sqrt{2}$ and

$$R(x) = \frac{1}{4}A^2\omega^2 - \frac{1}{4}A^2 - \lambda\left(\sqrt{1+\frac{1}{2}A^2} - \sqrt{1+A^2}\right).$$

Now, we set $R(x) = 0$. The positive solution of the resulting quadratic equation

$$\omega^2 - 1 - \frac{4\lambda}{A^2}\left(\sqrt{1+\frac{1}{2}A^2} - \sqrt{1+A^2}\right) = 0$$

is

$$\omega_C^{(1)} = \sqrt{1 + \frac{4\lambda}{A^2}\left(\sqrt{1+\frac{1}{2}A^2} - \sqrt{1+A^2}\right)}. \tag{3.54}$$

For the corresponding period, we obtain

$$T_C^{(1)} = \frac{2\pi}{\omega} = \frac{2\pi}{\sqrt{1 + \frac{4\lambda}{A^2}\left(\sqrt{1+\frac{1}{2}A^2} - \sqrt{1+A^2}\right)}}. \tag{3.55}$$

Thus, we get the following first-order approximation of the exact solution of the IVP (3.52)

$$y_C^{(1)}(x) = A\cos\left(\sqrt{1 + \frac{4\lambda}{A^2}\left(\sqrt{1+\frac{1}{2}A^2} - \sqrt{1+A^2}\right)}\,x\right). \tag{3.56}$$

In comparison, the exact frequency is (see [118])

$$\omega_{\text{ex}} = \frac{2\pi}{T_{\text{ex}}}, \tag{3.57}$$

where

$$T_{\text{ex}} = 4\int_0^{\pi/2}\left[1 - \frac{2\lambda}{\sqrt{1+A^2\sin(x)^2} + \sqrt{1+A^2}}\right]^{-1/2} dx. \tag{3.58}$$

In Table 3.6, for different values of λ and A the approximate frequency $\omega_C^{(1)}$ and the exact frequency ω_{ex} are given. $\qquad\square$

Example 3.10 Determine a second-order approximation of the exact solution of the IVP (3.52) by the EBM using the collocation technique.

Table 3.6 Values of $\omega_C^{(1)}$ and ω_{ex} determined for different parameter sets

No.	λ	A	ω_{ex}	$\omega_C^{(1)}$
(1)	0.2	2	0.9467768213	0.94825975661
(2)	0.5	2	0.8604467698	0.8648649692
(3)	0.7	1	0.6816273394	0.6851916996
(4)	0.9	1	0.5566810704	0.5638374876

Solution. As described in Example 3.6, we use the ansatz (3.42), i.e.,

$$y^{(2)}(x) = A_1 \cos(\omega x) + (A - A_1) \cos(3\omega x).$$

Substituting (3.42) into the residual (3.53), we obtain

$$R(x) = \frac{\omega^2}{2} \left(A_1 \sin(\omega x) + 3(A - A_1) \sin(3\omega x) \right)^2$$
$$+ \frac{1}{2} \left(A_1 \cos(\omega x) + (A - A_1) \cos(3\omega x) \right)^2$$
$$- \lambda \sqrt{1 + \left(A_1 \cos(\omega x) + (A - A_1) \cos(3\omega x) \right)^2}$$
$$- \frac{1}{2} A^2 + \lambda \sqrt{1 + A^2}.$$

Now, we use the two collocation points $\omega x = \pi/4$ and $\omega x = \pi/2$, and set

$$R\left(\frac{\pi}{4\omega}\right) = 0, \qquad R\left(\frac{\pi}{2\omega}\right) = 0.$$

It results the following system of two nonlinear algebraic equations for the two unknown ω and A_1

$$0 = 4\lambda \left(\sqrt{1 + A^2} - \sqrt{1 + \frac{A^2}{2} - 2AA_1 + 2A_1^2} \right)$$
$$+ \omega^2 \left(9A^2 + 4A_1^2 - 12AA_1 \right) + 4A_1 (A_1 - A) - A^2,$$

$$0 = 2\lambda \left(\sqrt{1 + A^2} - 1 \right) + \omega^2 (3A - 4A_1)^2 - A^2. \tag{3.59}$$

This system has 16 solutions. For 8 of these solutions, the ω-component is complex-valued. Thus, there remain 8 solutions with a real-valued ω-component. We have chosen the solution whose component $\omega_C^{(2)}$ is nearest to $\omega_C^{(1)}$. For the parameter sets (1)–(4) presented in Table 3.6, the corresponding solutions are given in Table 3.7. A comparison with the results in Table 3.6 shows that the second-order approximation $\omega_C^{(2)}$ is slightly better than the first-order approximation $\omega_C^{(1)}$. □

Table 3.7 Values of $\omega_C^{(2)}$ and $(A_1)_C^{(2)}$ determined for different parameter sets

No.	λ	A	$\omega_C^{(2)}$	$(A_1)_C^{(2)}$
(1)	0.2	2	0.9481247885	1.9936896281
(2)	0.5	2	0.8637698322	1.9811778812
(3)	0.7	1	0.6832313703	0.9871641681
(4)	0.9	1	0.5581704732	0.9759149690

Example 3.11 Determine a first-order approximation of the exact solution of the IVP (3.52) by the EBM using the Ritz–Galerkin technique.

Solution. We start with the ansatz (3.28), i.e.,

$$y^{(1)} = A\cos(\omega x),$$

and obtain

$$R(x) = \frac{1}{2}A^2\omega^2\sin(\omega x)^2 + \frac{1}{2}A^2\cos(\omega x)^2$$
$$- \lambda\sqrt{1 + A^2\cos(\omega x)^2} - \frac{1}{2}A^2 + \lambda\sqrt{1 + A^2}.$$

The next step is to set the weighted integral of the residual to zero, i.e.,

$$\int_0^{T/4} R(x)\cos(\omega x)dx \doteq 0, \quad T = \frac{2\pi}{\omega}.$$

This yields

$$0 = \frac{1}{2}A^2\omega^2\frac{1}{3\omega} + \frac{1}{2}A^2\frac{2}{3\omega} - \lambda\frac{(1 + A^2)\arcsin\left(\dfrac{A}{\sqrt{1 + A^2}}\right) + A}{2A\omega}$$
$$- \frac{1}{2}A^2\frac{1}{\omega} + \lambda\sqrt{1 + A^2}\frac{1}{\omega}$$
$$= \omega^2 - 1 - \frac{3\lambda}{A^3}\left((1 + A^2)\arcsin\left(\frac{A}{\sqrt{1 + A^2}}\right) + A\right)$$
$$+ \frac{6\lambda}{A^2}\sqrt{1 + A^2}.$$

The positive solution of this equation is

$$\omega_R^{(1)} = \sqrt{1 - \frac{6\lambda}{A^2}\sqrt{1 + A^2} + \frac{3\lambda}{A^3}\left((1 + A^2)\arcsin\left(\frac{A}{\sqrt{1 + A^2}}\right) + A\right)}. \quad (3.60)$$

Table 3.8 Values of $\omega_R^{(1)}$ and ω_{ex} determined for different parameter sets

No.	λ	A	ω_{ex}	$\omega_R^{(1)}$
(1)	0.2	2	0.9467768213	0.9457062842
(2)	0.5	2	0.8604467698	0.8578466878
(3)	0.7	1	0.6816273394	0.6774771762
(4)	0.9	1	0.5566810704	0.5517217102

Thus,

$$y_R^{(1)}(x) = A \cos\left(\omega_R^{(1)}x\right). \tag{3.61}$$

In Table 3.8, for the parameter sets (1)–(4) of the previous two tables the approximate frequency $\omega_R^{(1)}$ and the exact frequency ω_{ex} are given. $\square$

Example 3.12 Determine a second-order approximation of the exact solution of the IVP (3.52) by the EBM using the Ritz-Galerkin technique.

Solution. Substituting the ansatz

$$y^{(2)} = A_1 \cos(\omega x) + (A - A_1)\cos(3\omega x)$$

into the residual (3.53), we obtain

$$\begin{aligned}
R(x) = &\frac{\omega^2}{2}\left(A_1 \sin(\omega x) + 3(A - A_1)\sin(3\omega x)\right)^2 \\
&+ \frac{1}{2}\left(A_1 \cos(\omega x) + (A - A_1)\cos(3\omega x)\right)^2 \\
&- \lambda\sqrt{1 + \left(A_1 \cos(\omega x) + (A - A_1)\cos(3\omega x)\right)^2} \\
&- \frac{1}{2}A^2 + \lambda\sqrt{1 + A^2}.
\end{aligned}$$

Now, from (3.34) we get the following two nonlinear algebraic equations for the two unknowns A_1 and ω:

$$\begin{aligned}
\int_0^{T/4} R(x)\cos(\omega x)dx &= 0, \\
\int_0^{T/4} R(x)\cos(3\omega x)dx &= 0,
\end{aligned} \tag{3.62}$$

where $T = 2\pi/\omega$.

In Table 3.9, for the parameter sets (1)–(4) presented in the previous three tables, the numerically determined appropriate solutions of the nonlinear algebraic equations (3.62) are given. $\square$

Table 3.9 Values of $\omega_R^{(2)}$ and $(A_1)_R^{(2)}$ determined for different parameter sets

No.	λ	A	$\omega_R^{(2)}$	$(A_1)_R^{(2)}$
(1)	0.2	2	0.9474385576	1.9952077413
(2)	0.5	2	0.8621619690	1.9854583530
(3)	0.7	1	0.6824425348	0.9882819868
(4)	0.9	1	0.5575093394	0.9772737444

3.3 Hamiltonian Approach

3.3.1 Theoretical Background

As we mentioned before, many oscillation problems are nonlinear, and in most cases it is difficult to solve them analytically. In the previous section, we have introduced the energy balance method using the collocation technique and the Ritz technique. Both strategies are based on the Hamiltonian. Although this approach is simple, it strongly depends upon the location points (collocation technique) and the weight functions $w_j(x)$ (Ritz technique) that are chosen. Recently, He (see e.g. [52, 53]) has proposed the so-called Hamiltonian approach to overcome the shortcomings of the EBM.

In order to describe the Hamiltonian approach (HA), let us consider once more the IVP (3.23) of the general oscillator, i.e.,

$$y''(x) + f(y(x)) = 0, \quad y(0) = A, \quad y'(0) = 0.$$

The corresponding variational and the Hamiltonian of the oscillator are given in (3.24) and (3.25), respectively. As shown in the previous section, the Hamiltonian must be a constant value. Thus, we can write

$$H(y(x)) = \frac{1}{2}y'(x)^2 + F(y(x)) \doteq H_0, \qquad (3.63)$$

where H_0 is a real constant. Now, it is assumed that the solution can be expressed as

$$y(x) = A\cos(\omega x). \qquad (3.64)$$

Substituting (3.64) into (3.63) yields

$$H(y(x)) = \frac{1}{2}A^2\omega^2\sin(\omega x)^2 + F(A\cos(\omega x)) = H_0. \qquad (3.65)$$

Differentiating both sides of the identity (3.65) gives

$$\frac{\partial H(y)}{\partial A} = 0. \tag{3.66}$$

The next step is to introduce a new function $\bar{H}(y)$ defined as

$$\begin{aligned}
\bar{H}(y) &\equiv \int_0^{T/4} \left\{ \frac{1}{2} y'(x)^2 + F(y(x)) \right\} dx \\
&= \int_0^{T/4} H_0 \, dx = \frac{T}{4} H_0 = \frac{T}{4} H(y).
\end{aligned} \tag{3.67}$$

It holds

$$\frac{\partial \bar{H}(y)}{\partial T} = \frac{1}{4} H(y).$$

Thus, Eq. (3.66) is equivalent to

$$\frac{\partial^2 \bar{H}(y)}{\partial A \, \partial T} = 0,$$

or

$$\frac{\partial^2 \bar{H}(y)}{\partial A \, \partial(1/\omega)} = 0. \tag{3.68}$$

This formula can be used to determine an approximate frequency of the nonlinear oscillator. Substituting this approximate frequency into the ansatz (3.64), a first-order approximation of the solution of the IVP (3.23) is obtained.

3.3.2 Application of the Hamiltonian Approach

In this section, by means of two instructive examples we will show the reliability of the Hamiltonian approach.

Example 3.13 Let us consider once again the special case (3.36) of the Duffing ODE (3.35)

$$y''(x) + \alpha y(x) + \beta y(x)^3 = 0, \quad y(0) = A, \quad y'(0) = 0. \tag{3.69}$$

Determine an analytical approximation of the solution of the IVP (3.69) by the Hamiltonian approach.

Solution. For this problem, we have

$$f(y) = \alpha y + \beta y^3.$$

Thus,

$$F(y) = \frac{1}{2}\alpha y^2 + \frac{1}{4}\beta y^4.$$

The corresponding Hamiltonian is

$$H(y(x)) = \frac{1}{2}y'(x)^2 + \frac{1}{2}\alpha y(x)^2 + \frac{1}{4}\beta y(x)^4,$$

and

$$\bar{H}(y) = \int_0^{T/4} \left(\frac{1}{2}y'(x)^2 + \frac{1}{2}\alpha y(x)^2 + \frac{1}{4}\beta y(x)^4 \right) dx. \tag{3.70}$$

Now, we use the following ansatz for an approximate solution of the IVP (3.69)

$$y(x) = A\cos(\omega x).$$

Substituting this ansatz into (3.70), we obtain

$$\bar{H}(y) = \int_0^{T/4} \left(\frac{1}{2}A^2\omega^2 \sin(\omega x)^2 + \frac{1}{2}\alpha A^2 \cos(\omega x)^2 + \frac{1}{4}\beta A^4 \cos(\omega x)^4 \right) dx$$

$$= \int_0^{\frac{\pi}{2\omega}} \left(\frac{1}{2}A^2\omega^2 \sin(\omega x)^2 + \frac{1}{2}\alpha A^2 \cos(\omega x)^2 + \frac{1}{4}\beta A^4 \cos(\omega x)^4 \right) dx$$

$$= \frac{1}{2}A^2\omega^2 \cdot \frac{1}{4}\frac{\pi}{\omega} + \frac{1}{2}\alpha A^2 \cdot \frac{1}{4}\frac{\pi}{\omega} + \frac{1}{4}\beta A^4 \cdot \frac{3}{16}\frac{\pi}{\omega}$$

$$= \frac{1}{8}A^2\omega\pi + \frac{1}{\omega}\left(\frac{1}{8}\alpha A^2\pi + \frac{3}{64}\beta A^4\pi \right).$$

It follows

$$\frac{\partial \bar{H}(y)}{\partial(1/\omega)} = -\frac{1}{8}A^2\omega^2\pi + \frac{1}{8}\alpha A^2\pi + \frac{3}{64}\beta A^4\pi,$$

and

$$\frac{\partial^2 \bar{H}(y)}{\partial A \, \partial(1/\omega)} = -\frac{1}{4}A\omega^2\pi + \frac{1}{4}\alpha A\pi + \frac{3}{16}\beta A^3\pi.$$

We set

$$\frac{\partial^2 \bar{H}(y)}{\partial A \, \partial(1/\omega)} \doteq 0,$$

i.e.,

$$-\omega^2 + \alpha + \frac{3}{4}\beta A^2 = 0.$$

The positive root of this quadratic equation is a first-order approximation of the frequency of the nonlinear oscillator:

$$\omega_{HA}^{(1)} = \sqrt{\alpha + \frac{3}{4}\beta A^2}. \tag{3.71}$$

A comparison with formula (3.39) shows that this is the same approximation as that obtained by the energy balance method using the collocation point $x = \pi/(4\omega)$. Obviously, the corresponding approximation of the solution of the IVP (3.69) is

$$y_{HA}^{(1)}(x) = y_{C}^{(1)}(x) = A\cos\left(\sqrt{\alpha + \frac{3}{4}\beta A^2}\, x\right). \tag{3.72}$$

□

Note, that the Hamiltonian approach has required less computational work than the energy balance method.

Example 3.14 As a second example, let us consider the IVP of the Duffing equation with a fifth-order nonlinearity given by

$$y''(x) + y(x) + \alpha y(x)^5 = 0, \quad y'(0) = A, \quad y'(0) = 0, \tag{3.73}$$

where $\alpha \geq 0$ is a real constant. Determine an approximative solution of the IVP (3.73) by the Hamiltonian approach.

Solution. For this problem, we have

$$f(y) = y + \alpha y^5.$$

Thus,

$$F(y) = \frac{1}{2}y^2 + \frac{1}{6}\alpha y^6.$$

The corresponding Hamiltonian is

$$H(y(x)) = \frac{1}{2}y'(x)^2 + \frac{1}{2}y(x)^2 + \frac{1}{6}\alpha y(x)^6.$$

Integrating this Hamiltonian w.r.t. x from 0 to $T/4$, gives

$$\bar{H}(y) = \int_0^{T/4}\left(\frac{1}{2}y'(x)^2 + \frac{1}{2}y(x)^2 + \frac{1}{6}\alpha y(x)^6\right)dx. \tag{3.74}$$

Using the ansatz

$$y(x) = A\cos(\omega x)$$

for the approximate solution of the IVP (3.73), and substituting it into $\bar{H}(y)$, we obtain

$$\bar{H}(y) = \int_0^{\frac{\pi}{2\omega}} \left(\frac{1}{2}A^2\omega^2 \sin(\omega x)^2 + \frac{1}{2}A^2 \cos(\omega x)^2 + \frac{1}{6}\alpha A^6 \cos(\omega x)^6 \right) dx$$

$$= \frac{1}{2}A^2\omega^2 \cdot \frac{\pi}{4\omega} + \frac{1}{2}A^2 \cdot \frac{\pi}{4\omega} + \frac{1}{6}\alpha A^6 \cdot \frac{5\pi}{32\omega}$$

$$= \frac{1}{8}A^2\pi\omega + \frac{1}{\omega}\left(\frac{1}{8}A^2\pi + \frac{5}{192}\alpha A^6\pi \right).$$

Now, we compute

$$\frac{\partial \bar{H}(y)}{\partial(1/\omega)} = -\frac{1}{8}A^2\pi\omega^2 + \frac{1}{8}A^2\pi + \frac{5}{192}\alpha A^6\pi,$$

and

$$\frac{\partial^2 \bar{H}(y)}{\partial A\, \partial(1/\omega)} = -\frac{1}{4}A\pi\omega^2 + \frac{1}{4}A\pi + \frac{5}{32}\alpha A^5\pi.$$

We set

$$\frac{\partial^2 \bar{H}(y)}{\partial A\, \partial(1/\omega)} \doteq 0,$$

i.e.

$$-\omega^2 + 1 + \frac{5}{8}\alpha A^4 = 0.$$

The positive root of this quadratic equation is a first-order approximation of the frequency of the nonlinear oscillator:

$$\omega_{HA}^{(1)} = \sqrt{1 + \frac{5}{8}\alpha A^4}.$$

The corresponding approximation of the solution of the IVP (3.73) is

$$y_{HA}^{(1)}(x) = A \cos\left(\sqrt{1 + \frac{5}{8}\alpha A^4}\, x \right). \tag{3.75}$$

In Fig. 3.9 the approximate solution (3.75) is compared with the exact solution for $\alpha = 1.0$ and different values of A. It can be seen that the approximate solution gets worse if the parameter A is increased. $\qquad\square$

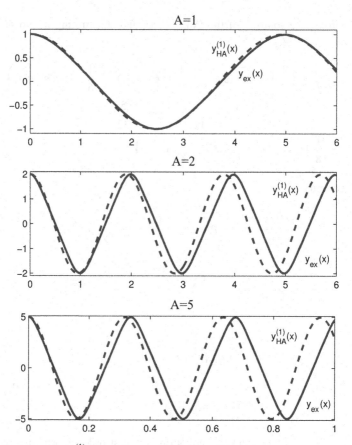

Fig. 3.9 Comparison of $y_{HA}^{(1)}(x)$ with the exact solution for $\alpha = 1.0$ and $A = 1, 2, 5$; *Note* for the third partial figure we have chosen the x-interval $[0, 1]$ instead of $[0, 6]$

3.4   Homotopy Analysis Method

3.4.1   Theoretical Background

The *homotopy analysis method* (HAM) is a semi-analytical technique to solve nonlinear ODEs. It uses the concept of the homotopy from topology to generate a convergent series solution for nonlinear problems. This is enabled by employing a homotopy-McLaurin series to deal with the nonlinearities in the ODE.

The HAM was first developed in 1992 by Liao [75] and further modified in [76–79]. The advantages of the HAM are

- It is a series expansion method that is not directly dependent on small or large physical parameters. Thus, it is applicable for strongly nonlinear problems.

- It has a greater generality than the other methods presented in this book, i.e., it often allows for strong convergence of the solution over larger spacial and parameter domains.
- It provides great freedom to choose the basis functions of the desired solution and the corresponding auxiliary linear operator of the homotopy.
- It gives a simple way to ensure the convergence of the solution series.

The essential idea of the HAM is to introduce an embedding parameter $p \in [0, 1]$, and a nonzero auxiliary parameter $\hbar$ referred to as the *convergence-control parameter*, to verify and enforce the convergence of the solution series. For $p = 0$, the original problem is usually reduced to a simplified problem that can be solved rather easily. If p is gradually increased toward 1, the problem goes through a sequence of deformations, and the solution at each stage is close to that at the previous stage of deformation. In most cases, for $p = 1$ the problem takes the original form and the final stage of the deformation gives the desired solution. In other words, as p increases from zero to one, the initial guess approaches a good analytical approximation of the exact solution of the given nonlinear problem.

In [77], Liao has shown that the HAM is a unified method for some perturbation and non-perturbation methods. For example, the ADM is a special case of the HAM and for certain values of $\hbar$, the VIM and the HAM are equivalent.

We now turn to the general approach used by the HAM to solve the nonlinear ODE

$$\mathcal{N}[y(x)] = 0, \quad x \in I, \tag{3.76}$$

subject to some initial or boundary conditions. Here, $\mathcal{N}$ is a nonlinear operator, and $y(x)$ is an unknown function defined over the interval I. The first step in the HAM solution of this ODE is to construct the homotopy

$$\begin{aligned} &H\left[\phi(x; p); y_0(x), \mathcal{H}(x), \hbar, p\right] \\ &\equiv (1 - p)\mathcal{L}[\phi(x; p) - y_0(x)] - p\,\hbar\,\mathcal{H}(x)\mathcal{N}[\phi(x; p)], \end{aligned} \tag{3.77}$$

where $\hbar \neq 0$ is an auxiliary parameter, $\mathcal{H}(x) \neq 0$ is an auxiliary function, $p \in [0, 1]$ is an embedding parameter, $y_0(x)$ is an initial approximation to the solution that satisfies the given initial or boundary conditions, $\phi(x; p)$ satisfies the initial or boundary conditions, and $\mathcal{L}$ is some linear operator. The linear operator $\mathcal{L}$ should normally be of the same order as the nonlinear operator $\mathcal{N}$.

We now set the homotopy (3.77) equal to zero so that

$$(1 - p)\mathcal{L}[\phi(x; p) - y_0(x)] = p\,\hbar\,\mathcal{H}(x)\mathcal{N}[\phi(x; p)]. \tag{3.78}$$

The Eq. (3.78) is called *zeroth-order deformation equation*. By setting $p = 0$ in this equation, we get

$$\mathcal{L}[\phi(x; 0) - y_0(x)] = 0.$$

From the definitions of $\mathcal{L}$, $\phi(x)$, and $y_0(x)$, it follows

$$\phi(x; 0) = y_0(x). \tag{3.79}$$

If we set $p = 1$ in the Eq. (3.78), we obtain

$$\mathcal{N}[\phi(x; 1)] = 0. \tag{3.80}$$

Under the assumption that $\phi(x; p)$ satisfies the initial or boundary conditions, formula (3.80) implies

$$\phi(x; 1) = y(x). \tag{3.81}$$

It is clear from (3.79) and (3.81) that $\phi(x; p)$ varies continuously from the initial approximation $y_0(x)$ to the required solution $y(x)$ as p increases from 0 to 1. In the topology, ϕ is called a *deformation*.

Next, let us define the terms

$$y_n(x) = \frac{1}{n!} \left. \frac{\partial^n \phi(x; p)}{\partial p^n} \right|_{p=0}. \tag{3.82}$$

By Taylor's theorem we can write

$$\phi(x; p) = \phi(x; 0) + \sum_{n=1}^{\infty} \frac{1}{n!} \left. \frac{\partial^n \phi(x; p)}{\partial p^n} \right|_{p=0} p^n = y_0(x) + \sum_{n=1}^{\infty} y_n(x) \, p^n, \tag{3.83}$$

In order to calculate the functions $y_n(x)$ in the solution series, we first differentiate equation (3.78) w.r.t. p. It follows

$$(1 - p)\mathcal{L} \left[\frac{\partial \phi(x; p)}{\partial p} \right] - \mathcal{L}[\phi(x; p) - y_0(x)]$$
$$= \hbar \mathcal{H}(x) \mathcal{N}[\phi(x; p)] + p \, \hbar \mathcal{H}(x) \frac{\partial \mathcal{N}[\phi(x; p)]}{\partial p}. \tag{3.84}$$

Setting $p = 0$ in this equation and applying the results of (3.79) and (3.81), we get

$$\mathcal{L}[y_1(x)] = \hbar \mathcal{H}(x) \mathcal{N}[y_0(x)]. \tag{3.85}$$

Now, we will prove by induction that the following relation is true

$$(1 - p)\mathcal{L} \left[\frac{\partial^n \phi(x; p)}{\partial p^n} \right] - n \, \mathcal{L} \left[\frac{\partial^{n-1} \phi(x; p)}{\partial p^{n-1}} \right]$$
$$= n \, \hbar \mathcal{H}(x) \frac{\partial^{n-1} \mathcal{N}[\phi(x; p)]}{\partial p^{n-1}} + p \, \hbar \mathcal{H}(x) \frac{\partial^n \mathcal{N}[\phi(x; p)]}{\partial p^n}, \tag{3.86}$$

where $n \geq 2$. Let us start with $n = 2$. Differentiating (3.84) w.r.t. p, gives

$$
(1-p)\mathcal{L}\left[\frac{\partial^2 \phi(x;p)}{\partial p^2}\right] - 2\mathcal{L}\left[\frac{\partial \phi(x;p)}{\partial p}\right]
$$
$$
= 2\hbar\mathcal{H}(x)\frac{\partial \mathcal{N}[\phi(x;p)]}{\partial p} + p\hbar\mathcal{H}(x)\frac{\partial^2 \mathcal{N}[\phi(x;p)]}{\partial p^2},
\tag{3.87}
$$

which shows that Eq. (3.86) is indeed true for $n = 2$.

Next, we assume that (3.86) is true for some value of n and show that it must also be true when n is replaced by $n + 1$. Since (3.86) is true by assumption, we can differentiate it w.r.t. p to obtain

$$
(1-p)\mathcal{L}\left[\frac{\partial^{n+1} \phi(x;p)}{\partial p^{n+1}}\right] - (n+1)\mathcal{L}\left[\frac{\partial^n \phi(x;p)}{\partial p^n}\right]
$$
$$
= (n+1)\hbar\mathcal{H}(x)\frac{\partial^n \mathcal{N}[\phi(x;p)]}{\partial p^n} + p\hbar\mathcal{H}(x)\frac{\partial^{n+1} \mathcal{N}[\phi(x;p)]}{\partial p^{n+1}},
\tag{3.88}
$$

which shows that (3.86) is true if n is replaced by $n + 1$. Thus, it can be concluded that Eq. (3.86) is valid for all $n \geq 2$.

Dividing (3.86) by $n!$ and setting $p = 0$, we get

$$
\mathcal{L}\left[\frac{1}{n!}\frac{\partial^n \phi(x;p)}{\partial p^n} - \frac{1}{(n-1)!}\frac{\partial^{n-1} \phi(x;p)}{\partial p^{n-1}}\right]\bigg|_{p=0}
$$
$$
= \hbar\mathcal{H}(x)\frac{1}{(n-1)!}\frac{\partial^{n-1} \mathcal{N}[\phi(x;p)]}{\partial p^{n-1}}\bigg|_{p=0}.
\tag{3.89}
$$

Using this equation and the functions $y_n(x)$ defined in (3.82), we obtain the relation

$$
\mathcal{L}[y_n(x) - y_{n-1}(x)] = \hbar\mathcal{H}(x)\frac{1}{(n-1)!}\frac{\partial^{n-1} \mathcal{N}[\phi(x;p)]}{\partial p^{n-1}}\bigg|_{p=0},
\tag{3.90}
$$

which is valid for all $n \geq 2$.

Now, we define an additional function

$$
\chi(n) \equiv \begin{cases} 0, & n \leq 1, \\ 1, & \text{else.} \end{cases}
\tag{3.91}
$$

Then, the Eqs. (3.85) and (3.90) can be written in the form

$$
\mathcal{L}[y_n(x) - \chi(n)y_{n-1}(x)] = \hbar\mathcal{H}(x)\frac{1}{(n-1)!}\frac{\partial^{n-1} \mathcal{N}[\phi(x;p)]}{\partial p^{n-1}}\bigg|_{p=0}.
\tag{3.92}
$$

The *linear* ODE (3.92) is valid for all $n \geq 1$. It is called the *nth-order deformation equation*.

Often the abbreviation

$$R_n(y_0(x), \ldots, y_{n-1}(x)) \equiv \frac{1}{(n-1)!} \left. \frac{\partial^{n-1} \mathcal{N}[\phi(x; p)]}{\partial p^{n-1}} \right|_{p=0} \tag{3.93}$$

is used to write the Eq. (3.92) in the short form

$$\mathcal{L}[y_n(x) - \chi(n)y_{n-1}(x)] = \hbar \mathcal{H}(x) R_n(y_0(x), \ldots, y_{n-1}(x)). \tag{3.94}$$

Obviously, the terms $y_n(x)$ can be obtained in order of increasing n by solving the linear deformation equations in succession.

Since the right-hand side of (3.94) depends only upon the known results $y_0(x)$, $y_1(x), \ldots, y_{n-1}(x)$, it can be obtained easily using computer algebra software, such as Maple, MathLab, and/or Mathematica.

The solution of the nth-order deformation equation (3.94) can be written as

$$y_n(x) = y_n^h(x) + y_n^c(x), \tag{3.95}$$

where $y_n^h(x)$ is the general solution of the linear homogeneous ODE

$$\mathcal{L}[y_n^h(x)] = 0, \tag{3.96}$$

and $y_n^c(x)$ is a particular solution of (3.94).

It is also possible to express $y_n^c(x)$ as

$$y_n^c(x) = \chi(n) y_{n-1}(x) + \mathcal{L}^{-1}\left(\hbar \mathcal{H}(x) R_n(y_0(x), \ldots, y_{n-1}(x))\right), \tag{3.97}$$

where $\mathcal{L}^{-1}$ is the inverse of the linear operator $\mathcal{L}$.

Finally, we define the nth partial sum of the terms $y_n(x)$ as

$$y^{[n]}(x) \equiv \sum_{k=0}^{n} y_k(x). \tag{3.98}$$

The solution of the Eq. (3.76) can be written as

$$y(x) = \phi(x; 1) = \sum_{k=0}^{\infty} y_k(x) = \lim_{n \to \infty} y^{[n]}(x). \tag{3.99}$$

This solution will be valid wherever the series converges.

For a better understanding of this method, let us consider a simple example.

Example 3.15 Solve the following IVP of a first-order nonlinear ODE by the HAM

$$y'(x) = \sqrt{1 - y(x)^2}, \quad y(0) = 0. \tag{3.100}$$

Solution. It is not difficult to show that the exact solution is $y(x) = \sin(x)$.

As an initial guess that satisfies the initial condition, we use $y_0(x) = 0$. Moreover, we set $\mathcal{L}[y(x)] \equiv y'(x)$ and $\mathcal{H}(x) \equiv 1$. Thus, (3.94) reads

$$(y_n(x) - \chi(n)\, y_{n-1}(x))' = \hbar \mathcal{H}(x)\, R_n(y_0(x), y_1(x), \ldots, y_{n-1}(x)).$$

Here, we have $y_n^h(x) = c_n$, where c_n is a constant, and

$$y_n^c(x) = \chi(n)\, y_{n-1}(x) + \hbar \int_0^x \mathcal{H}(t)\, R_n\big(y_0(t), \ldots, y_{n-1}(t)\big)\, dt. \tag{3.101}$$

Thus, $y_n(x) = c_n + y_n^c(x)$. It is

$$\mathcal{N}[y(x)] = y'(x) - \sqrt{1 - y(x)^2},$$

and

$$\mathcal{N}[\phi(x; p)] = y_0'(x) + \sum_{n=1}^{\infty} y_n'(x) p^n - \sqrt{1 - \left(y_0(x) + \sum_{n=1}^{\infty} y_n(x) p^n\right)^2}.$$

Looking at (3.93), we calculate

$$R_1(y_0(x)) = y_0'(x) - \sqrt{1 - y_0(x)^2} = -1.$$

Using formula (3.101), we obtain

$$y_1^c(x) = -\hbar \int_0^x dt = -\hbar x,$$

and $y_1(x) = y_1^h(x) + y_1^c(x) = c_1 - \hbar x$. Since $y_1(x)$ must satisfy the given initial condition, it follows $c_1 = 0$. Thus

$$y_1(x) = -\hbar x. \tag{3.102}$$

In the next step, we determine

$$R_2(y_0(x), y_1(x)) = y_1'(x) + \frac{y_0(x) y_1(x)}{\sqrt{1 - y_0(x)^2}} = -\hbar$$

and

$$y_2^c(x) = -\hbar x + \hbar \int_0^x (-\hbar) dt = -\hbar(1 + \hbar)x.$$

Thus, $y_2(x) = c_2 - \hbar(1 + \hbar)x$. Since the given initial condition yields $c_2 = 0$, we obtain

$$y_2(x) = -\hbar(1 + \hbar)x. \tag{3.103}$$

Finally, we present the result of the third step. We have

$$R_3(y_0(x), y_1(x), y_2(x)) = y_2'(x) + \frac{y_0(x)^2 y_1(x)^2}{\sqrt{(1 - y_0(x)^2)^3}}$$

$$+ \frac{y_1(x)^2}{2\sqrt{1 - y_0(x)^2}} + \frac{y_0(x) y_2(x)}{\sqrt{1 - y_0(x)^2}}$$

$$= -\hbar(1 + \hbar) + \frac{\hbar^2}{2} x^2,$$

and

$$y_3^c(x) = -\hbar(1 + \hbar)x + \hbar \int_0^x \left(-\hbar(1 + \hbar) + \frac{\hbar^2}{2} t^2 \right) dt$$

$$= -\hbar(1 + \hbar)^2 x + \frac{\hbar^3}{3!} x^3.$$

As before, we have $y_3(x) = c_3 - \hbar(1 + \hbar)^2 x + \frac{\hbar^3}{3!} x^3$. Since $y_3(x)$ must satisfy the given initial condition, it follows $c_3 = 0$. Thus,

$$y_3(x) = -\hbar(1 + \hbar)^2 x + \frac{\hbar^3}{3!} x^3. \tag{3.104}$$

If we set $\hbar = -1$, it can be shown that

$$y_{2n}(x) = 0, \qquad\qquad n \geq 0,$$
$$y_{2n-1}(x) = \frac{(-1)^{n-1}}{(2n-1)!} x^{2n-1}, \quad n \geq 1.$$

Substituting this result into (3.99), we get

$$y(x) = \frac{x}{1!} - \frac{x^3}{3!} + \frac{x^5}{5!} - \cdots$$

Obviously, the HAM generates the well-known Taylor series of the exact solution $y(x) = \sin(x)$ of the IVP (3.83). $\qquad\qquad\square$

As we have seen, the main aim of the HAM is to produce solutions that will converge in a much larger region than the solutions obtained with the traditional perturbation methods. The solutions obtained by this method depend on the choice of the linear operator $\mathcal{L}$, the auxiliary function $\mathcal{H}(x)$, the initial approximation $y_0(x)$, and the value of the auxiliary parameter $\hbar$. By varying these parameters, it is possible to adjust the region in which the series is convergent and the rate at which the series converges.

One of the important factors that influences the convergence of the solution series is the type of base functions used to express the solution. For example, we might try to express the solution as a polynomial or as a sum of exponential functions. It can be expected that base functions that more closely mimic the qualitative behavior of the actual solution should provide much better results than base functions whose behavior differs greatly from the qualitative behaviour of the actual solution. The choice of the linear operator $\mathcal{L}$, the auxiliary function $\mathcal{H}(x)$, and the initial approximation $y_0(x)$ often determines the base functions present in a solution.

Having selected $\mathcal{L}$, $\mathcal{H}(x)$ and $y_0(x)$, the deformation equations can be solved and a solution series can be determined. The solution obtained in this way will still contain the auxiliary parameter $\hbar$. But this solution should be valid for a range of values of $\hbar$. To determine the optimum value of $\hbar$, the so-called $\hbar$-curves are often plotted. These curves are obtained by plotting the partial sums $y^{[n]}(x)$ and/or their first few derivatives evaluated at a specific value $x = \bar{x}$ versus the parameter $\hbar$. As long as the given IVP or BVP has a unique (isolated) solution, the partial sums and their derivatives will converge to the exact solution for all values of $\hbar$ for which the approximate solution converges. This means that the $\hbar$-curves will be essentially horizontal over the range of $\hbar$ for which the solution converges. When $\hbar$ is chosen in this horizontal region, the approximate solution must converge to the exact solution of the given problem (see [40]). However, the drawback of this strategy is that the value $x = \bar{x}$ and/or the order of the derivative are not known a priori.

A more mathematically sophisticated strategy to determine the optimum value of $\hbar$ (or an appropriate $\hbar$-interval) is to substitute the analytical approximation

$$y^{[n]}(x; \hbar) \equiv \sum_{k=0}^{n} y_k(x)$$

into the left-hand side of (3.76). Obviously, this expression vanishes only for the exact solution. Now, the idea is to consider the norm of the residuum

$$F(\hbar) \equiv \sqrt{\int_{0}^{\infty} \left(\mathcal{N}\left(y^{[n]}(x; \hbar) \right) \right)^2 dx} \qquad (3.105)$$

and to compute the optimum value $\hbar_{\text{opt}}$ by solving the problem:

$$F(\hbar) \rightarrow \min. \qquad (3.106)$$

In our studies (see Examples 3.16 and 3.17) we have used the symbolic tool of the MATLAB to determine the optimum value $\hbar_{opt}$.

Note, the mentioned strategies for the determination of the auxiliary parameter $\hbar$ can only be used if the given problem is not ill-conditioned (see e.g. Example 3.17). This means that small errors in the problem data produce only small errors in the corresponding exact solution.

We have seen that the HAM has a high degree of freedom and a great flexibility to choose the equation-type of the higher-order deformation equation and the base functions of its solution. Now, we will consider the question how a linear operator, an auxiliary function, and an initial approximation can be chosen. Liao [77] gives several rules in order to ensure that these parameters are determined appropriately:

- *Rule of Solution Expression*
  The first step in calculating a HAM solution of a given problem is to select a set of base functions $\{e_n(x), n = 0, 1, 2, \ldots\}$ with which the solution of (3.76) should be expressed. The most suitable base functions can often be chosen by determining what types of functions most easily satisfy the initial or boundary conditions of a problem and by considering the physical interpretation and expected asymptotic behavior of the solution. Then, the solution can be represented in the form

$$y(x) = \sum_{n=0}^{\infty} c_n e_n(x),$$

  where c_n are real constants. From this the rule of solution expression is derived, which requires that every term in the solution is expressible in the form

$$y_n(x) = \sum_{n=0}^{M} a_n e_n(x),$$

  where the a_n are real constants, and M is a nonnegative integer depending on n. This rule together with the given initial or boundary conditions will guide the choice of the initial approximation $y_0(x)$.
  The rule of solution expression also guides the choice of the linear operator $\mathcal{L}$. From Eq. (3.95) we see that $y_n^h(x)$ should also obey the rule of solution expression. Then, from (3.96) we conclude that the linear operator $\mathcal{L}$ should be chosen such that the solution of the homogenous equation

$$\mathcal{L}[f(x)] = 0 \tag{3.107}$$

  is expressible as a linear combination of the base functions $\{e_n(x)\}$.

From (3.95) and the rule of solution expression, it is clear that the particular solutions $y_n^c(x)$ of the higher-order deformation equation must also be expressible as linear combinations of the chosen base functions. Since, according to (3.97), each $y_n^c(x)$ depends on the form of the auxiliary function $\mathcal{H}(x)$, the rule of solution expression restricts the choice of the auxiliary functions to those functions that will produce terms $y_n^c(x)$ having the desired form.

- *Rule of Coefficient Ergodicity*
  This rule guarantees the completeness of the solution. It says that each base $e_n(x)$ in the set of base functions should be present and modifiable in the series $\sum_{k=0}^m y_k(x)$ for $m \to \infty$. This rule further restricts the choice of the auxiliary function $\mathcal{H}(x)$ and, when combined with the rule of solution expression, sometimes uniquely determines the function $\mathcal{H}(x)$.

- *Rule of Solution Existence*
  This third rule requires that $\mathcal{L}$, $\mathcal{H}(x)$ and $y_0(x)$ should be chosen such that each one of the deformation equations (3.94) can be solved, preferably analytically.

In addition to ensuring that the above three rules are observed, the operator $\mathcal{L}$ should be chosen such that the solution of the higher-order deformation equation is not too difficult. In our experiments, we have seen that the following operators can be a good choice:

- $\mathcal{L}[y(x)] = y'(x),$
- $\mathcal{L}[y(x)] = y'(x) \pm y(x),$
- $\mathcal{L}[y(x)] = xy'(x) \pm y(x),$
- $\mathcal{L}[y(x)] = y''(x),$
- $\mathcal{L}[y(x)] = y''(x) \pm y(x).$

Now, let us come back to our Example 3.15 and apply the mentioned rules to the IVP (3.100). The given initial condition $y(0) = 0$ implies that the choice $y_0(x) = 0$ is appropriate. For simplicity, we express the solution by the following set of base functions

$$\{e_n(x) = x^n, n = 0, 1, 2, \ldots\}. \tag{3.108}$$

Thus, the solution can be represented in the form

$$y(x) = \sum_{n=0}^{\infty} c_n x^n. \tag{3.109}$$

Furthermore, it holds

$$y_n(x) = \sum_{n=0}^{M} a_n x^n. \tag{3.110}$$

According to the requirement (3.107), the choice $\mathcal{L}[y(x)] = y'(x)$ is suitable. This yields $y_n^h(x) = 0$. A look at formula (3.97) shows that $\mathcal{H}(x)$ must be of the form $\mathcal{H}(x) = x^k$, where k is a fixed integer. Therefore, this formula reads

$$y_n^c(x) = \chi_n \, y_{n-1}(x) + \hbar \int_0^x t^k R_n\big(y_0(t), \ldots, y_{n-1}(t)\big) \, dt.$$

It is obvious that, when $k \leq -1$, the term x^{-1} appears in the solution expression of $y_n(x)$, which is in contradiction with the rule of solution expression. If $k \geq 1$, the term x always disappears in the solution expression of the higher-order deformation equation. Thus, the coefficient of this term cannot be modified, even if the order of approximation tends to infinity. Taking into account the rule 1 and rule 2, we have to set $k = 0$, which uniquely determines the auxiliary function as $\mathcal{H}(x) = 1$. Until now, we still have freedom to choose a proper value of $\hbar$. Since the exact solution is known, we can deduce that $\hbar = -1$. However, if the exact solution is unknown, the strategy (3.105), (3.106) can be applied.

Example 3.16 Solve the following BVP of a second-order nonlinear ODE on the infinite interval $[0, \infty]$ by the HAM

$$y''(x) = -2y(x)y'(x), \quad y(0) = 0, \quad \lim_{x \to +\infty} y(x) = 1. \tag{3.111}$$

Solution. Here, we have $\mathcal{N}[y(x)] = y''(x) + 2y(x)y'(x)$. Since $y(+\infty)$ is finite, we express $y(x)$ by the set of base functions $\{e^{-nx}, \quad n = 1, 2, 3, \ldots\}$. As an initial guess that satisfies the given boundary conditions, we use $y_0(x) = 1 - e^{-x}$. Moreover, we set

$$\mathcal{L}[y(x)] \equiv y''(x) - y(x).$$

The solution of the homogeneous problem (3.96) is

$$y_n^h(x) = c_{n,1} e^x + c_{n,2} e^{-x}, \quad n \geq 1, \tag{3.112}$$

where the real constants $c_{n,1}$ and $c_{n,2}$ are determined by appropriate boundary conditions.

It holds

$$e^x \int e^{-2x} \int e^x \mathcal{L}[y(x)] dx \, dx$$

$$= e^x \int e^{-2x} \int e^x (y''(x) - y(x)) dx \, dx$$

$$= e^x \int e^{-2x} \left(e^x y'(x) - \int e^x y'(x) dx - \int e^x y(x) dx \right) dx$$

$$= e^x \left[\int e^{-x} y'(x) dx - \int e^{-2x} \left(\int e^x y'(x) dx + \int e^x y(x) dx \right) dx \right]$$

$$= e^x \left[\int e^{-x} y'(x) dx - \int e^{-2x} \left(\int e^x y'(x) dx + e^x y(x) \right. \right.$$

$$\left. \left. - \int e^x y'(x) dx \right) dx \right]$$

$$= e^x \left[\int e^{-x} y'(x) dx - \int e^{-x} y(x) dx \right]$$

$$= e^x \left[\int e^{-x} y'(x) dx + e^{-x} y(x) - \int e^{-x} y'(x) dx \right] = y(x).$$

Therefore, the inverse operator is

$$\mathcal{L}^{-1}[y(x)] = e^x \int e^{-2x} \int e^x y(x) dx \, dx.$$

Now, formula (3.97) can be applied, and we obtain

$$y_n^c(x) = \chi(n) y_{n-1}(x)$$
$$+ \hbar e^x \int_0^x e^{-2t} \int_0^t e^u \, \mathcal{H}(u) \, R_n(y_0(u), \dots, y_{n-1}(u)) du \, dt.$$

According to the rule of solution expression, the auxiliary function $\mathcal{H}(x)$ should be of the form $\mathcal{H}(x) = e^{-kx}$, where k is an integer. Setting $k = 1$, we get $\mathcal{H}(x) = e^{-x}$. Hence,

$$y_n^c(x) = \chi(n) y_{n-1}(x)$$
$$+ \hbar e^x \int_0^x e^{-2t} \int_0^t R_n(y_0(u), \dots, y_{n-1}(u)) du \, dt. \tag{3.113}$$

In order to compute $R_n(y_0(x), \dots, y_{n-1}(x))$, we have to substitute (3.83) into $\mathcal{N}[\phi(x; p)] = \phi''(x; p) + 2\phi(x; p)\phi'(x; p)$. We obtain

$$\mathcal{N}[\phi(x; p)] = -e^{-x} + \sum_{n=1}^{\infty} y_n''(x) p^n$$
$$+ 2\left(1 - e^{-x} + \sum_{n=1}^{\infty} y_n(x) p^n\right)\left(e^{-x} + \sum_{n=1}^{\infty} y_n'(x) p^n\right).$$

Now, the functions R_n are defined by (3.93). We compute

$$R_1(y_0(x)) = \mathcal{N}[\phi(x; p)]|_{p=0} = -e^{-x} + 2(1 - e^{-x})e^{-x} = e^{-x} - 2e^{-2x}.$$

Using formulas (3.112) and (3.113), we obtain

$$y_1(x) = y_1^h(x) + y_1^c(x)$$

$$= c_{1,1}e^x + c_{1,2}e^{-x} + \hbar e^x \int_0^x e^{-2t} \int_0^t \left(e^{-u} - 2e^{-2u}\right) du\, dt$$

$$= c_{1,1}e^x + c_{1,2}e^{-x} + \hbar e^x \int_0^x e^{-2t} \left(e^{-2t} - e^{-t}\right) dt$$

$$= c_{1,1}e^x + c_{1,2}e^{-x} + \hbar e^x \int_0^x \left(e^{-4t} - e^{-3t}\right) dt$$

$$= c_{1,1}e^x + c_{1,2}e^{-x} + \hbar e^x \left(\frac{1}{3}e^{-3x} - \frac{1}{4}e^{-4x} - \frac{1}{12}\right)$$

$$= c_{1,1}e^x + c_{1,2}e^{-x} + \hbar \left(-\frac{1}{4}e^{-3x} + \frac{1}{3}e^{-2x} - \frac{1}{12}e^x\right).$$

Since $y_0(x)$ satisfies the boundary conditions given in (3.111), in particular the second condition, the next iterates $y_n(x)$, $n \geq 1$, have to fulfill the boundary conditions

$$y_n(0) = 0, \quad \lim_{x \to +\infty} y_n(x) = 0. \tag{3.114}$$

By these boundary conditions the constants $c_{1,1}$ and $c_{1,2}$ are determined as $c_{1,1} = \hbar/12$ and $c_{1,2} = -\hbar/12$. Thus,

$$y_1(x) = \hbar \left(-\frac{1}{4}e^{-3x} + \frac{1}{3}e^{-2x} - \frac{1}{12}e^{-x}\right). \tag{3.115}$$

In the next step, we determine

$$R_2(y_0(x), y_1(x)) = \left.\frac{\partial \mathcal{N}[\phi(x; p)]}{\partial p}\right|_{p=0}$$

$$= y_1''(x) + 2e^{-x}y_1(x) + 2(1 - e^{-x})y_1'(x)$$

$$= \hbar \left(-\frac{9}{4}e^{-3x} + \frac{4}{3}e^{-2x} - \frac{1}{12}e^{-x}\right)$$

$$+ 2\hbar \left(-\frac{1}{4}e^{-4x} + \frac{1}{3}e^{-3x} - \frac{1}{12}e^{-2x}\right)$$

$$+ 2\hbar \left(\frac{3}{4}e^{-3x} - \frac{2}{3}e^{-2x} + \frac{1}{12}e^{-x}\right)$$

$$- 2\hbar \left(\frac{3}{4}e^{-4x} - \frac{2}{3}e^{-3x} + \frac{1}{12}e^{-2x}\right)$$

$$= \hbar \left(-2e^{-4x} + \frac{5}{4}e^{-3x} - \frac{1}{3}e^{-2x} + \frac{1}{12}e^{-x}\right),$$

and

$$y_2(x) = y_2^h(x) + y_2^c(x)$$

$$= c_{2,1}e^x + c_{2,2}e^{-x} + \hbar \left(-\frac{1}{4}e^{-3x} + \frac{1}{3}e^{-2x} - \frac{1}{12}e^{-x} \right)$$

$$+ \hbar^2 e^x \int_0^x e^{-2t} \int_0^t \left(-2e^{-4u} + \frac{5}{4}e^{-3u} - \frac{1}{3}e^{-2u} + \frac{1}{12}e^{-u} \right) du \, dt$$

$$= c_{2,1}e^x + c_{2,2}e^{-x} + \hbar \left(-\frac{1}{4}e^{-3x} + \frac{1}{3}e^{-2x} - \frac{1}{12}e^{-x} \right)$$

$$+ \hbar^2 e^x \int_0^x e^{-2t} \left(\frac{1}{2}e^{-4t} - \frac{5}{12}e^{-3t} + \frac{1}{6}e^{-2t} - \frac{1}{12}e^{-t} - \frac{1}{6} \right) dt$$

$$= c_{2,1}e^x + c_{2,2}e^{-x} + \hbar \left(-\frac{1}{4}e^{-3x} + \frac{1}{3}e^{-2x} - \frac{1}{12}e^{-x} \right)$$

$$+ \hbar^2 e^x \int_0^x \left(\frac{1}{2}e^{-6t} - \frac{5}{12}e^{-5t} + \frac{1}{6}e^{-4t} - \frac{1}{12}e^{-3t} - \frac{1}{6}e^{-2t} \right) dt$$

$$= c_{2,1}e^x + c_{2,2}e^{-x} + \hbar \left(-\frac{1}{4}e^{-3x} + \frac{1}{3}e^{-2x} - \frac{1}{12}e^{-x} \right)$$

$$+ \hbar^2 e^x \left(-\frac{1}{12}e^{-6x} + \frac{1}{12}e^{-5x} - \frac{1}{24}e^{-4x} + \frac{1}{36}e^{-3x} \right.$$

$$\left. + \frac{1}{12}e^{-2x} - \frac{5}{72} \right)$$

$$= c_{2,1}e^x + c_{2,2}e^{-x} + \hbar \left(-\frac{1}{4}e^{-3x} + \frac{1}{3}e^{-2x} - \frac{1}{12}e^{-x} \right)$$

$$+ \frac{\hbar^2}{12} \left(-e^{-5x} + e^{-4x} - \frac{1}{2}e^{-3x} + \frac{1}{3}e^{-2x} + e^{-x} - \frac{5}{6}e^x \right).$$

The function $y_2(x)$ satisfies the boundary conditions (3.114) if we set

$$c_{2,1} = \frac{5}{72}\hbar^2, \quad c_{2,2} = -\frac{5}{72}\hbar^2.$$

Then,

$$y_2(x) = \hbar \left(-\frac{1}{4}e^{-3x} + \frac{1}{3}e^{-2x} - \frac{1}{12}e^{-x} \right)$$

$$+ \frac{\hbar^2}{12} \left(-e^{-5x} + e^{-4x} - \frac{1}{2}e^{-3x} + \frac{1}{3}e^{-2x} + \frac{1}{6}e^{-x} \right).$$

(3.116)

Substituting the results into (3.99), we obtain the following approximation

$$
\begin{aligned}
y^{[2]}(x;\hbar) =\ &1 - e^{-x} + 2\hbar\left(-\frac{1}{4}e^{-3x} + \frac{1}{3}e^{-2x} - \frac{1}{12}e^{-x}\right) \\
&+ \frac{\hbar^2}{12}\left(-e^{-5x} + e^{-4x} - \frac{1}{2}e^{-3x} + \frac{1}{3}e^{-2x} + \frac{1}{6}e^{-x}\right).
\end{aligned}
\tag{3.117}
$$

Note, to indicate that the approximate solution $y^{[2]}$ still depends on the convergence-control parameter $\hbar$, we have used it as a second argument.

It can be easily shown that the exact solution of the BVP (3.112) is the function $y(x) = \tanh(x)$.

The next task is to determine values of the parameter $\hbar$ for which the series (3.99) converges to the exact solution. Here, we have used the strategy (3.105), (3.106). For a better understanding we have determined additional terms $y_3(x), \ldots, y_{20}(x)$ using the symbolic tool of the MATLAB. Then we have solved the optimization problem without constraints (3.106) for the nth partial sum of the terms $y_n(x), n = 1, \ldots, 20$. In the Fig. 3.10, for each $y^{[n]}$ the function $F(\hbar)$ is plotted versus $\hbar$. It can be easily seen that a region for appropriate values of $\hbar$ is $[-2, 0)$. Moreover, in Table 3.10 we present the corresponding optimum values $\hbar_{\text{opt}}^{(n)}, n = 1, \ldots, 20$.

To determine an approximation of $\hbar_{\text{opt}}^{(n)}$ if n tends to ∞, we have computed a fourth-degree polynomial, which fits the points $n = 3, \ldots, 20$ given in Table 3.10; see Fig. 3.11. Then, we have extrapolated to $F(\hbar) = 0$. The resulting value is $\hbar^* = -1.835$.

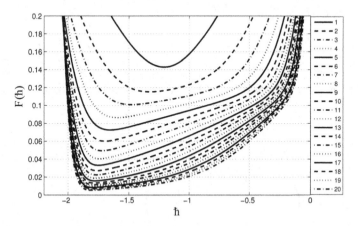

Fig. 3.10 $F(\hbar)$ versus $\hbar$ for $y^{[n]}, n = 1, \ldots, 20$

Table 3.10 Values of $\hbar_{opt}^{(n)}$ and $F(\hbar_{opt}^{(n)})$ if the number n of terms in (3.98) is increased

n	$\hbar_{opt}^{(n)}$	$F(\hbar_{opt}^{(n)})$	n	$\hbar_{opt}^{(n)}$	$F(\hbar_{opt}^{(n)})$
1	-1.207	1.4e-1	11	-1.765	2.3e-2
2	-1.325	1.2e-1	12	-1.773	1.9e-2
3	-1.480	1.0e-1	13	-1.776	1.6e-2
4	-1.592	8.6e-2	14	-1.783	1.3e-2
5	-1.653	7.3e-2	15	-1.786	1.1e-2
6	-1.695	6.0e-2	16	-1.792	9.8e-3
7	-1.719	4.9e-2	17	-1.796	8.5e-3
8	-1.738	4.0e-2	18	-1.801	7.3e-3
9	-1.749	3.3e-2	19	-1.804	6.3e-3
10	-1.760	2.7e-2	20	-1.808	5.4e-3

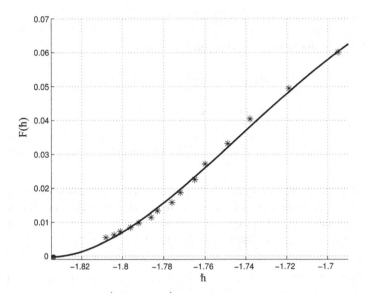

Fig. 3.11 $*$: from *right* to *left*: $\left(\hbar_{opt}^{(n)}, F(\hbar_{opt}^{(n)})\right)$, $n = 3, \ldots, 20$; $\circ$: extrapolated point ($\hbar^*$ such that $F(\hbar^*) = 0$)

3.4.2 Application of the Homotopy Analysis Method

In this section we apply the Homotopy Analysis Method to two examples, which are often discussed in related papers and books.

Example 3.17 Once again, let us consider Bratu's problem (see Examples 1.25 and 2.4):

$$y''(x) = -\lambda\, e^{y(x)}, \quad y(0) = y(1) = 0. \tag{3.118}$$

Find an approximate solution of this equation and compare the result with the data presented in Example 2.4.

Solution. Here, we have $\mathcal{N}[y(x)] = y''(x) + \lambda\, e^{y(x)}$. According to the ODE and the boundary conditions, we express the solution by the following set of base functions

$$\{e_n(x) = x^n, n = 0, 1, 2, \ldots\}.$$

Thus, the solution can be represented in the form

$$y(x) = \sum_{n=0}^{\infty} a_n\, x^n.$$

We start with the function $y_0(x) \equiv 0$, which satisfies the boundary conditions. Moreover, we set

$$\mathcal{L}[y(x)] \equiv y''(x).$$

The solution of the homogeneous problem (3.96) is

$$y_n^h(x) \equiv c_{n,1} + c_{n,2}x, \quad n \geq 1, \tag{3.119}$$

Now, formula (3.97) can be applied, and we obtain

$$y_n^c(x) = \chi(n)y_{n-1}(x) + \hbar \int_0^x \int_0^t \mathcal{H}(u)\, R_n(y_0(u), \ldots, y_{n-1}(u))du\, dt.$$

The auxiliary function $\mathcal{H}(x)$ should be of the form $\mathcal{H}(x) = e^{kx}$, where k is an integer. For $k \neq 0$, exponential terms will appear in the solution expression of $y_n^c(x)$, which disobeys the rule of solution expression. Hence, $k = 0$, which uniquely determines the corresponding auxiliary function as $\mathcal{H}(x) \equiv 1$. Thus,

$$y_n^c(x) = \chi(n)y_{n-1}(x) + \hbar \int_0^x \int_0^t R_n(y_0(u), \ldots, y_{n-1}(u))du\, dt. \tag{3.120}$$

In order to compute $R_n(y_0(x), \ldots, y_{n-1}(x))$, we have to substitute (3.83) into $\mathcal{N}[\phi(x; p)] = \phi(x; p)''(x) + \lambda\, e^{\phi(x;p)}$. We obtain

$$\mathcal{N}[\phi(x; p)] = \sum_{n=0}^{\infty} y_n''(x)p^n + \lambda \exp\left(\sum_{n=0}^{\infty} y_n(x)p^n\right).$$

Now, the functions R_n are defined by (3.93). We compute

$$R_1(y_0(x)) = \mathcal{N}[\phi(x; p)]|_{p=0} = \lambda.$$

Using formulas (3.112) and (3.120), we obtain

$$y_1(x) = y_1^h(x) + y_1^c(x)$$

$$= c_{1,1} + c_{1,2}x + \hbar\lambda \int_0^x \int_0^t du\, dt$$

$$= c_{1,1} + c_{1,2}x + \hbar\lambda \frac{x^2}{2}.$$

Since $y_1(x)$ must satisfy the given boundary conditions, we obtain

$$c_{1,1} = 0, \quad c_{1,2} = -\frac{1}{2}\hbar\lambda.$$

Thus,

$$y_1(x) = -\frac{\hbar\lambda}{2}x + \frac{\hbar\lambda}{2}x^2. \tag{3.121}$$

In the next step, we determine

$$R_2(y_0(x), y_1(x)) = \left.\frac{\partial \mathcal{N}[\phi(x; p)]}{\partial p}\right|_{p=0}$$

$$= y_1''(x) + \lambda\, y_1(x)$$

$$= \hbar\lambda - \frac{\hbar\lambda^2}{2}x + \frac{\hbar\lambda^2}{2}x^2.$$

and

$$y_2(x) = y_2^h(x) + y_2^c(x)$$

$$= c_{2,1} + c_{2,2}x$$

$$\quad + y_1(x) + \hbar^2\lambda \int_0^x \int_0^t du\, dt + \frac{\hbar^2\lambda^2}{2} \int_0^x \int_0^t (u^2 - u)du\, dt$$

$$= c_{2,1} + c_{2,2}x$$

$$\quad - \frac{\hbar\lambda}{2}x + \frac{\hbar\lambda}{2}x^2 + \frac{\hbar^2\lambda}{2}x^2 + \frac{\hbar^2\lambda^2}{2}\left(\frac{x^4}{12} - \frac{x^3}{6}\right)$$

$$= c_{2,1} + c_{2,2}x - \frac{\hbar\lambda}{2}x + \frac{1}{2}(\hbar\lambda + \hbar^2\lambda)x^2 - \frac{\hbar^2\lambda^2}{12}x^3 + \frac{\hbar^2\lambda^2}{24}x^4.$$

Considering the given boundary conditions, we get

$$c_{2,1} = 0, \quad c_{2,2} = \frac{\hbar^2\lambda^2}{24} - \frac{\hbar^2\lambda}{2}.$$

Thus,

$$
\begin{aligned}
y_2(x) = \left(\frac{\hbar^2\lambda^2}{24} - \frac{\hbar^2\lambda}{2} - \frac{\hbar\lambda}{2}\right) x + \frac{1}{2}(\hbar\lambda + \hbar^2\lambda)x^2 \\
- \frac{\hbar^2\lambda^2}{12}x^3 + \frac{\hbar^2\lambda^2}{24}x^4.
\end{aligned}
\tag{3.122}
$$

Now, we determine

$$
R_3(y_0(x), y_1(x), y_2(x)) = \frac{1}{2} \left.\frac{\partial^2 \mathcal{N}[\phi(x; p)]}{\partial p^2}\right|_{p=0}.
$$

This yields

$$
\begin{aligned}
R_3(\cdot) &= y_2''(x) + \lambda\left(y_2(x) + \frac{1}{2}y_1(x)^2\right) \\
&= (\hbar\lambda + \hbar^2\lambda) + \left(\frac{\hbar^2\lambda^3}{24} - \frac{\hbar\lambda^2}{2} - \hbar^2\lambda^2\right) x \\
&\quad + \left(\hbar^2\lambda^2 + \frac{1}{2}\hbar\lambda^2 + \frac{1}{8}\hbar^2\lambda^3\right) x^2 - \frac{1}{3}\hbar^2\lambda^3 x^3 + \frac{1}{6}\hbar^2\lambda^3 x^4,
\end{aligned}
$$

and

$$
\begin{aligned}
y_3(x) &= c_{3,1} + c_{3,2}x \\
&\quad + y_2(x) + \hbar\int_0^x \int_0^t R_3(y_0(u), y_1(u), y_2(u))du\, dt \\
&= c_{3,1} + c_{3,2}x \\
&\quad + \left(\frac{1}{24}\hbar^2\lambda^2 - \frac{1}{2}\hbar^2\lambda - \frac{1}{2}\hbar\lambda\right) x + \frac{1}{2}(\hbar\lambda + \hbar^2\lambda)x^2 \\
&\quad - \frac{1}{12}\hbar^2\lambda^2 x^3 + \frac{1}{24}\hbar^2\lambda^2 x^4 \\
&\quad + \frac{1}{2}\left(\hbar^2\lambda + \hbar^3\lambda\right) x^2 + \left(\frac{1}{144}\hbar^3\lambda^3 - \frac{1}{12}\hbar^2\lambda^2 - \frac{1}{6}\hbar^3\lambda^2\right) x^3 \\
&\quad + \left(\frac{1}{12}\hbar^3\lambda^2 + \frac{1}{24}\hbar^2\lambda^2 + \frac{1}{96}\hbar^3\lambda^3\right) x^4 \\
&\quad - \frac{1}{60}\hbar^3\lambda^3 x^5 + \frac{1}{180}\hbar^3\lambda^3 x^6 \\
&= c_{3,1} + c_{3,2}x \\
&\quad + \left(\frac{1}{24}\hbar^2\lambda^2 - \frac{1}{2}\hbar^2\lambda - \frac{1}{2}\hbar\lambda\right) x + \left(\hbar^2\lambda + \frac{1}{2}\hbar\lambda + \frac{1}{2}\hbar^3\lambda\right) x^2 \\
&\quad + \left(-\frac{1}{6}\hbar^2\lambda^2 + \frac{1}{144}\hbar^3\lambda^3 - \frac{1}{6}\hbar^3\lambda^2\right) x^3
\end{aligned}
$$

$$+ \left(\frac{1}{12}\hbar^2\lambda^2 + \frac{1}{12}\hbar^3\lambda^2 + \frac{1}{96}\hbar^3\lambda^3 \right) x^4$$

$$- \frac{1}{60}\hbar^3\lambda^3 x^5 + \frac{1}{180}\hbar^3\lambda^3 x^6.$$

Since $y_3(x)$ must satisfy the given boundary conditions, we obtain

$$c_{3,1} = 0,$$

$$c_{3,2} = \frac{1}{24}\hbar^2\lambda^2 - \frac{1}{2}\hbar^2\lambda - \frac{1}{2}\hbar^3\lambda - \frac{1}{160}\hbar^3\lambda^3 + \frac{1}{12}\hbar^3\lambda^2.$$

Thus,

$$
\begin{aligned}
y_3(x) = {} & \left(\frac{1}{12}\hbar^2\lambda^2 - \hbar^2\lambda - \frac{1}{2}\hbar\lambda - \frac{1}{2}\hbar^3\lambda - \frac{1}{160}\hbar^3\lambda^3 + \frac{1}{12}\hbar^3\lambda^2 \right) x \\
& + \left(\hbar^2\lambda + \frac{1}{2}\hbar\lambda + \frac{1}{2}\hbar^3\lambda \right) x^2 \\
& + \left(-\frac{1}{6}\hbar^2\lambda^2 + \frac{1}{144}\hbar^3\lambda^3 - \frac{1}{6}\hbar^3\lambda^2 \right) x^3 \\
& + \left(\frac{1}{12}\hbar^2\lambda^2 + \frac{1}{12}\hbar^3\lambda^2 + \frac{1}{96}\hbar^3\lambda^3 \right) x^4 \\
& - \frac{1}{60}\hbar^3\lambda^3 x^5 + \frac{1}{180}\hbar^3\lambda^3 x^6.
\end{aligned}
\tag{3.123}
$$

Now, we substitute our results into the representation (3.98), and obtain

$$y^{[3]}(x; \hbar; \lambda) = y_0(x) + y_1(x) + y_2(x) + y_3(x). \tag{3.124}$$

As before, the next step is to adjust $\hbar$ such that (3.124) is a good approximation of the solution $y(x)$ of the BVP (3.118). In order to obtain a more exact result, we have used the symbolic tool of the MATLAB to compute additional terms $y_4(x), \ldots, y_7(x)$. Then, we have solved the optimization problem without constraints (3.106) for the nth partial sum of the terms $y_n(x)$, $n = 1, \ldots, 7$. In the Fig. 3.12, the results for $\lambda_1 = 1$, $\lambda_2 = 2$, $\lambda_3 = 3$, and $\lambda_4 = 3.513$ are represented. For each $y^{[n]}$, the function $F(\hbar)$ is plotted versus $\hbar$. Moreover, in Table 3.11, for the mentioned values of λ, the corresponding optimum values $\hbar_{\text{opt}}^{(n)}$, $n = 1, \ldots, 7$, are given. Obviously, the optimum values $\hbar_{\text{opt}}^{(n)}$ change with the parameter λ.

In the Tables 3.12 and 3.13, we compare $y^{[3]}(x; \hbar; \lambda)$ with the exact solution of Bratu's problem for $\lambda = 1$ and $\lambda = 2$, respectively. In both cases we have used the optimum $\hbar$, which is given in Table 3.11, i.e., $\hbar_{\text{opt}}^{(\lambda=1)} = -1.11$ and $\hbar_{\text{opt}}^{(\lambda=2)} = -1.25$. It can be seen that for these values the HAM gives similar good results as the VIM presented in Sect. 2.2 (see Example 2.4).

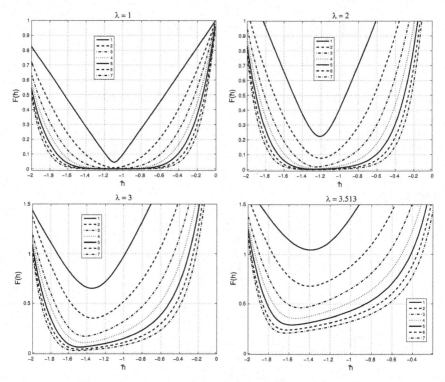

Fig. 3.12 For $y^{[n]}$, $n = 1, \ldots, 7$, and $\lambda = 1, 2, 3, 3.513$: $F(\hbar)$ versus $\hbar$

Table 3.11 For $\lambda = 1, 2, 3, 3.513$: optimum values $\hbar_{opt}^{(n)}$, $n = 1, \ldots, 7$

n	$\lambda_1 = 1$	$\lambda_2 = 2$	$\lambda_3 = 3$	$\lambda_4 = 3.513$
1	-1.090	-1.210	-1.350	-1.390
2	-1.090	-1.200	-1.330	-1.400
3	-1.110	-1.250	-1.420	-1.490
4	-1.110	-1.250	-1.430	-1.540
5	-1.120	-1.280	-1.480	-1.570
6	-1.120	-1.280	-1.490	-1.620
7	-1.130	-1.300	-1.520	-1.630

In Example 5.21, we will show how the solution curve of Bratu's BVP can be computed by appropriate numerical techniques. This curve possesses a simple turning point at $\lambda_0 = 3.51383071912$. Turning points are singular points of the given problem, i.e. problem (3.118) is ill-conditioned in the neighborhood of λ_0. It is therefore to be expected that the HAM fails or produces inaccurate results. This is confirmed by the results presented in Table 3.14, where we have used $\lambda = 3.513$.

Table 3.12 Results computed by the HAM for Bratu's problem with $\lambda = 1$; $\theta = 1.5171645990507543685218444212962$

x	$y^{[3]}(x; -1.11; 1)$	$y(x)$	Relative error
0.1	0.04983221	0.04984679	2.93 e-04
0.2	0.08915608	0.08918993	3.80 e-04
0.3	0.11755708	0.11760910	4.42 e-04
0.4	0.13472542	0.13479025	4.81 e-04
0.5	0.14046976	0.14053921	4.94 e-04
0.6	0.13472542	0.13479025	4.81 e-04
0.7	0.11755708	0.11760910	4.42 e-04
0.8	0.08915608	0.08918993	3.80 e-04
0.9	0.04983221	0.04984679	2.93 e-04

Table 3.13 Results computed by the HAM for Bratu's problem with $\lambda = 2$; $\theta = 2.3575510538774020425939799885899$

x	$y^{[3]}(x; -1.25; 2)$	$y(x)$	Relative error
0.1	0.11382305	0.11441074	5.14 e-03
0.2	0.20514722	0.20641912	6.16 e-03
0.3	0.27198555	0.27387931	6.91 e-03
0.4	0.31276250	0.31508936	7.38 e-03
0.5	0.32647027	0.32895242	7.55 e-03
0.6	0.31276250	0.31508936	7.38 e-03
0.7	0.27198555	0.27387931	6.91 e-03
0.8	0.20514722	0.20641912	6.16 e-03
0.9	0.11382305	0.11441074	5.14 e-03

Table 3.14 Results computed by the HAM for Bratu's problem with $\lambda = 3.513$; $\theta = 4.7374700066634551362743387948882$

x	$y^{[3]}(x; -1.49; 3.513)$	$y(x)$	Relative error
0.1	0.26743526	0.37261471	2.80 e-01
0.2	0.48796610	0.69393980	2.95 e-01
0.3	0.65320219	0.94487744	3.07 e-01
0.4	0.75571630	1.10579685	3.15 e-01
0.5	0.79047843	1.16138892	3.18 e-01
0.6	0.75571630	1.10579685	3.15 e-01
0.7	0.65320219	0.94487744	3.07 e-01
0.8	0.48796610	0.69393980	2.95 e-01
0.9	0.26743526	0.37261471	2.80 e-01

Moreover, the numerically determined solution curve shows: Bratu's BVP has no solution for $\lambda > \lambda_0$, one (non-isolated) solution for $\lambda = \lambda_0$, and two solutions for $\lambda < \lambda_0$. In [4] an analytical shooting method is proposed to determine for a fixed value $\lambda < \lambda_0$ both solutions by the HAM.   □

Example 3.18 In [3] the following IVP is considered, which describes the cooling of a lumped system with variable specific heat

$$\left(1 + \varepsilon y(x)\right)y'(x) + y(x) = 0, \quad y(0) = 1. \tag{3.125}$$

Determine an approximate solution of this IVP.

Solution. Here, we have $\mathcal{N}[y(x)] = \left(1 + \varepsilon y(x)\right)y'(x) + y(x)$. According to the governing ODE and the initial condition (3.125), the solution can be expressed by the following set of base functions

$$\{e_n(x) = e^{-nx}, \quad n = 1, 2, 3 \ldots\},$$

in the form

$$y(x) = \sum_{n=1}^{\infty} d_n e^{-nx}, \tag{3.126}$$

where the real coefficients d_n have to be determined. Now, the rule of solution expression says that the solution of (3.125) must be expressed in the same form as (3.126) and the other expressions such as $x^m e^{-nx}$ must be avoided. We start with the function $y_0(x) = e^{-x}$, which satisfies the initial condition. Moreover, we set

$$\mathcal{L}[y(x)] \equiv y'(x) + y(x).$$

The general solution of the linear homogeneous problem $\mathcal{L}[y(x)] = 0$ is

$$y_n^h(x) = c_n e^{-x}, \tag{3.127}$$

where c_n is a real constant.
    Since

$$e^{-x} \int e^x \mathcal{L}[y(x)] dx = e^{-x} \int e^x \left(y'(x) + y(x)\right) dx = y(x),$$

it holds

$$\mathcal{L}^{-1}[y(x)] = e^{-x} \int e^x y(x) dx.$$

Thus, for $n \geq 1$ the solution of the nth-order deformation equation becomes

$$y_n^c(x) = \chi(n)y_{n-1}(x) + \hbar e^{-x} \int_0^x e^t \mathcal{H}(t) R_n(y_0(t), \ldots, y_{n-1}(t)) dt.$$

According to the rule of solution expression denoted by (3.126) and from (3.94), the auxiliary function $\mathcal{H}(x)$ should be in the form $\mathcal{H}(x) = e^{-ax}$, where a is an integer. It is found that, when $a \leq -1$, the solution of the high-order deformation (3.94) contains the term xe^{-x}, which incidentally disobeys the rule of solution expression. When $a \geq 1$, the base e^{-2x} always disappears in the solution expression of the high-order deformation, so that the coefficient of the term e^{-2x} cannot be modified even if the order of approximation tends to infinity. Thus, we have to set $a = 0$, which uniquely determines the corresponding auxiliary function as $\mathcal{H}(x) \equiv 1$. Hence, we have

$$y_n^c(x) = \chi(n)y_{n-1}(x) + \hbar e^{-x} \int_0^x e^t R_n(y_0(t), \ldots, y_{n-1}(t)) dt. \tag{3.128}$$

In order to compute $R_n(y_0(x), \ldots, y_{n-1}(x))$, we have to substitute (3.83) into $\mathcal{N}[\phi(x; p)] = \big(1 + \varepsilon\phi(x; p)\big)\phi'(x; p) + \phi(x; p)$. We obtain

$$\mathcal{N}[\phi(x; p)] = \left(1 + \varepsilon\left(y_0(x) + \sum_{n=1}^\infty y_n(x)p^n\right)\right)\left(y_0'(x) + \sum_{n=1}^\infty y_n'(x)p^n\right)$$
$$+ y_0(x) + \sum_{n=1}^\infty y_n(x)p^n.$$

Now, the functions R_n are defined by (3.93). We compute

$$R_1(y_0(x)) = \mathcal{N}[\phi(x; p)]|_{p=0} = -\varepsilon e^{-2x}.$$

Using formula (3.128), we obtain

$$y_1^c(x) = -\varepsilon\hbar e^{-x} \int_0^x e^t e^{-2t} dt = -\varepsilon\hbar e^{-x} \int_0^x e^{-t} dt = -\varepsilon\hbar e^{-x} + \varepsilon\hbar e^{-2x}.$$

Using formulas (3.95) and (3.127), it follows

$$y_1(x) = c_1 e^{-x} - \varepsilon\hbar e^{-x} + \varepsilon\hbar e^{-2x}.$$

Since $y_1(x)$ must satisfy the given initial condition, we get $c_1 = 0$, and

$$y_1(x) = -\varepsilon\hbar e^{-x} + \varepsilon\hbar e^{-2x}. \tag{3.129}$$

In the next step, we determine

$$
\begin{aligned}
R_2(y_0(x), y_1(x)) &= \left. \frac{\partial \mathcal{N}[\phi(x; p)]}{\partial p} \right|_{p=0} \\
&= \varepsilon y_1(x) y_0'(x) + (1 + \varepsilon y_0(x)) y_1'(x) + y_1(x) \\
&= \left(-\varepsilon^2 \hbar e^{-x} + \varepsilon^2 \hbar e^{-2x} \right) \left(-e^{-x} \right) \\
&\quad + \left(1 + \varepsilon e^{-x} \right) \left(\varepsilon \hbar e^{-x} - 2 \varepsilon \hbar e^{-2x} \right) + \left(-\varepsilon \hbar e^{-x} + \varepsilon \hbar e^{-2x} \right) \\
&= \left(2 \varepsilon^2 \hbar - \varepsilon \hbar \right) e^{-2x} - 3 \varepsilon^2 \hbar e^{-3x},
\end{aligned}
$$

and

$$
\begin{aligned}
y_2^c(x) &= -\varepsilon \hbar e^{-x} + \varepsilon \hbar e^{-2x} + \hbar e^{-x} \int_0^x \left((2\varepsilon^2 \hbar - \varepsilon \hbar) e^{-t} - 3\varepsilon^2 \hbar e^{-2t} \right) dt \\
&= -\varepsilon \hbar e^{-x} + \varepsilon \hbar e^{-2x} \\
&\quad + (2\varepsilon^2 \hbar^2 - \varepsilon \hbar^2) \left(e^{-x} - e^{-2x} \right) - \frac{3}{2} \varepsilon^2 \hbar^2 \left(e^{-x} - e^{-3x} \right) \\
&= \left(\frac{1}{2} \varepsilon^2 \hbar^2 - \varepsilon \hbar (1 + \hbar) \right) e^{-x} + (\varepsilon \hbar (1 + \hbar) - 2\varepsilon^2 \hbar^2) e^{-2x} + \frac{3}{2} \varepsilon^2 \hbar^2 e^{-3x}.
\end{aligned}
$$

Thus, we obtain

$$
\begin{aligned}
y_2(x) &= c_2 e^{-x} + \left(\frac{1}{2} \varepsilon^2 \hbar^2 - \varepsilon \hbar (1 + \hbar) \right) e^{-x} \\
&\quad + (\varepsilon \hbar (1 + \hbar) - 2\varepsilon^2 \hbar^2) e^{-2x} + \frac{3}{2} \varepsilon^2 \hbar^2 e^{-3x}.
\end{aligned}
$$

As before, the function $y_2(x)$ must satisfy the given initial condition. This implies $c_2 = 0$, and we have

$$
\begin{aligned}
y_2(x) &= \left(\frac{1}{2} \varepsilon^2 \hbar^2 - \varepsilon \hbar (1 + \hbar) \right) e^{-x} + (\varepsilon \hbar (1 + \hbar) - 2\varepsilon^2 \hbar^2) e^{-2x} \\
&\quad + \frac{3}{2} \varepsilon^2 \hbar^2 e^{-3x}.
\end{aligned}
\tag{3.130}
$$

Now, we substitute our results into the representation (3.98), and we obtain the following HAM approximation

$$
y^{[2]}(x; \hbar; \varepsilon) = y_0(x) + y_1(x) + y_2(x),
\tag{3.131}
$$

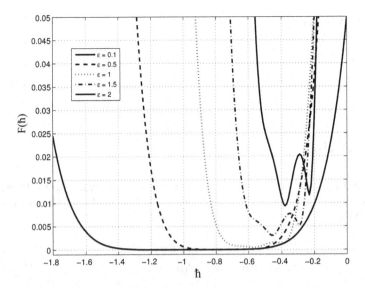

Fig. 3.13 For $y^{[2]}$ and different values of ε: $F(\hbar)$ versus $\hbar$

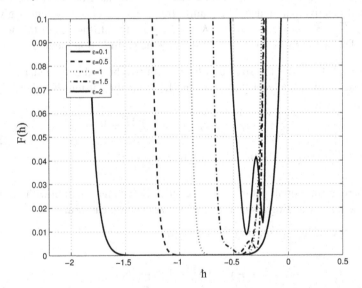

Fig. 3.14 For $y^{[7]}$ and different values of ε: $F(\hbar)$ versus $\hbar$

which depends on the external parameter ε and the auxiliary parameter $\hbar$. In Fig. 3.13, for different values of ε and the three-term approximation $y^{[2]}$, the function $F(\hbar)$ is plotted versus $\hbar$. In Fig. 3.14 the same is shown for $y^{[7]}$, where this approximation has been computed with the symbolic tool of the MATLAB. Finally, in Table 3.15 we present the optimum values $\hbar_{\text{opt}}$ for different values of ε.

Table 3.15 Values of $\hbar_{\text{opt}}^{(2)}$ and $\hbar_{\text{opt}}^{(7)}$ for different values of ε

ε	0.1	0.5	1	1.5	2
$\hbar_{\text{opt}}^{(2)}$	−0.95	−0.78	−0.62	−0.51	−0.43
$\hbar_{\text{opt}}^{(7)}$	−0.96	−0.74	−0.57	−0.46	−0.38

3.5 Exercises

Exercise 3.1 Let us consider the free vibration of a mass m connected to two non-linear springs on a frictionless contact surface, with a small nonlinear perturbation. The IVP for this system is

$$y''(x) + \omega_n^2 y(x) + \varepsilon y(x)^3 = 0,$$
$$y(0) = A, \quad y'(0) = 0.$$

Here, $y(x)$ is the displacement at time x, $y''(x)$ is the acceleration and ε ($\varepsilon \ll 1$) is a small positive parameter. The natural frequency ω_n of the system is given by $\omega_n = \sqrt{K_1 + K_2/m}$ (rad/s), where K_1 and K_2 are the stiffnesses of the nonlinear springs. Discuss the solution of this system by the perturbation method.

Exercise 3.2 Consider the structural beam with two hinged ends under a concentrated load at the middle of the span. The BVP for this system has the following form:

$$y''(x) - \frac{Fx}{2EI} \left(1 + y'(x)^2\right)^{\frac{3}{2}} = 0,$$
$$y(0) = 0, \quad y'(0) = \frac{FL^2}{16EI} \tag{3.132}$$

Note, there is not a small parameter in the ODE. Solve this equation with the perturbation method by adding a small parameter to the nonlinear term.

Exercise 3.3 Solve the BVP (3.132) by setting

$$\left(1 + y'(x)^2\right)^{\frac{3}{2}} \approx 1 + \frac{3}{2} y'(x)^2,$$

and compare your results with those obtained in Exercise 3.2.

Exercise 3.4 Consider the following IVP for a nonlinear oscillator with a discontinuity

$$y''(x) + \omega_n^2 y(x) + \varepsilon y(x)^3 \text{sign}(y(x)) = 0,$$
$$y(0) = A, \quad y'(0) = 0.$$

Discuss the solution of this IVP by the energy balance method.

Exercise 3.5 Consider the motion of a rigid rod rocking back on the circular surface without slipping. The governing IVP of this motion is

$$y''(x) + \frac{3}{4}y(x)^2 y''(x) + \frac{3}{4}y(x)y'(x)^2 + 3\frac{g}{l}y(x)\cos(y(x)) = 0,$$

$$y(0) = A, \quad y'(0) = 0.$$

Here, g and l are positive constants. Discuss the solution of this system by the energy balance method.

Exercise 3.6 Consider the free vibrations of an autonomous conservative oscillator with inertia and static type fifth-order nonlinear ties. The IVP of this dynamical system is

$$y''(x) + \varepsilon_1 y(x)^2 y''(x) + \varepsilon_2 y(x)^4 y''(x) + \varepsilon_1 y(x)y'(x)^2$$
$$+ 2\varepsilon_2 y(x)^3 y'(x)^2 + \lambda y(x) + \varepsilon_3 y(x)^3 + \varepsilon_4 y(x)^5 = 0, \qquad (3.133)$$
$$y(0) = A, \quad y'(0) = 0.$$

Here, $\varepsilon_1, \ldots, \varepsilon_4$ are positive parameters, and λ is an integer which may take the values $-1, 0, 1$. Discuss the solution of this system by the energy balance method.

Exercise 3.7 Consider another nonlinear oscillator with a discontinuity

$$y''(x) + y(x) + y(x)^3 = 0,$$
$$y(0) = A, \quad y'(0) = 0.$$

Discuss the solution of this system by the Hamiltonian approach.

Exercise 3.8 Consider the free vibration of a tapered beam. In dimensionless form, the governing IVP corresponding to the fundamental vibration mode is

$$y''(x) + \varepsilon_1 \left(y(x)^2 y''(x) + y(x)y'(x)^2\right) + y(x) + \varepsilon_2 y(x)^3 = 0,$$
$$y(0) = A, \quad y'(0) = 0, \qquad (3.134)$$

where ε_1 and ε_2 are arbitrary real constants, and $y(x)$ is the displacement at time x. Discuss the solution of this system by the Hamiltonian approach.

Exercise 3.9 Consider the motion of a particle on a rotating parabola. The IVP that describes this motion is

$$y''(x) + 4q^2 y(x)^2 y''(x) + 4q^2 y(x)y'(x)^2 + \Delta y(x) = 0,$$
$$y(0) = A, \quad y'(0) = 0.$$

Here, $y''(x)$ is the acceleration, $y'(x)$ is the velocity, q and Δ are positive constants. Discuss the solution of this IVP by the Hamiltonian approach.

Exercise 3.10 As we know, the vibration of the simple pendulum can be modeled mathematically by the following IVP

$$y''(x) + \omega_n^2 \sin(y(x)) = 0,$$
$$y(0) = A, \quad y'(0) = 0,$$

where $\omega_n = \sqrt{g/l}$ is the natural frequency of the pendulum. Solve this IVP by the homotopy analysis method.

Exercise 3.11 Consider the forced vibration of a mass m connected to a nonlinear spring on a frictionless contact surface. The IVP of this dynamical system is

$$y''(x) + \omega_n^2 y(x) + \varepsilon y(x)^3 = \frac{kb \cos \omega x}{m},$$
$$y(0) = A, \quad y'(0) = 0.$$

In this ODE, ω_n is the natural frequency of the system which is given by $\omega_n = \sqrt{k/m}$ (rad/s), $\varepsilon \ll 1$ is a positive real parameter, and k is the stiffness of the nonlinear spring. Solve this equation by the homotopy analysis method.

Exercise 3.12 Solve the IVP (3.133) given in Example 3.6 by the Hamiltonian approach and compare the results.

Exercise 3.13 Solve the IVP (3.134) given in Example 3.8 by the energy balance method and compare the results.

Exercise 3.14 In the theory of the plane jet, which is studied in the hydrodynamics, we are encountered with the following IVP (see [3])

$$ay'''(x) + y(x)y''(x) + y'(x)^2 = 0,$$
$$y(0) = y''(x) = 0, \quad \lim_{x \to +\infty} y'(x) = 0.$$

Here, $a > 0$ is a real parameter. Discuss the solution of this IVP by the homotopy analysis method.

Exercise 3.15 Consider the first extension of Bratu's problem (see [128])

$$y'(x) + e^{y(x)} + e^{2y(x)} = 0,$$
$$y(0) = y(1) = 0.$$

Solve this BVP by the homotopy analysis method.

Chapter 4
Nonlinear Two-Point Boundary Value Problems

4.1 Introduction

Let us consider, the following system of n first-order ordinary differential equations (ODEs)

$$
\begin{aligned}
y_1'(x) &= f_1(x, y_1(x), y_2(x), \ldots, y_n(x)), \\
y_2'(x) &= f_2(x, y_1(x), y_2(x), \ldots, y_n(x)), \\
&\cdots \\
y_n'(x) &= f_n(x, y_1(x), y_2(x), \ldots, y_n(x)),
\end{aligned}
\tag{4.1}
$$

where $y_i'(x)$ denotes the derivative of the function $y_i(x)$ w.r.t. x, $i = 1, \ldots, n$.

Setting $\mathbf{y}(x) \equiv (y_1(x), \ldots, y_n(x))^T$ and $\mathbf{f}(x, \mathbf{y}) \equiv (f_1(x, \mathbf{y}), \ldots, f_n(x, \mathbf{y}))^T$, this system can be formulated in vector notation as

$$
\mathbf{y}'(x) \equiv \frac{d}{dx} \mathbf{y}(x) = \mathbf{f}(x, \mathbf{y}(x)), \quad x \in [a, b], \quad \mathbf{y}(x) \in \Omega \subset \mathbb{R}^n.
\tag{4.2}
$$

We assume that the function $\mathbf{f} : [a, b] \times \Omega \to \mathbb{R}^n$ is sufficiently smooth.

In the applications higher order ODEs are often encountered. Such a higher order equation

$$
u^{(n)}(x) = G\left(x, u(x), u'(x), u''(x), \ldots, u^{(n-1)}(x)\right)
\tag{4.3}
$$

can be transformed into the system (4.1) by setting

$$
\mathbf{y} = (y_1, y_2, \ldots, y_n)^T \equiv \left(u, u', u'', \ldots, u^{(n-1)}\right)^T.
$$

Obviously, the first-order system reads now

$$
y_1' = y_2, \quad y_2' = y_3, \ldots, y_{n-1}' = y_n, \quad y_n' = G(x, y_1, y_2, \ldots, y_n).
\tag{4.4}
$$

© Springer India 2016
M. Hermann and M. Saravi, *Nonlinear Ordinary Differential Equations*,
DOI 10.1007/978-81-322-2812-7_4

Example 4.1 Given the following ODE, which models the buckling of a thin rod

$$u''(t) + \lambda \sin(u(t)) = 0. \tag{4.5}$$

Transform this second-order ODE into a first-order system.

Solution: We obtain the following system of two first-order ODEs

$$y_1'(x) = y_2(x), \quad y_2'(x) = -\lambda \sin(y_1(t)). \qquad \square$$

In the monograph [63], the ODEs (4.1) have been subjected to initial and linear two-point boundary conditions. Here, we add *nonlinear* two-point boundary conditions

$$
\begin{aligned}
g_1(y_1(a), \ldots, y_n(a), y_1(b), \ldots, y_n(b)) &= 0, \\
g_2(y_1(a), \ldots, y_n(a), y_1(b), \ldots, y_n(b)) &= 0, \\
&\vdots \\
g_n(y_1(a), \ldots, y_n(a), y_1(b), \ldots, y_n(b)) &= 0.
\end{aligned}
\tag{4.6}
$$

to (4.1). The n algebraic equations (4.6) can be written in vector notation in the form

$$g(y(a), y(b)) = 0, \tag{4.7}$$

and we assume that the function $g : \Omega_1 \times \Omega_2 \to \mathbb{R}^n$, $\Omega_1 \subset \mathbb{R}^n$ and $\Omega_1 \subset \mathbb{R}^n$, is sufficiently smooth.

The combination of (4.2) and (4.7), i.e.,

$$
\begin{aligned}
y'(x) &= f(x, y(x)), \quad a \le x \le b, \\
g(y(a), y(b)) &= 0,
\end{aligned}
\tag{4.8}
$$

is called *two-point boundary value problem* (BVP). If components of the solution $y(x)$ are prescribed at more than two points, we have a *multipoint boundary value problem*. Initial value problems (IVPs) are typical for evolution problems, and x represents the time. In contrast, boundary value problems are used to model static problems (e.g., elasto-mechanical problems) since the solution depends on the state of the system in the past (at the point a) as well as on the state of the system in the future (at the point b). In general, the variable x in a BVP does not describe the time, i.e., no dynamics is inherent in the given problem.

In the applications, the function g is often linear. In that case, the boundary conditions (4.7) can be written in the form

$$B_a y(a) + B_b y(b) = \beta, \tag{4.9}$$

where B_a, $B_b \in \mathbb{R}^{n \times n}$ and $\beta \in \mathbb{R}^n$. We speak of a linear BVP, if the ODE and the boundary conditions are linear, i.e.,

$$y'(x) = A(x)\, y(x) + r(x), \quad a \leq x \leq b,$$
$$B_a y(a) + B_b y(b) = \beta,$$

(4.10)

where $A(x)$, B_a, $B_b \in \mathbb{R}^{n \times n}$, $r(x)$, $\beta \in \mathbb{R}^n$.

The proof of the existence and uniqueness of solutions for nonlinear BVPs is extremely difficult and is seldom as successful as for linear BVPs (*see*, e.g., [57]). Therefore, we always assume that the BVP (4.8) has a locally unique (isolated) solution $y(x)$, which has to be approximated by numerical methods.

The most important numerical methods for (4.8) can be subdivided into three classes:

- shooting methods,
- finite difference methods, and
- collocation methods.

In this book, we shall consider only the class of shooting methods, since an important component of these methods are the sophisticated numerical techniques for IVPs. In particular, this means the use of step-size controlled IVP-solvers, which we have described in our text [63].

4.2 Simple Shooting Method

The idea of the simple shooting method is to assign an IVP

$$u'(x) = f(x, u(x)), \quad u(a) = s \in \mathbb{R}^n,$$

(4.11)

to the nonlinear BVP (4.8). Obviously, this IVP contains the still undetermined initial vector s, which can be considered as a free parameter. Therefore, we denote the solution of (4.11) with $u \equiv u(x; s)$.

The essential requirement on the IVP, which brings the IVP in relation to the BVP, is that the IVP possesses the same solution as the BVP. Since in (4.8) and (4.11) the differential equations are identical, the vector s in (4.11) has to be determined such that the corresponding solution trajectory $u(x; s)$ satisfies the boundary conditions in (4.8), too. This means that the unknown n-dimensional vector s must satisfy the n-dimensional system of nonlinear algebraic equations

$$F(s) \equiv g(s, u(b; s)) = 0.$$

(4.12)

The following relations between the solutions of the algebraic system (4.12) and the solutions of the BVP (4.8) can be shown:

- if $s = s^*$ is a root of (4.12), then $y(x) \equiv u(x; s^*)$ is a solution of the BVP (4.8),
- if $y(x)$ is a solution of the BVP (4.8), then $s = y(a)$ is a root of (4.12),
- for each isolated solution of the BVP (4.8), there is a simple root of the algebraic equations (4.12).

Thus, the nonlinear BVP (4.8), which is defined in appropriate (infinitely dimensional) function spaces, is transformed into the finite dimensional nonlinear algebraic problem (4.12)—and this in an exact way, i.e., without any approximation error. This is a significant advantage of the shooting methods, in comparison with the other classes of BVP-solvers mentioned above.

As with all nonlinear problems, the numerical treatment of (4.12) requires knowledge of the approximate position of the root s^* and an associated neighborhood U^* where no other solution exists. For an arbitrary nonlinear BVP this information can rarely be formulated in theory-based statements. Instead, extensive numerical experiments are necessary to find good starting values. In many cases, the information about the practical problem, which is modeled by (4.8), can be helpful. If nothing is known about the given problem, the so-called *trial-and-error* technique is often used.

Numerical tools to globalize the locally convergent methods for nonlinear BVPs are *continuation* and *embedding techniques*, which are also referred to as *homotopy methods*. Here, the BVP (4.8) is embedded into a one-parametric family of BVPs. Then, an attempt is made to increase successively the parameter value until the original BVP (4.8) is reached. The idea of this strategy is to use the approximate solution of the previous problem as starting trajectory for the subsequent problem.

If it is possible to separate a linear term in the ODE and/or the boundary conditions, i.e., if the BVP can be splitted as follows:

$$
\begin{aligned}
y'(x) &= f(x, y(x)) = A(x)\, y(x) + r(x) + f_1(x, y(x)), \\
g(y(a), y(b)) &= B_a y(a) + B_b y(b) - \beta + g_1(y(a), y(b)) = 0,
\end{aligned}
\tag{4.13}
$$

then the nonlinear part is multiplied by the embedding parameter λ:

$$
\begin{aligned}
y'(x) &= A(x)\, y(x) + r(x) + \lambda\, f_1(x, y(x)), \\
B_a y(a) &+ B_b y(b) - \beta + \lambda\, g_1(y(a), y(b)) = 0.
\end{aligned}
\tag{4.14}
$$

Obviously, for $\lambda = 0$ we have a simpler (linear) problem, and for $\lambda = 1$ the original problem. The continuation method, which is often called *killing of the nonlinearity*, starts with $\lambda = 0$, i.e., with the numerical solution of the linear problem. Then, the approximate solution of the linear problem is used as starting trajectory to solve the problem (4.14) for a slightly increased value of λ, say λ_1. If it is not possible to compute a solution of the new problem, λ_1 must be decreased. Otherwise, λ_1 is increased. This process is executed step-by-step until a solution of the original BVP ($\lambda = 1$) is determined.

Another preferred continuation method for BVPs uses the interval length $b - a$ as the embedding parameter λ; see, for example, [69, 102].

Further information on continuation methods for the numerical treatment of systems of nonlinear equations, such as (4.12), can be found in the standard reference works on nonlinear equations [93, 101, 110]. These books also contain the basic algorithms for the numerical treatment of (4.12).

The most common methods for the numerical solution of (4.12) are Newton-like techniques. To ensure the feasibility of these methods and a sufficiently fast convergence of the iterates toward the desired solution s^*, it is usually assumed that the Jacobian of F, i.e., the matrix $M(s^*) \equiv \partial F(s^*)/\partial s$, is nonsingular. Under this hypothesis the solution $y^* \equiv u(x; s^*)$ of the BVP (4.8) is isolated. The presence of non-isolated solutions can lead, inter alia, to the phenomenon of branching solutions. The study of branching solutions is the objective of the bifurcation theory, which we will consider in the next chapter.

Let us now assume that we use the multidimensional Newton method (see, e.g., [58, 93]) to solve (4.12). Starting with a vector s_0, which is sufficiently close to the solution s^*, i.e.,

$$s_0 \in S(s^*) \equiv \{s \in \mathbb{R}^n : \|s - s^*\| \leq \varepsilon\},$$

a sequence of iterates $\{s_k\}_{k=0}^{\infty}$ is computed by the formula

$$s_{k+1} = s_k - c_k, \quad k = 0, 1, \ldots, \tag{4.15}$$

where the increment c_k is determined by the following n-dimensional system of linear equations

$$M_k c_k = q_k. \tag{4.16}$$

For the right-hand side of (4.16), it holds $q_k \equiv F(s_k)$. The system matrix $M_k \equiv M(s_k)$ is the Jacobian $M(s)$ of $F(s)$ at $s = s_k$, i.e., $M_k = \partial F(s_k)/\partial s$.

To simplify the representation, let $g(v, w) = 0$ be the abbreviation of the boundary conditions in (4.8). Using the chain rule, we obtain for the Jacobian $M(s)$ the following expression

$$M(s) = \frac{\partial}{\partial s} g(s, u(b; s)) = \left[\frac{\partial}{\partial v} g(v, w) + \frac{\partial}{\partial w} g(v, w) \frac{\partial}{\partial s} u(b; s) \right]\Bigg|_{\substack{v = s \\ w = u(b; s)}}.$$

Considering the IVP (4.11), the function $\partial u(b; s)/\partial s$ can be computed by the IVP

$$\frac{d}{dx} \frac{\partial}{\partial s} u(x; s) = \frac{\partial}{\partial u} f(x, u)\Bigg|_{u=u(x;s)} \cdot \frac{\partial}{\partial s} u(x; s), \quad \frac{\partial}{\partial s} u(a; s) = I. \tag{4.17}$$

Now, if we introduce the following notation

$$B_{a,k} \equiv B_a(s_k) \equiv \frac{\partial}{\partial \boldsymbol{v}} \boldsymbol{g}(\boldsymbol{v}, \boldsymbol{w}) \Big|_{\substack{s=s_k \\ \boldsymbol{v}=s \\ \boldsymbol{w}=\boldsymbol{u}(b;s)}} \quad,$$

$$B_{b,k} \equiv B_b(s_k) \equiv \frac{\partial}{\partial \boldsymbol{w}} \boldsymbol{g}(\boldsymbol{v}, \boldsymbol{w}) \Big|_{\substack{s=s_k \\ \boldsymbol{v}=s \\ \boldsymbol{w}=\boldsymbol{u}(b;s)}} \quad,$$

$$X(t; s_k) \equiv \frac{\partial}{\partial s} \boldsymbol{u}(x; s) \Big|_{s=s_k}, \quad A(x; s_k) \equiv \frac{\partial}{\partial \boldsymbol{u}} \boldsymbol{f}(x, \boldsymbol{u}) \Big|_{\substack{s=s_k \\ \boldsymbol{u}=\boldsymbol{u}(x;s)}} \quad,$$

then M_k and $\boldsymbol{q}_k$ can be written in the form

$$M_k = B_{a,k} + B_{b,k} X_k^e, \qquad \boldsymbol{q}_k = \boldsymbol{F}(s_k), \tag{4.18}$$

where $X_k^e \equiv X(b; s_k)$. Note, that with our notation the matrix M_k is similarly struc-
tured as the matrix $M = B_a + B_b X^e$ of the standard simple shooting method for
linear BVPs (see [57, 63]).

According to the formula (4.17), the matrix $X(x) \equiv X(x; s_k)$ is the solution of
the matrix IVP

$$\begin{aligned} X'(x) &= A(x; s_k) \, X(x), \quad a \le x \le b, \\ X(a) &= I. \end{aligned} \tag{4.19}$$

Problem (4.19) is also known as the *variational problem*.

In practice, it is often very difficult and time-consuming to calculate analytically the matrices $A(x; s)$. Therefore, in the existing implementations of the simple
shooting method the matrix $X(x; s_k)$ is approximated by forward-difference approximations. For this purpose, $\boldsymbol{u}(x; s_k)$ is determined for $s = s_k$ from the IVP (4.11).
In addition, n further functions $\boldsymbol{u}^{(j)} \equiv \boldsymbol{u}^{(j)}(x; s_k)$, $j = 1, \ldots, n$, are computed from
the slightly modified IVPs

$$\begin{aligned} \frac{d\boldsymbol{u}^{(j)}}{dx}(x) &= \boldsymbol{f}(x, \boldsymbol{u}^{(j)}(x)), \quad a \le x \le b, \\ \boldsymbol{u}^{(j)}(a) &= s_k + h \, \boldsymbol{e}^{(j)}, \quad j = 1, \ldots, n, \end{aligned} \tag{4.20}$$

where $\boldsymbol{e}^{(j)} \equiv (0, \ldots, 1, \ldots, 0)^T$ denotes the jth unit vector in $\mathbb{R}^n$, and $h \ne 0$ is a
small positive number with $h \approx \sqrt{\nu}$. Here, ν is the relative machine accuracy (see,
e.g., [58]). Obviously, if we define

$$\hat{\boldsymbol{u}}^{(j)}(x; s_k) \equiv (\boldsymbol{u}^{(j)}(x; s_k) - \boldsymbol{u}(x; s_k))/h,$$

then

$$\hat{X}(x; s_k) \equiv [\hat{u}^{(1)}(x; s_k), \hat{u}^{(2)}(x; s_k), \dots, \hat{u}^{(n)}(x; s_k)] \qquad (4.21)$$

is a sufficiently accurate difference approximation of $X(x; s_k)$.

With regard to the convergence of the Newton method (4.15), (4.16), the following statement applies.

Theorem 4.2 (THEOREM OF NEWTON AND KONTOROVICH)
Let $F(s)$ be continuously differentiable in the ball

$$S(s_0, \rho) \equiv \{s \in \mathbb{R}^n : \|s - s_0\| \le \rho\}.$$

Assume that the inverse of the Jacobian $M_0 \equiv M(s_0)$ exists and there are three constants $\alpha_1, \alpha_2, \alpha_3 \ge 0$ such that

$$\|M_0^{-1} F(s_0)\| \le \alpha_1, \quad \|M_0^{-1}\| \le \alpha_2,$$
$$\|M_1 - M_2\| \le \alpha_3 \|s_1 - s_2\| \text{ for } s_1, s_2 \in S(s_0, \rho).$$

Let $\sigma \equiv \alpha_1 \alpha_2 \alpha_3$. If

$$\sigma \le 1/2 \text{ and } \rho \ge \rho_- \equiv (1 - \sqrt{1 - 2\sigma})/(\alpha_2 \alpha_3)$$

then it holds

1. *there exists a unique root $s \in S(s_0, \rho_-)$, and*
2. *the sequence of Newton iterates $\{s_k\}_{k=0}^{\infty}$ converges to s^*.*

If $\sigma < 1/2$, then it holds

3. *s^* is the unique solution in $S(s_0, \min\{\rho, \rho_+\})$, where*
   $\rho_+ \equiv (1 + \sqrt{1 - 2\sigma})/(\alpha_2 \alpha_3)$, and
4. *$\|s_k - s^*\| \le (2\sigma)^{2^k}/(\alpha_2 \alpha_3), k = 0, 1, \dots$*

Proof See, e.g., [32]. ■

It can be inferred from this theorem that the condition $\sigma \le 1/2$ will only be satisfied if the starting vector s_0 is located in a sufficiently small neighborhood of the exact solution s^* (this follows from the fact that in this case $\|F(s_0)\|$ is very small). Thus, the Newton method is only a locally convergent method (see also [58]).

Let us assume that the right-hand side f of the differential equation in (4.8) is uniformly Lipschitz continuous on $[a, b] \times \Omega$, i.e., there exists a constant $L \ge 0$ such that

$$\|f(x, y_1) - f(x, y_2)\| \le L \|y_1 - y_2\|$$

for all $(x, y_i) \in [a, b] \times \Omega, i = 1, 2$.

For the next claim we need Gronwall's Lemma.

Lemma 4.3 (GRONWALL'S LEMMA)
*Let $u(x)$ and $v(x)$ be two nonnegative continuous functions on the interval $[a, b]$.
Assume that $C \geq 0$ is a constant and it holds*

$$v(x) \leq C + \int_a^x v(\tau)u(\tau)d\tau, \quad \text{for } a \leq x \leq b.$$

Then

$$v(x) \leq C \exp \int_a^x u(\tau)\, d\tau, \quad \text{for } a \leq x \leq b.$$

If, in particular $C = 0$ then $v(x) \equiv 0$.

Proof See, e.g., [119].   ∎

Using the integral representation of the differential equation (4.11) and Gronwall's Lemma, we obtain

$$\|u(b; s_1) - u(b; s_2)\| \leq \|s_1 - s_2\| + \int_a^b L \|u(x; s_1) - u(x; s_2)\|\, dx$$
$$\leq e^{L(b-a)}\|s_1 - s_2\|.$$

Obviously, the matrix $X(x; s)$ depends similarly sensitive on the initial vector s as $u(x; s)$. Thus, it can be expected that the constant α_3 (see Theorem 4.2) is of the magnitude $e^{L(b-a)}$. In other words, it must be assumed that the radius of convergence satisfies $\rho = O(e^{-L(b-a)})$. This is a serious limitation, if $L(b - a)$ is not sufficiently small. In particular, we have to expect considerable difficulties when the simple shooting method is used over a large interval $[a, b]$.

We will now introduce the following well-known test problem to demonstrate the above mentioned difficulties.

Example 4.4 (TROESCH'S PROBLEM)
Consider the following nonlinear BVP (see, e.g., [56, 117, 120, 122])

$$y''(x) = \lambda \sinh(\lambda y(x)), \quad 0 \leq x \leq 1,$$
$$y(0) = 0, \quad y(1) = 1,$$
$$(4.22)$$

where λ is a positive constant. The larger the constant λ is, the more difficult it is to determine the solution of (4.22) with the simple shooting method. The solution is almost constant for $x \geq 0$, i.e., $y(x) \approx 0$, and near to $x = 1$ it increases very rapidly to fulfill the prescribed boundary condition $y(1) = 1$. In Fig. 4.1 the graph of the exact solution $y(x)$ is represented for $\lambda = 10$.

Let us use the simple shooting method to determine an approximation of the unknown exact initial value $y'(0) = s^*$. Then, the solution of the BVP (4.22) can be computed from the associated IVP. About this initial value is known that it holds $s^* > 0$, but s^* is very small. If we are trying to solve the IVP

$$u''(x) = \lambda \sinh(\lambda u(x)), \quad u(0) = 0, \quad u'(0) = s, \tag{4.23}$$

for $s > s^*$ with a numerical IVP-solver, we must recognize that the integration step-size h tends to zero and the integration stops before the right boundary $x = 1$ of the given interval is reached. The reason for this behavior is a singularity in the solution $u(x; s)$ of the IVP (4.22) at

$$x_s = \frac{1}{\lambda} \int_0^\infty \frac{d\xi}{\sqrt{s^2 + 2\cosh(\xi) - 2}} \approx \frac{1}{\lambda} \ln\left(\frac{8}{|s|}\right).$$

In order to be able to perform the integration of the IVP over the complete interval $[a, b]$, the initial value s must be chosen such that the value x_s is located behind $x = 1$. This leads to the condition

$$|s| \le s_b \equiv 8 e^{-\lambda}, \tag{4.24}$$

which is a very strong restriction on the initial value, and thus also on the starting value for the Newton method.

The exact solution of Troesch's problem can be represented in the form

$$y(x) = u(x; s^*) = \frac{2}{\lambda} \operatorname{arcsinh}\left(\frac{s^* \operatorname{sn}(\lambda x, k)}{2 \operatorname{cn}(\lambda x, k)}\right), \quad k^2 = 1 - \frac{(s^*)^2}{4},$$

where $\operatorname{sn}(\lambda x, k)$ and $\operatorname{cn}(\lambda x, k)$ are the Jacobian elliptic functions (see [6]) and the parameter s^* is the solution of the equation

$$\frac{2}{\lambda} \operatorname{arcsinh}\left(\frac{s \operatorname{sn}(\lambda, k)}{2 \operatorname{cn}(\lambda, k)}\right) = 1.$$

In Table 4.1, the numerical results for Troesch's problem obtained with the code RWPM (parameter INITX $\doteq$ simple shooting) are presented (see [59]). The IVPs have been integrated with the IVP-solver RKEX78 (see [98, 113]) using an error tolerance of 10^{-9}. For the numerical solution of the nonlinear algebraic equations we have used a damped Newton method with an error tolerance of 10^{-7}.

Here, s^* is the numerically determined approximation for the exact initial slope $u'(0)$. Moreover, it denotes the number of Newton steps and NFUN is the number of calls of the right-hand side of the ODE (4.22). For the parameter value $\lambda = 9$, the simple shooting method provided an error code ierr=8, which means that the BVP could not be solved within the required accuracy.

In Fig. 4.1, the solutions of the IVP (4.23) for $\lambda = 8$ and different values of s are plotted. In that case, the bound in (4.24) is $s_b = 2.68370102\ldots \times 10^{-3}$.

Table 4.1 Results of the simple shooting method for Troesch's problem

λ	it	NFUN	s^*
1	3	734	8.4520269×10^{-1}
2	4	1626	5.1862122×10^{-1}
3	5	2858	2.5560422×10^{-1}
4	5	3868	1.1188017×10^{-1}
5	11	9350	4.5750461×10^{-2}
6	11	10815	1.7950949×10^{-2}
7	10	10761	6.8675097×10^{-3}
8	11	13789	2.5871694×10^{-3}

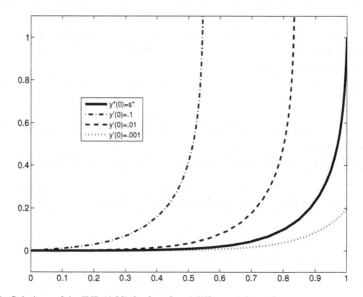

Fig. 4.1 Solutions of the IVP (4.23) for $\lambda = 8$ and different values of s

Now, we have the following situation. If $s > s_b$, the singularity x_s is within the interval $[0, 1]$, and thus, the curves do not reach the right end of the interval. Moreover, if $s < s^* = 2.5871694 \times 10^{-3}$, then $y(1) < 1$. $\square$

At the end of this section, we will explain the name of the method. In former times the cannoneers had to solve the following problem with the use of a cannon. In Fig. 4.2 the picture of a cannon is displayed. It can be shown that the ballistic trajectory of the cannonball is the solution of a second-order ODE

$$y''(x) = f(x, y(x)), \quad a \leq x \leq b.$$

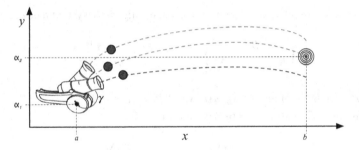

Fig. 4.2 The origin of the shooting method

The associated initial conditions are the position of the cannon $y(a) = \alpha_1$ and the unknown slope $y'(a) = \gamma$ of the cannon's barrel. Now, the cannoneer must determine the slope (i.e., the missing initial condition of the corresponding IVP) such that at $x = b$ the trajectory of the cannonball hits the given target $y(b) = \alpha_2$ (i.e., the trajectory of the IVP satisfies this boundary condition, too). In practice, the cannoneer has solved this problem by trial-and-error. For those who do not like the military terminology, the name *shooting method* can be replaced by *Manneken Pis method* in reference to the well-known sculpture and emblem of Brussels.

4.3 Method of Complementary Functions

In [63], we have considered the method of complementary functions for linear partially separated BVPs. Before we study nonlinear BVPs with (completely) nonlinear partially separated boundary conditions of the form

$$
\begin{aligned}
&y'(x) = f(x, y(x)), \quad a \le x \le b, \\
&g_p(y(a)) = 0, \\
&g_q(y(a), y(b)) = 0,
\end{aligned}
\tag{4.25}
$$

we will assume that the separated part of the boundary conditions is linear, i.e.,

$$
\begin{aligned}
&y'(x) = f(x, y(x)), \quad a \le x \le b, \\
&g_p(y(a)) \equiv B_a^{(1)} y(a) - \beta^{(1)} = 0, \\
&g_q(y(a), y(b)) = 0,
\end{aligned}
\tag{4.26}
$$

where $n = p + q$, $f : \Omega_1 \to \mathbb{R}^n$, $\Omega_1 \subset (a, b) \times \mathbb{R}^n$, $g_q : \Omega_2 \to \mathbb{R}^q$, $\Omega_2 \subset \mathbb{R}^n \times \mathbb{R}^n$, $B_a^{(1)} \in \mathbb{R}^{p \times n}$, and $\beta^{(1)} \in \mathbb{R}^p$. Let us assume $\mathrm{rank}(B_a^{(1)}) = p$.

If the simple shooting method is used to compute a solution of the BVP (4.26), in each step of the Newton method the linear algebraic system (4.16) has to be solved.

For the special boundary conditions in (4.26) this algebraic system reads

$$
\begin{pmatrix} B_a^{(1)} \\ B_{a,k}^{(2)} + B_{b,k}^{(2)} X_k^e \end{pmatrix} c_k = \begin{pmatrix} B_a^{(1)} s_k - \beta^{(1)} \\ g_q(s_k, u(b; s_k)) \end{pmatrix},
\tag{4.27}
$$

where $B_{i,k}^{(2)} = B_i^{(2)}(s_k) \equiv \partial g_q(s_k, u(b; s_k))/\partial y(i)$, $i = a, b$.

Using the QR factorization of the matrix $(B_a^{(1)})^T$,

$$
(B_a^{(1)})^T = \mathcal{Q} \begin{pmatrix} \mathcal{U} \\ 0 \end{pmatrix} = (\mathcal{Q}^{(1)} | \mathcal{Q}^{(2)}) \begin{pmatrix} \mathcal{U} \\ 0 \end{pmatrix} = \mathcal{Q}^{(1)} \mathcal{U},
\tag{4.28}
$$

the vector c_k is transformed as follows

$$
c_k = \mathcal{Q}\mathcal{Q}^T c_k = \mathcal{Q} \begin{pmatrix} w_k \\ z_k \end{pmatrix} = (\mathcal{Q}^{(1)} | \mathcal{Q}^{(2)}) \begin{pmatrix} w_k \\ z_k \end{pmatrix},
\tag{4.29}
$$

where $w_k \in \mathbb{R}^p$, $z_k \in \mathbb{R}^q$, $\mathcal{Q}^{(1)} \in \mathbb{R}^{n \times p}$, $\mathcal{Q}^{(2)} \in \mathbb{R}^{n \times q}$ und $\mathcal{U} \in \mathbb{R}^{p \times p}$.

Substituting (4.29) into (4.27), the left-hand side of (4.27) turns into

$$
\begin{bmatrix} \mathcal{U}^T (\mathcal{Q}^{(1)})^T \\ B_{a,k}^{(2)} + B_{b,k}^{(2)} X_k^e \end{bmatrix} \underbrace{(\mathcal{Q}^{(1)} | \mathcal{Q}^{(2)}) \begin{bmatrix} w_k \\ z_k \end{bmatrix}}_{c_k}
$$

$$
= \begin{bmatrix} \mathcal{U}^T \underbrace{(\mathcal{Q}^{(1)})^T \mathcal{Q}^{(1)}}_{I_p} & \mathcal{U}^T \underbrace{(\mathcal{Q}^{(1)})^T \mathcal{Q}^{(2)}}_{0_{p \times q}} \\ [B_{a,k}^{(2)} + B_{b,k}^{(2)} X_k^e] \mathcal{Q}^{(1)} & [B_{a,k}^{(2)} + B_{b,k}^{(2)} X_k^e] \mathcal{Q}^{(2)} \end{bmatrix} \begin{bmatrix} w_k \\ z_k \end{bmatrix}
$$

and the linear algebraic system reads

$$
\begin{bmatrix} \mathcal{U}^T & 0 \\ [B_{a,k}^{(2)} + B_{b,k}^{(2)} X_k^e] \mathcal{Q}^{(1)} & [B_{a,k}^{(2)} + B_{b,k}^{(2)} X_k^e] \mathcal{Q}^{(2)} \end{bmatrix} \begin{bmatrix} w_k \\ z_k \end{bmatrix} = \begin{bmatrix} B_a^{(1)} s_k - \beta^{(1)} \\ g_q(s_k, u(b; s_k)) \end{bmatrix}.
\tag{4.30}
$$

Now, using the forward substitution the unknown w_k can be computed numerically very fast from the following p-dimensional quadratic system with a lower triangular system matrix

$$
\mathcal{U}^T w_k = B_a^{(1)} s_k - \beta^{(1)}.
$$

For the numerical treatment of the second (block-) row of (4.30), in principle, $u(b; s_k)$ and X_k^e have to be determined as solutions of the $n + 1$ IVPs (4.11) and (4.19). However, in (4.30) the matrix X_k^e occurs only in combination with the matrix $\mathcal{Q}^{(2)}$ and with the vector $\mathcal{Q}^{(1)} w_k$. The idea is now to approximate the expressions $X_k^e \mathcal{Q}^{(2)}$ and $X_k^e \mathcal{Q}^{(1)} w_k$ by discretized directional derivatives. This can be realized as follows. Assuming that the solution $u(x_1; x_0, s)$ of the IVP $u' = f(x, u)$, $u(x_0) = s$, has already been computed, then Algorithm 1.1 generates an approximation $XR \in \mathbb{R}^{n \times j}$ of $X(x_1; x_0, s)R$, with $X(x_1; x_0, s) \equiv \partial u(x_1; x_0, s)/\partial s \in \mathbb{R}^{n \times n}$ und $R \in \mathbb{R}^{n \times j}$.

Algorithm 1.1 Approximation of directional derivatives
INPUT: dimension j, interval limits x_0 and x_1,
           point s, directions R, step-size h,
           subroutine for the computation of $f(x, y)$
SET: $\varepsilon = (\|s\| + 1)h$
For $i = 1 : j$ Do
           EVALUATE: $u(x_1; x_0, s + \varepsilon R e^{(i)})$
           {as solution of the IVP
             $u' = f(x, u), \ u(x_0) = s + \varepsilon R e^{(i)}$}
           SET: $XR(:, i) = (u(x_1; x_0, s + \varepsilon R e^{(i)}) - u(x_1; x_0, s))/\varepsilon$
Endfor
OUTPUT: XR

In Algorithm 1.1, $h > 0$ is a small constant and $e^{(i)}$ denotes the ith unit vector in $\mathbb{R}^j$.

The Algorithm 1.1 can now be used to determine approximations of $X_k^e Q^{(2)}$ and $X_k^e Q^{(1)} w_k$. To do this, in the first case we set

$$x_0 = a, \quad x_1 = b, \quad j = q, \quad s = s_k, \quad R = Q^{(2)},$$

and in the second case

$$x_0 = a, \quad x_1 = b, \quad j = 1, \quad s = s_k, \quad R = Q^{(1)} w_k.$$

It can easily be seen that this strategy needs only the integration of $q + 2$ IVPs. More precisely, the vectors $u(b; s_k)$, $u\left(b; s_k + \varepsilon Q^{(2)} e^{(i)}\right)$, $i = 1, \ldots, q$, and $u(b; s_k + \varepsilon Q^{(1)} w_k)$ have to be computed. The shooting method described above (including Algorithm 1.1) is called the *Newton form* of the nonlinear method of complementary functions.

In the nonlinear method of complementary functions the total number of integrations as well as the numerical effort for the algebraic operations can be reduced even further. To demonstrate this, let us return to the algebraic system (4.30). First, we modify the vector s_k such that the relation

$$B_a^{(1)} s_k = \beta^{(1)} \tag{4.31}$$

is satisfied. The general solution $\hat{s}_k$ of this underdetermined system can be written in the form

$$\hat{s}_k = (B_a^{(1)})^+ \beta^{(1)} + \left[I - (B_a^{(1)})^+ B_a^{(1)}\right] \omega, \quad \omega \in \mathbb{R}^n, \tag{4.32}$$

where $(B_a^{(1)})^+$ denotes the Moore–Penrose pseudoinverse of $B_a^{(1)}$ (see, e.g., [42]).

Since $\text{rank}(B_a^{(1)}) = p$, it holds

$$\hat{s}_k = \mathcal{Q}^{(1)}\mathcal{U}^{-T}\beta^{(1)} + \mathcal{Q}^{(2)}(\mathcal{Q}^{(2)})^T\omega. \qquad (4.33)$$

If we set $\omega = s_k$, formula (4.33) transforms the vector s_k into a vector $\hat{s}_k$, which satisfies the condition (4.31). Now, in the algebraic equations (4.30) we use the transformed vector $\hat{s}_k$ instead of s_k. Obviously, we obtain $w_k = 0$. This implies in turn that the n-dimensional linear algebraic system (4.30) is reduced to the following q-dimensional linear algebraic system for the vector $z_k \in \mathbb{R}^q$:

$$\hat{M}_k z_k = \hat{q}_k, \qquad (4.34)$$

where

$$\hat{M}_k \equiv [B_a^{(2)}(\hat{s}_k) + B_b^{(2)}(\hat{s}_k)X(b;\hat{s}_k)]\mathcal{Q}^{(2)} \quad \text{and} \quad \hat{q}_k \equiv g_q(\hat{s}_k, u(b;\hat{s}_k)).$$

Taking into account formula (4.29), the new iterate is determined according to the formula

$$\hat{s}_{k+1} = \hat{s}_k - \mathcal{Q}^{(2)}z_k. \qquad (4.35)$$

Note that each further iterate satisfies the condition (4.31) automatically, i.e., it holds

$$B_a^{(1)}\hat{s}_{k+1} = \underbrace{B_a^{(1)}\hat{s}_k}_{\beta^{(1)}} - B_a^{(1)}\mathcal{Q}^{(2)}z_k = \beta^{(1)} - \mathcal{U}^T\underbrace{(\mathcal{Q}^{(1)})^T\mathcal{Q}^{(2)}}_{0_{p\times q}}z_k = \beta^{(1)}.$$

Thus, in exact arithmetic, it is only necessary to apply the transformation (4.33) to the initial guess s_0. However, since in numerical computations rounding errors are unavoidable, the transformation (4.33) should be realized in each νth iteration step ($\nu \approx 10$) to counteract numerical instabilities.

As in the Newton form of the method of complementary functions, the matrix $X(b;\hat{s}_k)\mathcal{Q}^{(2)}$ (see formula (4.34)) can be computed by discretized directional derivatives. To do this, we set

$$x_0 \equiv a, \quad x_1 \equiv b, \quad j \equiv q, \quad s \equiv \hat{s}_k, \quad R \equiv \mathcal{Q}^{(2)},$$

and apply Algorithm 1.1. For the realization of this approximation $q + 1$ IVPs have to be solved, namely one integration for $u(b;\hat{s}_k)$ and q integrations for $u\left(b;\hat{s}_k + \varepsilon\mathcal{Q}^{(2)}e^{(i)}\right)$, $i = 1,\dots,q$. Thus, the number of integrations is reduced from $n + 1$ (simple shooting method) to $q + 1$ (method of complementary functions). The shooting technique, which is based on the formulas (4.28), (4.32)–(4.35) (including Algorithm 1.1), is called the *standard form* of the (nonlinear) method of complementary functions. To be consistent with the *linear* method of complementary functions we also denote the columns of $X(b;s)\mathcal{Q}^{(2)}$ as *complementary functions*.

Table 4.2 Results of the method of complementary functions for Troesch's problem

λ	it	NFUN	s^*
1	3	558	8.4520269×10^{-1}
2	4	1082	5.1862122×10^{-1}
3	5	2092	2.5560422×10^{-1}
4	5	3663	1.1188017×10^{-1}
5	11	6440	4.5750462×10^{-2}
6	12	8310	1.7950949×10^{-2}
7	12	9545	6.8675097×10^{-3}
8	16	15321	2.5871694×10^{-3}

Example 4.5 (once again: TROESCH'S PROBLEM)
We want to look once again at Troesch's problem, which we have considered in
Example 4.4. If the standard form of the method of complementary functions (program RWPM with the input parameter INITX $\doteq$ method of complementary functions)
is applied to the BVP (4.22), the results presented in Table 4.2 are produced.

Here, s^* is the numerically determined approximation for the exact initial slope
$u'(0)$. Moreover, it denotes the number of Newton steps and NFUN is the number
of calls of the right-hand side of the ODE (4.22). For the parameter value $\lambda = 9$ the
method of complementary functions provided an error code ierr=3, which means
that a descent direction could not be determined.

A comparison with the results of the simple shooting method (see Table 4.1) shows
that the theoretically predicted savings in computational costs could be achieved for
parameter values up to $\lambda = 5$. For parameter values larger than 5 the method of
complementary functions requires more Newton steps to guarantee the prescribed
accuracy. Therefore, the number of IVPs, which have to be solved, increases (see the
column NFUN).   □

Both forms of the nonlinear method of complementary functions can also be used
to solve *linear* BVPs with partially separated boundary conditions. Here, only one
iteration step is theoretically necessary. If additional iteration steps are performed the
accuracy of the results is increased compared with the linear method of complementary functions. These additional steps can be considered as an iterative refinement of
s_1, which is often used in the course of the numerical solution of systems of linear
algebraic equations.

4.4 Multiple Shooting Method

The basic idea of the multiple shooting method is to subdivide the given interval
$[a, b]$ into m subintervals

$$a = \tau_0 < \tau_1 < \cdots < \tau_{m-1} < \tau_m = b. \tag{4.36}$$

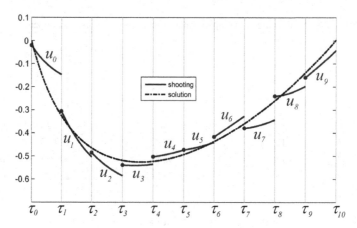

Fig. 4.3 Example of an intermediate step of the multiple shooting method and the exact solution $y(x)$

The grid points τ_k are called *shooting points*. Now, on each subinterval $[\tau_j, \tau_{j+1}]$, $0 \leq j \leq m - 1$, an IVP will be defined

$$
\begin{aligned}
u'_j(x) &= f(x, u_j(x)), \quad \tau_j \leq x \leq \tau_{j+1}, \\
u_j(\tau_j) &= s_j \in \mathbb{R}^n.
\end{aligned}
\tag{4.37}
$$

Let us denote the solution of (4.37) as $u_j \equiv u_j(x; s_j)$. By a suitable choice of the unknown vectors s_j in (4.37), the solutions $u_j(x; s_j)$ on the subintervals $[\tau_j, \tau_{j+1}]$, $j = 0, \ldots, m - 1$, are composed such that a continuous function results, which is continuous on the entire interval $[a, b]$ and satisfies not only the differential equation, but also the boundary conditions in (4.8); see Fig. 4.3.

Since we require the continuity of the solution at intermediate grid points, the following $m - 1$ matching conditions must be satisfied:

$$
u_j(\tau_{j+1}; s_j) - s_{j+1} = 0, \quad j = 0, \ldots, m - 2.
\tag{4.38}
$$

In addition, the solution must fulfill the boundary conditions, i.e.,

$$
g(s_0, u_{m-1}(b; s_{m-1})) = 0.
\tag{4.39}
$$

Writing the boundary conditions (4.39) first, and then the matching conditions (4.38), we obtain the following mn-dimensional system of nonlinear algebraic equations for the unknown initial vectors $s_j \in \mathbb{R}^n$ in (4.37)

$$
F^{(m)}(s^{(m)}) = 0,
\tag{4.40}
$$

where

$$s^{(m)} \equiv \begin{pmatrix} s_0 \\ s_1 \\ s_2 \\ \vdots \\ s_{m-1} \end{pmatrix}, \quad F^{(m)}(s^{(m)}) \equiv \begin{pmatrix} g(s_0, u_{m-1}(b; s_{m-1})) \\ u_0(\tau_1; s_0) - s_1 \\ u_1(\tau_2; s_1) - s_2 \\ \vdots \\ u_{m-2}(\tau_{m-1}; s_{m-2}) - s_{m-1} \end{pmatrix}.$$

The numerical algorithm defined by the formulas (4.36)–(4.40) is called *multiple shooting method*. Sometimes it is also denoted as *parallel shooting method*. This name makes reference to the fact that on the individual subintervals of the grid (4.36) the IVPs (4.37) can be integrated in parallel, i.e., independently of one another. Thus, the multiple shooting method is the prototype of numerical parallel algorithm.

If $y^*(t)$ denotes the exact (isolated) solution of the BVP (4.8), whose existence we assume here, the following claim can be shown. Under the assumption that f is sufficiently smooth there exists for each $s_j^* \equiv y^*(\tau_j)$ a n-dimensional neighborhood U_j^* such that the IVP (4.37) has a solution $u_j(t; s_j)$, which exists on the entire interval $[\tau_j, \tau_{j+1}]$. Thus, there is also a mn-dimensional neighborhood

$$\hat{U}^* \equiv U_0^* \times U_1^* \times \cdots \times U_{m-1}^*$$

of the vector $s_*^{(m)} \equiv (s_0^*, \ldots, s_{m-1}^*)^T \in \mathbb{R}^{mn}$ where $F^{(m)} : \hat{U}^* \to \mathbb{R}^{mn}$ is defined and continuously differentiable.

For the solution of the system of nonlinear algebraic equations (4.40) a (multidimensional) Newton-like method is commonly used. In the kth iteration step of the Newton method the following system of mn linear algebraic equations has to be solved

$$M_k^{(m)} c_k^{(m)} = q_k^{(m)}, \tag{4.41}$$

where the Jacobian $M_k^{(m)} \equiv M^{(m)}(s_k^{(m)}) \in \mathbb{R}^{mn \times mn}$ and the vector of the right-hand side $q_k^{(m)} \equiv q^{(m)}(s_k^{(m)}) \in \mathbb{R}^{mn}$ are defined as

$$M_k^{(m)} \equiv \begin{pmatrix} B_{a,k} & & & & B_{b,k} X_{m-1,k}^e \\ X_{0,k}^e & -I & & & \\ & X_{1,k}^e & -I & & \\ & & \ddots & \ddots & \\ & & & X_{m-2,k}^e & -I \end{pmatrix} \quad \text{and} \quad q_k^{(m)} \equiv F^{(m)}(s_k^{(m)}),$$

with

$$X_{j,k}^e \equiv \frac{\partial u_j(\tau_{j+1}; s_{j,k})}{\partial s_j} \in \mathbb{R}^{n \times n}, \quad j = 0, \ldots, m-1,$$

and

$$B_{i,k} \equiv B_i(s_k^{(m)}) \equiv \frac{\partial g(s_{0,k}, u_{m-1}(b; s_{m-1,k}))}{\partial y(i)} \in \mathbb{R}^{n \times n}, \quad i = a, b.$$

The new iterate is then calculated as

$$s_{k+1}^{(m)} = s_k^{(m)} - c_k^{(m)}.$$

The system matrix $M_k^{(m)}$ has the same structure as the matrix $M^{(m)}$ of the multiple shooting method for linear BVPs (see, e.g., [57, 63]). But the difference is that the matrices $X_{j,k}^e$, $j = 0, \ldots, m-1$, and $B_{a,k}$, $B_{b,k}$ now depend on the initial vectors $s_{j,k}$.

It can be shown that $M_k^{(m)}$ is nonsingular if and only if the matrix

$$M(s_{0,k}, s_{m-1,k}) \equiv B_{a,k} + B_{b,k} X_{m-1,k}^e$$

is nonsingular (see [57]).

For $s_{0,k} = s_k$, $s_{m-1,k} = u(\tau_{m-1}; s_k)$, the matrix $M(s_{0,k}, s_{m-1,k})$ coincides with the matrix M_k of the simple shooting method (see formula (4.18)). Thus, if $y^*(x) = u(x; s^*)$ is an isolated solution of (4.8), then the matrix $M(s^*, u(\tau_{m-1}; s^*))$ is nonsingular. This applies also to the matrix $M^{(m)}(s_*^{(m)})$ with

$$s_*^{(m)} \equiv (s^*, u(\tau_1; s^*), \ldots, u(\tau_{m-1}; s^*))^T.$$

Setting

$$A_j(x; s_{j,k}) \equiv \frac{\partial f}{\partial u_j}(x, u_j(x; s_{j,k})), \quad j = 0, \ldots, m-1,$$

and using the notation $Z_j(x; s_{j,k})$ for the solution of the (matrix) IVP

$$\begin{aligned} \dot{Z}_j(x) &= A_j(x; s_{j,k}) Z_j(x), \quad \tau_j \le x \le \tau_{j+1}, \\ Z_j(\tau_j) &= I, \end{aligned} \tag{4.42}$$

we have $X_{j,k}^e = Z_j(\tau_{j+1}; s_{j,k})$. This shows that the matrix $X_{j,k}^e$ can be computed by integrating the n IVPs (4.42). The analytical determination of the matrices $A_j(x; s_{j,k})$ that appear on the right-hand side of (4.42) is very time-consuming and sometimes not possible. Analogous to the simple shooting method (see formulas (4.20) and (4.21)) it is more efficient to approximate the matrices $X_{j,k}^e$, $j = 0, \ldots, m-1$, by finite differences. For this purpose, at each iteration step, mn additional IVPs must be integrated. Thus, per step, a total of $m(n+1)$ IVPs have to be solved numerically.

There is still the problem that the user has to supply the matrices $B_{a,k}$ and $B_{b,k}$. In general, these matrices are generated by analytical differentiation. In practice, BVPs are frequently encountered with linear boundary conditions (4.9). In this case, we have $B_{a,k} = B_a$ and $B_{b,k} = B_b$.

Essentially, there are two numerical techniques to solve the system of linear algebraic equations (4.41). The first method is based on the so-called *compactification* (see [33, 34, 117]). The idea is to transform the system (4.41) into an equivalent linear system

$$\bar{M}_k^{(m)} c_k^{(m)} = \bar{q}_k^{(m)}, \tag{4.43}$$

whose system matrix has a lower block-diagonal structure. Let the bordered matrix of the original system be

$$[M_k^{(m)} | q_k^{(m)}] \equiv \begin{pmatrix} B_{a,k} & & & & B_{b,k} X_{m-1,k}^e & q_{0,k}^{(m)} \\ X_{0,k}^e & -I & & & & q_{1,k}^{(m)} \\ & X_{1,k}^e & -I & & & q_{2,k}^{(m)} \\ & & \ddots & \ddots & & \vdots \\ & & & X_{m-2,k}^e & -I & q_{m-1,k}^{(m)} \end{pmatrix}. \tag{4.44}$$

In a first step, the block Gaussian elimination is used to eliminate the last block-element in the first row of $M_k^{(m)}$. To do this, the mth row of $[M_k^{(m)} | q_k^{(m)}]$ is multiplied from the left by $B_{b,k} X_{m-1,k}^e$, and the result is added to the first row. In the second step, the $(m-1)$th row is multiplied from the left by $B_{b,k} X_{m-1,k}^e X_{m-2,k}^e$, and the result is added again to the first row. This procedure is continued for $i = m-2, \ldots, 2$ by applying an analogous elimination step to the ith row. At the end of this block elimination strategy the following equivalent bordered matrix is obtained

$$[\bar{M}_k^{(m)} | \bar{q}_k^{(m)}] \equiv \begin{pmatrix} S_k & & & & \bar{q}_{0,k}^{(m)} \\ X_{0,k}^e & -I & & & q_{1,k}^{(m)} \\ & X_{1,k}^e & -I & & q_{2,k}^{(m)} \\ & & \ddots & \ddots & \vdots \\ & & & X_{m-2,k}^e & -I & q_{m-1,k}^{(m)} \end{pmatrix}, \tag{4.45}$$

with

$$S_k = B_{a,k} + B_{b,k} X_{m-1,k}^e X_{m-2,k}^e X_{m-3,k}^e \cdots X_{1,k}^e X_{0,k}^e,$$

$$\bar{q}_{0,k}^{(m)} = q_{0,k}^{(m)} + B_{b,k} X_{m-1,k}^e \left(q_{m-1,k}^{(m)} + X_{m-2,k}^e q_{m-2,k}^{(m)} \right.$$

$$\left. + X_{m-2,k}^e X_{m-3,k}^e q_{m-3,k}^{(m)} + \cdots + X_{m-2,k}^e X_{m-3,k}^e \cdots X_{2,k}^e X_{1,k}^e q_{1,k}^{(m)} \right).$$

The solution of the corresponding system of linear equations is realized as follows:

1. determination of the solution $c_{0,k}^{(m)} \in \mathbb{R}^n$ of the n-dimensional system

$$S_k c_{0,k}^{(m)} = \bar{q}_{0,k}^{(m)}, \quad \text{and} \tag{4.46}$$

2. performing the recursion

$$c_{j+1,k}^{(m)} = X_{j,k}^e \, c_{j,k}^{(m)} - q_{j+1,k}^{(m)}, \quad j = 0, 1, \ldots, m - 2. \tag{4.47}$$

Having determined the vector $c_k^{(m)} = (c_{0,k}^{(m)}, \ldots, c_{m-1,k}^{(m)})^T$, the new iterate of the Newton method is calculated according to the formula

$$s_{k+1}^{(m)} = s_k^{(m)} - c_k^{(m)}.$$

However, in [55] it is shown that the recursion (4.47) is a numerically instable process. But it is possible to reduce slightly the accumulation of rounding errors in this solution technique by a subsequent iterative refinement (see [34]).

A more appropriate treatment of the system of linear equations (4.41) is based on a LU factorization of the matrix $M_k^{(m)}$ with partial pivoting, scaling, and an iterative refinement. For general systems of linear algebraic equations it is shown in [115] that this strategy leads to a numerically stable algorithm. Without taking into account any form of packed storage, the LU factorization of $M_k^{(m)}$ requires a storage space for $m \times [n^2 m]$ real numbers. The implementation RWPM of the multiple shooting method (see [59, 60, 61]) is based on this linear equation solver. However, in RWPM a packed storage is used where only a maximum of $4 \times [n^2(m - 1)]$ real numbers have to be stored.

It is also possible to use general techniques for sparse matrices (various methods are available in the MATLAB package) to solve the linear shooting equations. Our tests with the MATLB package have shown that the storage for the LU factorization using sparse matrix techniques is also below the limit $4 \times [n^2(m - 1)]$ mentioned above. For the case $m = 10$ and $n = 10$ a typical situation for the fill-in is graphically presented in Fig. 4.4. The value nz is the number of nonzero elements in the corresponding matrices.

The convergence properties of the iterates $\{s_k^{(m)}\}_{k=0}^{\infty}$ can be derived directly from the application of Theorem 4.2 to the system of nonlinear algebraic equations (4.40).

To show that the radius of convergence of the Newton method will increase if the multiple shooting method is used instead of the simple shooting method, we represent the multiple shooting method as the simple shooting method for an (artificially) enlarged BVP. For this, we define a new variable τ on the subinterval $[\tau_j, \tau_{j+1}]$ as

$$\tau \equiv \frac{t - \tau_j}{\Delta_j}, \quad \Delta_j \equiv \tau_{j+1} - \tau_j, \quad j = 0, \ldots, m - 1. \tag{4.48}$$

With $y_j(\tau) \equiv y(\tau_j + \tau \Delta_j)$, and

$$f_j(\tau, y_j(\tau)) \equiv \Delta_j f(\tau_j + \tau \Delta_j, y_j(\tau)), \quad j = 0, \ldots, m - 1, \tag{4.49}$$

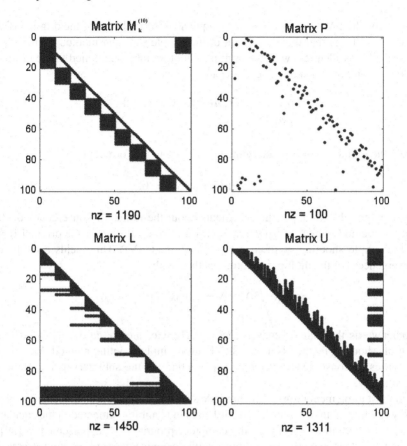

Fig. 4.4 Example for the fill-in the LU factorization

the differential equation in (4.8) can be written in the form

$$\bar{y}'(\tau) = \bar{f}(\tau, \bar{y}(\tau)), \quad 0 \le \tau \le 1, \quad \bar{y} \equiv \begin{pmatrix} y_0 \\ \vdots \\ y_{m-1} \end{pmatrix}, \quad \bar{f} \equiv \begin{pmatrix} f_0 \\ \vdots \\ f_{m-1} \end{pmatrix}. \quad (4.50)$$

The boundary conditions and the matching conditions are

$$\bar{g}(\bar{y}(0), \bar{y}(1)) \equiv \begin{pmatrix} g(y_0(0), y_{m-1}(1)) \\ y_1(0) - y_0(1) \\ \vdots \\ y_{m-1}(0) - y_{m-2}(1) \end{pmatrix} = 0. \quad (4.51)$$

Obviously, the BVP (4.50), (4.51) is an equivalent formulation of the original BVP
(4.8). This means that the application of the simple shooting method to the transformed BVP is identical with the multiple shooting method applied to the original
BVP. We can now formulate the associated IVP as

$$\bar{u}'(\tau) = \bar{f}(\tau, \bar{u}(\tau)), \quad 0 \le \tau \le 1,$$
$$\bar{u}(0) = \bar{s},$$

(4.52)

where the initial vector $\bar{s}$ is determined according to the formula (4.51) by

$$\bar{F}(\bar{s}) \equiv \bar{g}(\bar{s}, \bar{u}(1; \bar{s})) = 0.$$

(4.53)

It is now possible to apply the statements about the radius of convergence of the
Newton method, which are given in Sect. 4.2 for the solution of the original BVP
by the simple shooting method to the transformed BVP. This yields a radius of
convergence $\bar{\varrho}$ of the multiple shooting method with

$$\bar{\varrho} \equiv O(e^{-L\Delta}), \quad \Delta \equiv \max_{0 \le j \le m-1} (\tau_{j+1} - \tau_j).$$

In other words, the radius of convergence of the Newton method can be increased if the
multiple shooting method is used instead of the simple shooting method. Moreover,
the radius of convergence increases exponentially if the subintervals are reduced
successively.

In most implementations of the multiple shooting method, the Newton method or
a Newton-like method with discretized Jacobian is not implemented in the standard
form described above (see, e.g., the computer program RWPM presented in [59]).
Instead, the Newton method is combined with *damping* and *regularization strategies*
(see [58]) to force global convergence of the nonlinear equation solver. Furthermore, it makes sense not to compute the very costly finite difference approximation
of the Jacobian in each iteration step. This becomes possible by using the rank-1
modification formula of Sherman and Morrison, which leads to Broyden's method
(see [58]).

In the above-mentioned multiple shooting code RWPM a damped Quasi-Newton
method is used, which is transformed automatically into a regularized Quasi-Newton
method if the descending behavior is insufficient or the Jacobian is nearly singular.
The kth iteration step is of the form

$$s_{k+1}^{(m)} = s_k^{(m)} - \gamma_k \left[(1 - \lambda_k) D^k + \lambda_k M_k^{(m)} \right]^{-1} F^{(m)}(s_k^{(m)}).$$

(4.54)

Here, λ_k is the regularization parameter with $0 < \lambda_k \le 1$. If $\lambda_k = 1$ we have a damped
Newton method. The parameter γ_k is the damping parameter with $0 < \gamma_k \le 1$. The
matrix $D^k \in \mathbb{R}^{nm \times nm}$ is a diagonal matrix with $D^k = \text{diag}(\|M_k^{(m)} e^{(i)}\|_2 + 10^{-3})$,
where $e^{(i)}$ denotes the ith unit vector. The descending condition used is

$$\|\boldsymbol{F}^{(m)}(s_{k+1}^{(m)})\|_2^2 \leq (1 - 2\delta\gamma_k)\,\|\boldsymbol{F}^{(m)}(s_k^{(m)})\|_2^2, \quad \delta \in (0, 1/2). \tag{4.55}$$

A precise description of this Quasi-Newton method can be found in [57].

Up to now, we have made the assumption that the shooting points τ_j, $j = 0, \ldots, m$, are given by the user and they are not changed when the multiple shooting method is carried out. But this is not always an appropriate procedure. The multiple shooting method is based on two different types of grids, a *fine* inner grid, which is produced by the IVP-solver, and a coarse *outer* grid, which consists of the shooting points. The inner grid is always generated automatically and cannot be accessed by the user. In order to keep the computational costs low, besides both endpoints a and b, additional shooting points should be set only when this is necessary. Since the shape of the solution curve is not known a priori, the outer grid (4.36) must be generated automatically, too. The different implementations of the multiple shooting method each have their own strategy and heuristics for an automatic control of the shooting points. The practical experience of the program developer plays an important role in this context.

It seems reasonable that the user proposes a subdivision of the interval $[a, b]$, which can be modified by the algorithm. In principle, it must be possible that in the first iteration step the proposed shooting points are replaced by more appropriate points. Additional shooting points should be generated automatically during the solution process of the system of nonlinear algebraic equations (4.40) at two places. Namely, when $\boldsymbol{F}^{(m)}(s_0^{(m)})$ is evaluated, where $s_0^{(m)}$ is the starting vector, and when the Jacobian is approximated by a finite difference approximation $\delta\boldsymbol{F}^{(m)}(s_k^{(m)}; h_k)$, $k = 0, 1, \ldots$. For the selection of a shooting point three criteria are significant, which we want to describe in the following. Here, the two constants

$$c_1 \equiv 1/\sqrt{\text{KPMACH}}, \quad \text{KPMACH smallest positive machine number},$$

$$c_2 \equiv 1/\sqrt{\text{EPMACH}}, \quad \text{EPMACH relative machine accuracy},$$

play an important role. The criteria are:

1. Monitoring the growth of the solution $\boldsymbol{u}_j(x)$ of the IVP (4.37):
   the point $x = \tau^*$ will be added as a new shooting point, if it holds $\|\boldsymbol{u}_j(\tau^*)\| \geq \max\left(c_1, (\|s_j\| + 1)c_2\right)$,
2. Monitoring the partial derivatives $X_{j,k}^e$ ($j = 0, \ldots, m - 1$):
   a new shooting point τ^* will be placed in the subinterval $[\tau_j, \tau_{j+1}]$, if

$$\max_{i=1,\ldots,n} \|X_{j,k}^e\,\boldsymbol{e}^{(i)}\| / \min_{i=1,\ldots,n} \|X_{j,k}^e\,\boldsymbol{e}^{(i)}\| \geq c_2,$$

3. A new shooting point $\tau^* = \bar{x}$ will be added, if the IVP-solver is terminated irregularly at $x = \bar{x}$ because of:

   a. the required accuracy could not be obtained,
   b. $\boldsymbol{f}(\bar{x}, \boldsymbol{u})$ could not be evaluated,
   c. the upper bound for the number of integrations has been exceeded.

Table 4.3 Results of the multiple shooting method for Troesch's problem

λ	m	it	NFUN	s^*
5	10	13	8,568	4.5750462×10^{-2}
10	10	9	12,641	3.5833778×10^{-4}
15	10	11	26,616	2.4445130×10^{-6}
20	10	13	34,425	1.6487732×10^{-8}
25	11	21	83,873	$1.1110272 \times 10^{-10}$
30	13	16	78,774	$7.4860938 \times 10^{-13}$
35	13	24	143,829	$5.0440932 \times 10^{-15}$
40	14	25	169,703	$3.3986835 \times 10^{-17}$
45	15	16	273,208	$2.2900149 \times 10^{-19}$

Example 4.6 (once more again: TROESCH'S PROBLEM)
Let us look at Troesch's problem that was studied in Examples 4.4 and 4.5. Here, we present the results, which we have obtained with the multiple shooting method (program RWPM with the input parameter INITX $\doteq$ multiple shooting). For the parameter values, $\lambda = 5 : 5 : 40$ the IVPs have been integrated with the numerical IVP-solver RKEX78 (see [98, 113]). In order to guarantee the demanded accuracy for $\lambda = 45$, it was necessary to exchange the IVP-solver. We used as IVP-solver the Bulirsch-Gragg-Stoer extrapolation algorithm BGSEXP (see [66]). As start selection of shooting points we have chosen an equidistant net with $m = 10$, and the starting points were $y(\tau_i) = 0$, $i = 0, \ldots, 9$. In Table 4.3, we denote by m the number of shooting points. If $m > 10$ the code has added automatically new shooting points to the initial net. As before, s^* is the numerically determined approximation for the exact initial slope $u_0'(0)$, it denotes the number of Newton steps and NFUN is the number of calls of the right-hand side of the ODE (4.22).   □

Example 4.7 In Sect. 3.4 the following nonlinear BVP on an infinite interval is considered (see formula (3.111)):

$$y''(x) = -2y(x)y'(x), \quad y(0) = 0, \quad \lim_{x \to +\infty} y(x) = 1. \qquad (4.56)$$

To treat this problem by numerical techniques, we transform (4.56) into a series of BVPs on finite intervals

$$y''(x) = -2y(x)y'(x), \quad y(0) = 0, \quad y(b_i) = 1, \quad i = 1, 2, \ldots, \qquad (4.57)$$

where $\{b_i\}$ is a sequence of increasing positive real numbers. Now, (4.57) must be written in the standard form (4.2) as

$$\begin{pmatrix} y_1'(x) \\ y_2'(x) \end{pmatrix} = \begin{pmatrix} y_2(x) \\ -2y_1(x)y_2(x) \end{pmatrix},$$

$$\begin{pmatrix} 1 & 0 \\ 0 & 0 \end{pmatrix} \begin{pmatrix} y_1(0) \\ y_2(0) \end{pmatrix} + \begin{pmatrix} 0 & 0 \\ 1 & 0 \end{pmatrix} \begin{pmatrix} y_1(b_i) \\ y_2(b_i) \end{pmatrix} = \begin{pmatrix} 0 \\ 1 \end{pmatrix}, \qquad (4.58)$$

i.e.,

$$y'(x) = f(x, y), \quad B_a y(0) + B_b y(b_i) = \beta.$$

Let $y(x; b_i)$ be the solution of (4.58). The computation of $y(x; b_i)$, $i = 1, 2, \ldots$, is stopped if $y(x; b_N)$ satisfies $\triangle y_N < TOL$, where TOL is a given tolerance,

$$\triangle y_i \equiv \max_{j=0,\ldots m} \frac{|y_1(\tau_j; b_i) - y_1(\tau_j; b_{i-1})|}{|y_1(\tau_j; b_i)| + eps},$$

and eps is a constant prescribed by the MATLAB ($eps = 2.2 \ldots \times 10^{-16}$). Then, the approximate solution of the BVP (4.56) is

$$y(x) \equiv \begin{cases} y_1(x; b_N), & x \in [0, b_N] \\ 1, & x > b_N \end{cases}$$

In our numerical experiments, we have used the multiple shooting method with $m = 20$ to solve the BVPs (4.58). Moreover, we have set $TOL = 10^{-8}$. In Table 4.4, the corresponding results are presented. Here, $\varepsilon(y_i)$ compares the approximate solution with the exact solution, i.e.,

$$\varepsilon(y_i) \equiv \max_{j=0,\ldots,m} \frac{|\tanh(\tau_j) - y_1(\tau_j; b_i)|}{|\tanh(\tau_j)| + eps}.$$

The results show that the numerical solution for $b_6 = 9$ is a good approximation of the solution of the BVP (4.56). $\qquad \square$

Table 4.4 Numerical results for problem (4.57)

i	b_i	$\varepsilon(y_i)$	$\triangle y_i$
1	4	1.308525×10^{-3}	
2	5	1.777780×10^{-4}	1.130546×10^{-3}
3	6	2.407612×10^{-5}	1.536982×10^{-4}
4	7	3.258719×10^{-6}	2.081733×10^{-5}
5	8	4.410273×10^{-7}	2.817690×10^{-6}
6	9	5.968657×10^{-8}	3.813407×10^{-7}

4.5   Nonlinear Stabilized March Method

In Sect. 4.3, we have developed two nonlinear variants of the method of complementary functions for BVPs with nonlinear partially separated boundary conditions

$$
\begin{aligned}
y'(x) &= f(x, y(x)), \quad a \leq x \leq b, \\
g_p(y(a)) &\equiv B_a^{(1)} y(a) - \beta^{(1)} = 0, \\
g_q(y(a), y(b)) &= 0,
\end{aligned}
\tag{4.59}
$$

where $n = p+q$, $f : \Omega_1 \to \mathbb{R}^n$, $\Omega_1 \subset (a, b) \times \mathbb{R}^n$, $g_q : \Omega_2 \to \mathbb{R}^q$, $\Omega_2 \subset \mathbb{R}^n \times \mathbb{R}^n$, $B_a^{(1)} \in \mathbb{R}^{p \times n}$, and $\beta^{(1)} \in \mathbb{R}^p$. As before, let us assume $\mathrm{rank}(B_a^{(1)}) = p$.

Here, we will study a segmented version of the method of complementary functions, which solves BVPs of the form (4.59) with less numerical effort than the multiple shooting method. It is a generalization of the linear stabilized march method (see, e.g., [63]) to nonlinear BVPs.

The starting point for the following considerations are the equations (4.40)–(4.42) of the multiple shooting method. It should be recalled that in the kth iteration step of the multiple shooting method the system of linear algebraic equations (4.41) has to be solved. As we have seen this linear system is usually solved by Gaussian elimination and a special version of packed storage. If the boundary conditions are partially separated as described above the required storage space can be reduced significantly by special block elimination techniques. As a by-product the number of IVPs, which have to be integrated, is reduced, too. The latter result is much more important than the reduction of the storage space of the respective shooting method.

Without changing the solution, we can write the first block equation of system (4.41) as last block equation. Then, we substitute the special form (4.59) of the boundary conditions into this algebraic system and obtain

$$
\begin{pmatrix}
X_{0,k}^e & -I & & & \\
 & \ddots & \ddots & & \\
 & & X_{m-2,k}^e & -I & \\
\begin{pmatrix} B_a^{(1)} \\ B_{a,k}^{(2)} \end{pmatrix} & & & \begin{pmatrix} 0 \\ B_{b,k}^{(2)} \end{pmatrix} & X_{m-1,k}^e
\end{pmatrix}
\begin{pmatrix}
c_{0,k} \\ \vdots \\ c_{m-2,k} \\ c_{m-1,k}
\end{pmatrix}
$$

$$
=
\begin{pmatrix}
u_{0,k}^e - s_{1,k} \\
\vdots \\
u_{m-2,k}^e - s_{m-1,k} \\
B_a^{(1)} s_{0,k} - \beta^{(1)} \\
g_q(s_{0,k}, u_{m-1,k}^e)
\end{pmatrix},
\tag{4.60}
$$

where $\boldsymbol{u}_{j,k}^e \equiv \boldsymbol{u}_j(\tau_{j+1}; \boldsymbol{s}_{j,k})$. The following transformations are based on two orthogonal matrices Q_k and $\tilde{Q}_k^T$. They are constructed such that the matrix of the above system is transformed into a (2×2)-block matrix whose $(1,1)$-block is a lower triangular matrix and the $(1,2)$-block only consists of zeros. These matrices are

$$
Q_k = \begin{pmatrix} Q_{0,k}^{(1)} & & & Q_{0,k}^{(2)} & & \\ & \ddots & & & \ddots & \\ & & Q_{m-1,k}^{(1)} & & & Q_{m-1,k}^{(2)} \end{pmatrix}, \tag{4.61}
$$

$$
\tilde{Q}_k^T = \begin{pmatrix} & & & & & I_p \\ (Q_{1,k}^{(1)})^T & & & & & \\ & \ddots & & & & \\ & & (Q_{m-1,k}^{(1)})^T & & & \\ (Q_{1,k}^{(2)})^T & & & & & \\ & \ddots & & & & \\ & & (Q_{m-1,k}^{(2)})^T & & & \\ & & & & & I_q \end{pmatrix}, \tag{4.62}
$$

where the matrix elements $Q_{i,k}^{(1)} \in \mathbb{R}^{n \times p}$ and $Q_{i,k}^{(2)} \in \mathbb{R}^{n \times q}$ are determined recursively as follows:

- $i = 0$:
  compute the QR factorization (4.28) of $(B_a^{(1)})^T$ and set

$$
Q_{0,k} = (Q_{0,k}^{(1)} \mid Q_{0,k}^{(2)}) \equiv (Q^{(1)} \mid Q^{(2)}). \tag{4.63}
$$

- $i = 1, \ldots, m - 1$:
  compute the QR factorization of $X_{i-1,k}^e Q_{i-1,k}^{(2)}$

$$
X_{i-1,k}^e Q_{i-1,k}^{(2)} = (\tilde{Q}_{i-1,k}^{(2)} \mid \tilde{Q}_{i-1,k}^{(1)}) \begin{pmatrix} \tilde{\mathcal{U}}_{i-1,k} \\ 0 \end{pmatrix} \tag{4.64}
$$

and set

$$
Q_{i,k} = (Q_{i,k}^{(1)} \mid Q_{i,k}^{(2)}) \equiv (\tilde{Q}_{i-1,k}^{(1)} \mid \tilde{Q}_{i-1,k}^{(2)}). \tag{4.65}
$$

Now, using the matrix Q_k we carry out the coordinate transformation

$$
\boldsymbol{c}_k^{(m)} = Q_k \boldsymbol{d}_k^{(m)}, \quad \boldsymbol{d}_k^{(m)} \in \mathbb{R}^{mn}, \tag{4.66}
$$

and substitute (4.66) into the system (4.60). Then, the resulting equation is multiplied from the left by $\tilde{Q}_k^T$. This yields

$$(\tilde{Q}_k^T M_k^{(m)} Q_k) d_k^{(m)} = \tilde{Q}_k^T q_k^{(m)}, \qquad (4.67)$$

where the right-hand side of (4.60) is denoted with $q_k^{(m)}$. For the following considerations a detailed representation of this system of algebraic equations is important. However, due to the lack of space this can only be realized step-by-step. In a first step we form $T \equiv \tilde{Q}_k^T M_k^{(m)}$, i.e.,

$$
\begin{pmatrix}
(Q_{1,k}^{(1)})^T & & & & & I_p \\
& \ddots & & & & \\
& & (Q_{m-1,k}^{(1)})^T & & & \\
(Q_{1,k}^{(2)})^T & & & & & \\
& \ddots & & & & \\
& & (Q_{m-1,k}^{(2)})^T & & & \\
& & & & & I_q
\end{pmatrix}
\cdot
\begin{pmatrix}
X_{0,k}^e & -I & & \\
& \ddots & \ddots & \\
& & X_{m-2,k}^e & -I \\
\begin{pmatrix} B_a^{(1)} \\ B_{a,k}^{(2)} \end{pmatrix} & & \begin{pmatrix} 0 \\ B_{b,k}^{(2)} \end{pmatrix} X_{m-1,k}^e
\end{pmatrix}.
$$

Thus

$$
T =
\begin{pmatrix}
B_a^{(1)} & & & & \\
(Q_{1,k}^{(1)})^T X_{0,k}^e & -(Q_{1,k}^{(1)})^T & & & \\
& \ddots & & \ddots & \\
& & & (Q_{m-1,k}^{(1)})^T X_{m-2,k}^e & -(Q_{m-1,k}^{(1)})^T \\
(Q_{1,k}^{(2)})^T X_{0,k}^e & -(Q_{1,k}^{(2)})^T & & & \\
& \ddots & & \ddots & \\
& & & (Q_{m-1,k}^{(2)})^T X_{m-2,k}^e & -(Q_{m-1,k}^{(2)})^T \\
B_{a,k}^{(2)} & & & & B_{b,k}^{(2)} X_{m-1,k}^e
\end{pmatrix}.
$$

In the next step, we consider the relation $c_k^{(m)} = Q_k d_k^{(m)}$. Writing the vector $d_k^{(m)} \in \mathbb{R}^{mn}$ in the form

$$d_k^{(m)} = (d_{0,k}^{(1)}, \ldots, d_{m-1,k}^{(1)}, d_{0,k}^{(2)}, \ldots, d_{m-1,k}^{(2)})^T, \quad d_{i,k}^{(1)} \in \mathbb{R}^p, \quad d_{i,k}^{(2)} \in \mathbb{R}^q,$$

we obtain

$$c_{0,k} = Q_{0,k}^{(1)} d_{0,k}^{(1)} + Q_{0,k}^{(2)} d_{0,k}^{(2)},$$

$$\vdots$$

$$c_{m-1,k} = Q_{m-1,k}^{(1)} d_{m-1,k}^{(1)} + Q_{m-1,k}^{(2)} d_{m-1,k}^{(2)}.$$

On the other hand, using the abbreviations

$$
\begin{aligned}
A^{(1)} &= B_a^{(1)} Q_{0,k}^{(1)}, & A^{(2)} &= B_a^{(1)} Q_{0,k}^{(2)} \\
B_{i,k}^{(1)} &= (Q_{i,k}^{(1)})^T X_{i-1,k}^e Q_{i-1,k}^{(1)}, & C_{i,k}^{(1)} &= -(Q_{i,k}^{(1)})^T Q_{i,k}^{(1)}, & i &= 1, \ldots, m-1 \\
B_{i,k}^{(2)} &= (Q_{i,k}^{(1)})^T X_{i-1,k}^e Q_{i-1,k}^{(2)}, & C_{i,k}^{(2)} &= -(Q_{i,k}^{(1)})^T Q_{i,k}^{(2)}, & i &= 1, \ldots, m-1 \\
F_{i,k}^{(1)} &= (Q_{i,k}^{(2)})^T X_{i-1,k}^e Q_{i-1,k}^{(1)}, & G_{i,k}^{(1)} &= -(Q_{i,k}^{(2)})^T Q_{i,k}^{(1)}, & i &= 1, \ldots, m-1 \\
F_{i,k}^{(2)} &= (Q_{i,k}^{(2)})^T X_{i-1,k}^e Q_{i-1,k}^{(2)}, & G_{i,k}^{(2)} &= -(Q_{i,k}^{(2)})^T Q_{i,k}^{(2)}, & i &= 1, \ldots, m-1 \\
K^{(1)} &= B_{a,k}^{(2)} Q_{0,k}^{(1)}, & L^{(1)} &= B_{b,k}^{(2)} X_{m-1,k}^e Q_{m-1,k}^{(1)} \\
K^{(2)} &= B_{a,k}^{(2)} Q_{0,k}^{(2)}, & L^{(1)} &= B_{b,k}^{(2)} X_{m-1,k}^e Q_{m-1,k}^{(2)},
\end{aligned}
$$

the system matrix $\tilde{Q}_k^T M_k^{(m)} Q_k = T Q_k$ of the algebraic system (4.67) can be represented in the form

$$
T Q_k = \begin{pmatrix}
A^{(1)} & & & 0 & A^{(2)} & & & 0 \\
B_{1,k}^{(1)} & C_{1,k}^{(1)} & & & B_{1,k}^{(2)} & C_{1,k}^{(2)} & & \\
& \ddots & \ddots & & & \ddots & \ddots & \\
0 & & B_{m-1,k}^{(1)} & C_{m-1,k}^{(1)} & 0 & & B_{m-1,k}^{(2)} & C_{m-1,k}^{(2)} \\
F_{1,k}^{(1)} & G_{1,k}^{(1)} & & & F_{1,k}^{(2)} & G_{1,k}^{(2)} & & \\
& \ddots & \ddots & & & \ddots & \ddots & \\
0 & & F_{m-1,k}^{(1)} & G_{m-1,k}^{(1)} & 0 & & F_{m-1,k}^{(2)} & G_{m-1,k}^{(2)} \\
K^{(1)} & & & L^{(1)} & K^{(2)} & & & L^{(2)}
\end{pmatrix}.
$$

Let us study in more detail the entries of the matrix above. Taking into account the factorization (4.28), we get $A^{(1)} = \mathcal{U}^T$ and $A^{(2)} = 0$. Obviously, $C_{i,k}^{(1)} = -I_p$ und $C_{i,k}^{(2)} = 0$, $i = 1, \ldots, m-1$. The formulas (4.64) and (4.65) imply $B_{i,k}^{(2)} = 0$, $i = 1, \ldots, m-1$. The relations $G_{i,k}^{(1)} = 0$ and $G_{i,k}^{(2)} = -I_q$, $i = 1, \ldots, m-1$, are obvious again. Finally, from (4.64) and (4.65) we deduce that $F_{i,k}^{(2)} = \tilde{U}_{i-1,k}$, $i = 1, \ldots, m-1$.

Thus, using the new notation

$$
\begin{aligned}
U_{0,k} &\equiv \mathcal{U}^T, & U_{i,k} &\equiv \tilde{\mathcal{U}}_{i-1,k}, & i &= 1, \ldots, m-1, \\
Z_{0,k} &\equiv B_{a,k}^{(2)} Q_{0,k}^{(1)}, & Z_{i,k} &\equiv (Q_{i,k}^{(2)})^T X_{i-1}^e Q_{i-1,k}^{(1)}, & i &= 1, \ldots, m-1, \\
Z_{m,k} &\equiv B_{b,k}^{(2)} X_{m-1}^e Q_{m-1,k}^{(1)}, & V_{i,k} &\equiv (Q_{i,k}^{(1)})^T X_{i-1,k}^e Q_{i-1,k}^{(1)}, & i &= 1, \ldots, m-1, \\
R_{a,k} &\equiv B_{a,k}^{(2)} Q_{0,k}^{(2)}, & R_{b,k} &\equiv B_{b,k}^{(2)} X_{m-1,k}^e Q_{m-1,k}^{(2)} &&
\end{aligned}
$$

$$(4.68)$$

the system of linear equations (4.67) can be represented in the very simple form

$$
\begin{pmatrix}
U_{0,k} & & & & & & \\
V_{1,k} & -I_p & & & & 0 & \\
& \ddots & \ddots & & & & \\
& & V_{m-1,k} & -I_p & & & \\
- & - & - & - & - & - & - & - & - \\
Z_{1,k} & & & & U_{1,k} & -I_q & \\
& \ddots & & & & \ddots & \ddots \\
& Z_{m-1,k} & & & & U_{m-1,k} & -I_q \\
Z_{0,k} & & Z_{m,k} & R_{a,k} & & & R_{b,k}
\end{pmatrix}
\begin{pmatrix}
r_k^{(1)} \\
- \\
r_k^{(2)}
\end{pmatrix}
$$

$$
= \left(\boldsymbol{\eta}_k^{(1)}, \boldsymbol{\eta}_k^{(2)} \right)^T, \qquad (4.69)
$$

where

$$
\boldsymbol{r}_k^{(1)} \equiv \left(\boldsymbol{d}_{0,k}^{(1)}, \ldots, \boldsymbol{d}_{m-1,k}^{(1)} \right)^T, \qquad \boldsymbol{r}_k^{(2)} \equiv \left(\boldsymbol{d}_{0,k}^{(2)}, \ldots, \boldsymbol{d}_{m-1,k}^{(2)} \right)^T,
$$

$$
\boldsymbol{\eta}_k^{(1)} \equiv \Big(B_a^{(1)} \boldsymbol{s}_{0,k} - \boldsymbol{\beta}^{(1)}, (Q_{1,k}^{(1)})^T (\boldsymbol{u}_{0,k}^e - \boldsymbol{s}_{1,k}), \ldots,
$$

$$
(Q_{m-1,k}^{(1)})^T (\boldsymbol{u}_{m-2,k}^e - \boldsymbol{s}_{m-1,k}) \Big)^T,
$$

$$
\boldsymbol{\eta}_k^{(2)} \equiv \Big((Q_{1,k}^{(2)})^T (\boldsymbol{u}_{0,k}^e - \boldsymbol{s}_{1,k}), \ldots,
$$

$$
(Q_{m-1,k}^{(2)})^T (\boldsymbol{u}_{m-2,k}^e - \boldsymbol{s}_{m-1,k}), \boldsymbol{g}_q(\boldsymbol{s}_{0,k}, \boldsymbol{u}_{m-1,k}^e) \Big)^T.
$$

If the boundary conditions in (4.59) are *completely separated*, i.e.,

$$
\boldsymbol{g}_q = \boldsymbol{g}_q(\boldsymbol{y}(b)),
$$

we have $Z_{0,k} = 0$ and $R_{a,k} = 0$. Therefore, the linear algebraic system (4.69) can be solved by the following stable block elimination technique, which requires only minimum storage space. In the first step of this technique the unknown vector $\boldsymbol{r}_k^{(1)}$ is determined from the linear system, which has as system matrix the bidiagonal (1, 1)-block matrix of (4.69) and as right-hand side the vector $\boldsymbol{\eta}_k^{(1)}$. Then, the result is substituted into (4.69), and the second unknown vector $\boldsymbol{r}_k^{(2)}$ is computed from the lower block system. The corresponding system matrix is the bidiagonal (2, 2)-block matrix from (4.69). An appropriate implementation of this solution strategy (using an adequate form of compact storage) requires only $mq + n^2$, $m > 2$, memory locations. If in addition some steps of the iterative refinement are executed to improve the accuracy, the memory requirements are increased to $m(n^2 + 2n + q) + n^2 - 2q$ memory locations. By contrast, the application of the standard Gaussian elimination would require $3mn^2$ memory locations (without iterative refinement) and $(4m - 8)n^2 + 2mn$ memory locations (with iterative refinement). In general, the block elimination technique requires slightly more CPU time than the standard Gaussian elimination (depending on p and the orthogonalization technique used). However, this fact is relatively uninteresting since in a shooting method the linear equation solver requires only a very small percentage of the total CPU time. What really matters is the number of IVPs which have to be integrated.

Let us now come back to the BVP (4.59) with *partially separated* boundary conditions, and our aim to develop a nonlinear stabilized march method, which is a real analog of the linear stabilized march method. There are at least two possibilities to realize the modification of the stabilized march method. The resulting algorithms differ with respect to the numerical effort and the convergence properties.

To derive the first modification, we start again with the block elimination of the $(1,1)$-block system in (4.69), i.e., the first unknown vector $r_k^{(1)} = (d_{0,k}^{(1)}, \ldots, d_{m-1,k}^{(1)})^T$ is determined from the linear equations

$$
\begin{aligned}
U_{0,k} \, d_{0,k}^{(1)} &= B_a^{(1)} s_{0,k} - \beta^{(1)}, \\
d_{i,k}^{(1)} &= (Q_{i,k}^{(1)})^T (u_{i-1,k}^e - s_{i,k}) + (Q_{i,k}^{(1)})^T X_{i-1,k}^e Q_{i-1,k}^{(1)} d_{i-1,k}^{(1)}, \\
i &= 1, \ldots, m-1.
\end{aligned} \tag{4.70}
$$

This strategy is immediately plausible, if we look at the upper part of the system (4.69) and take notice of the special lower triangular form. Since the matrix $X_{i-1,k}^e$ appears in (4.70) only in combination with the vector $Q_{i-1,k}^{(1)} d_{i-1,k}^{(1)}$, the corresponding term can be approximated by a (discretized) directional derivative. To do this, we use Algorithm 1.1 with the parameters $x_0 \equiv \tau_{i-1}$, $x_1 \equiv \tau_i$, $j \equiv 1$, $s \equiv s_{i-1,k}$, $R \equiv Q_{i-1,k}^{(1)} d_{i-1,k}^{(1)}$, $i = 1, \ldots, m-1$. Obviously, this realization requires the integration of only $2(m-1)$ IVPs. Now, $r_k^{(1)}$ is substituted into (4.69) and the second unknown vector $r_k^{(2)} = (d_{0,k}^{(2)}, \ldots, d_{m-1,k}^{(2)})^T$ is the solution of the $(2,2)$-block system in (4.69). Thus, we have to solve the following system of algebraic equations

$$
\begin{pmatrix}
U_{1,k} & -I_q & & & \\
& \ddots & \ddots & & \\
& & U_{m-1,k} & -I_q & \\
B_{a,k}^{(2)} Q_{0,k}^{(2)} & & & B_{b,k}^{(2)} X_{m-1,k}^e Q_{m-1,k}^{(2)}
\end{pmatrix}
\begin{pmatrix}
d_{0,k}^{(2)} \\
\vdots \\
d_{m-2,k}^{(2)} \\
d_{m-1,k}^{(2)}
\end{pmatrix}
$$

$$
=
\begin{pmatrix}
(Q_{1,k}^{(2)})^T (u_{0,k}^e - s_{1,k}) - (Q_{1,k}^{(1)})^T X_{0,k}^e Q_{0,k}^{(1)} d_{0,k}^{(1)} \\
\vdots \\
(Q_{m-1,k}^{(2)})^T (u_{m-2,k}^e - s_{m-1,k}) - (Q_{m-2,k}^{(1)})^T X_{m-2,k}^e Q_{m-2,k}^{(1)} d_{m-2,k}^{(1)} \\
g_q(s_{0,k}, u_{m-1,k}^e) - B_{a,k}^{(2)} Q_{0,k}^{(1)} d_{0,k}^{(1)} - B_{b,k}^{(2)} X_{m-1,k}^e Q_{m-1,k}^{(1)} d_{m-1,k}^{(1)}
\end{pmatrix}. \tag{4.71}
$$

To ensure that this system is well defined, the following quantities are required:

$$u^e_{i,k},$$

$$X^e_{i,k} Q^{(1)}_{i,k} d^{(1)}_{i,k} \quad i = 0, \ldots, m-1,$$

$$U_{i,k} \equiv X^e_{i-1,k} Q^{(2)}_{i-1,k} \quad i = 1, \ldots, m-1,$$

$$X^e_{m-1,k} Q^{(2)}_{m-1,k}.$$

Note that $u^e_{i,k}$ and $X^e_{i,k} Q^{(1)}_{i,k} d^{(1)}_{i,k}$, $i = 0, \ldots, m-2$, have been already computed in the determination of $r^{(1)}_k$ (see formula (4.70)). To compute the vector $u^e_{m-1,k}$ and an approximation of $X^e_{m-1,k} Q^{(1)}_{m-1,k} d^{(1)}_{m-1,k}$ (on the basis of Algorithm 1.1), two additional IVPs have to be integrated. Up to this point, we have in total $2m$ integrations. Setting $x_0 \equiv \tau_{i-1}$, $x_1 \equiv \tau_i$, $j \equiv p$, $s \equiv s_{i-1,k}$, $R \equiv Q^{(2)}_{i-1,k}$, $i = 1, \ldots, m$, Algorithm 1.1 can be used again to approximate $X^e_{i-1,k} Q^{(2)}_{i-1,k}$, $i = 1, \ldots, m$, with mq integrations. Thus, in the kth step of our first variant of the nonlinear stabilized march method a total of $(q+2)m$ IVPs have to be integrated. We call this variant the *Newtonian form* of the nonlinear stabilized march method. Compared with the (nonlinear) multiple shooting method, this new method reduces the number of IVPs, which have to be integrated, by $(p-1)m$. As a by-product, the dimension of the corresponding system of linear algebraic equations is reduced from mn to mq.

Once an approximation of $d^{(m)}_k$ is computed with the technique described above, the new iterate $s^{(m)}_{k+1}$ is determined with the formula

$$s^{(m)}_{k+1} = s^{(m)}_k - Q_k d^{(m)}_k. \tag{4.72}$$

As for the conventional Newton method, it is of course also possible to integrate damping and regularization strategies into the iteration method described above. This leads to a more or less appropriate globalization of the locally convergent iteration.

Obviously, if the BVP is linear then only one step of the iteration (4.72) is theoretically necessary. However, this iteration step does not match with the result of the linear stabilized march method (see, e.g. ,[63]). The linear method requires only the integration of $(q+1)m$ IVPs.

If the Newtonian form of the nonlinear stabilized march method is executed on a computer, more than one iteration steps are needed to solve a linear BVP with the required accuracy. These additional iteration steps can be interpreted as a special form of the iterative refinement, which compensates the unavoidable rounding errors. The inherent iterative refinement is a significant advantage over the linear stabilized method.

To get a nonlinear stabilized march method, which solves (theoretically) linear BVPs in one iteration step and requires the same number of integrations as the linear stabilizedm march method, we now present a second possibility to modify the linear shooting technique. This new approach is based on a decoupling of the system (4.69) by introducing an intermediate step, which transforms the vectors $s_{0,k}, \ldots, s_{m-1,k}$ into vectors $\hat{s}_{0,k}, \ldots, \hat{s}_{m-1,k}$ such that the relations

$$
\begin{align}
&\text{(i)} \quad (Q_{i,k}^{(1)})^T (u_{i-1,k}^e - \hat{s}_{i,k}) = 0, \quad i = 1, \ldots, m-1, \\
&\text{(ii)} \quad B_a^{(1)} \hat{s}_{0,k} - \beta^{(1)} = 0
\end{align}
\tag{4.73}
$$

are fulfilled. Then, the regularity of the $(1,1)$-block matrix in (4.69) implies $r_k^{(1)} = 0$ and the linear algebraic system (4.69) is reduced to the following mq-dimensional linear system

$$
\hat{M}_k^{(m)} z_k^{(m)} = \hat{q}_k^{(m)},
\tag{4.74}
$$

where $\hat{M}_k^{(m)}$ denotes the $(2,2)$-block matrix of (4.69). Moreover, we have $z_k^{(m)} \equiv r_k^{(2)}$ and $\hat{q}_k^{(m)} \equiv \eta_k^{(2)}$.

The vectors $\hat{s}_{i,k}$, $i = 0, \ldots, m-1$, which satisfy the relations (4.73), can be determined in a similar manner as described in Sect. 4.3 for the standard form of the (nonlinear) method of complementary functions. In particular, the general solution of the first underdetermined system in (4.73) is

$$
\begin{align}
\hat{s}_{i,k} = &\{(Q_{i,k}^{(1)})^T\}^+ (Q_{i,k}^{(1)})^T u_{i-1,k}^e \\
&+ [I - \{(Q_{i,k}^{(1)})^T\}^+ (Q_{i,k}^{(1)})^T] \omega_{i,k}, \quad \omega_{i,k} \in \mathbb{R}^n.
\end{align}
\tag{4.75}
$$

Since the matrices $Q_{i,k}^{(1)}$ have full rank, the corresponding pseudoinverses can be determined very easily (see, e.g., [57]):

$$
\{(Q_{i,k}^{(1)})^T\}^+ = Q_{i,k}^{(1)} \left((Q_{i,k}^{(1)})^T Q_{i,k}^{(1)} \right)^{-1} = Q_{i,k}^{(1)}.
$$

Therefore, it is possible to present formula (4.75) in the form

$$
\hat{s}_{i,k} = Q_{i,k}^{(1)} (Q_{i,k}^{(1)})^T u_{i-1,k}^e + Q_{i,k}^{(2)} (Q_{i,k}^{(2)})^T \omega_{i,k}, \quad i = 1, \ldots, m-1.
\tag{4.76}
$$

The general solution of the second underdetermined system in (4.73) can be written analogously to (4.33) as follows

$$
\hat{s}_{0,k} = Q_{0,k}^{(1)} (U_{0,k}^{(1)})^{-1} \beta^{(1)} + Q_{0,k}^{(2)} (Q_{0,k}^{(2)})^T \omega_{0,k}.
\tag{4.77}
$$

If we set $\omega_{i,k} \equiv s_{i,k}$, $i = 0, \ldots, m-1$, the formulas (4.76) and (4.77) transform the vectors $s_{i,k}$ into vectors $\hat{s}_{i,k}$, which satisfy the conditions (4.73).

The main advantage of this strategy is that in addition to the reduction of the linear algebraic systems, the number of IVPs, which have to be integrated, can be reduced. Namely, in (4.74) the matrices $X_{i,k}^e$ occur only in the combination $X_{i,k}^e Q_{i,k}^{(2)}$. If these matrix products are approximated by discretized directional derivatives (using the

Algorithm 1.1), the computational effort can be reduced significantly: in each iteration step of the second variant of the nonlinear stabilized march method only $m(q+1)$
IVPs have to be solved. Note, the multiple shooting method requires the integration
of $m(n+1)$ IVPs. If the number p of separated boundary conditions is not too small
compared with the dimension q of the nonseparated boundary conditions, the computational effort for the additional algebraic manipulations (QR factorization, etc.)
is negligible.

Once the solution $z_k^{(m)}$ of the linear algebraic system (4.74) is computed, the new
iterate $s_{k+1}^{(m)}$ can be determined by the formula

$$s_{k+1}^{(m)} = \hat{s}_k^{(m)} - \text{diag}\left(Q_{0,k}^{(2)}, \ldots, Q_{m-1,k}^{(2)}\right) z_k^{(m)}. \tag{4.78}$$

We call the shooting method, which is defined by the formulas (4.74)–(4.78), the
standard form of the nonlinear stabilized march method.

Example 4.8 Let us consider the following parametrized BVP (see, e.g., [13, 57, 60,
111])

$$y_1'(x) = \lambda \frac{(y_3(x) - y_1(x))y_1(x)}{y_2(x)}, \quad y_2'(x) = -\lambda(y_3(x) - y_1(x)),$$

$$y_3'(x) = \frac{0.9 - 10^3(y_3(x) - y_5(x)) - \lambda(y_3(x) - y_1(x))y_3(x)}{y_4(x)},$$

$$y_4'(x) = \lambda(y_3(x) - y_1(x)), \quad y_5'(x) = -100(y_5(x) - y_3(x)), \quad 0 \le x \le 1,$$

$$y_1(0) = y_2(0) = y_3(0) = 1, \quad y_4(0) = -10, \quad y_3(1) = y_5(1).$$

$$\tag{4.79}$$

Here, $n = 5$, $q = 1$, $p = 4$, and λ is a real parameter. Our aim was to determine
the solution of (4.79) for the parameter value $\lambda = 10,000$ by the multiple shooting
method and the standard form of the nonlinear stabilized march method. To find a
starting trajectory we have used the following strategy. In a first step, we have solved
the simplified BVP with $\lambda = 0$, using the starting trajectory

$$y_1(x) = y_2(x) = y_3(x) \equiv 1, \quad y_4(x) \equiv -10, \quad y_5(x) \equiv 1.$$

Then, the obtained result was taken as starting trajectory for the solution of the
problem with $\lambda = 100$. In a second step, we started with the result for $\lambda = 100$ to
determine the solution of the BVP for $\lambda = 10,000$. The corresponding IVPs were
solved by the semi-implicit extrapolation method SIMPRS [17].

In the Tables 4.5 and 4.6, we present the results for a grid with 11 and 21 equidistributed shooting points, respectively. Here, we use the following abbreviations: MS
= multiple shooting method, SM = stabilized march method, it = number of iteration steps, NIVP = number of IVP-solver calls, and NFUN = number of calls of the
right-hand side of the ODE (4.79).   □

Table 4.5 Using a grid with 11 equidistant shooting points

Code	$\lambda = 0$			$\lambda = 100$			$\lambda = 10,000$		
	it	NIVP	NFUN	it	NIVP	NFUN	it	NIVP	NFUN
MS	1	70	3,430	6	370	90,194	4	250	44,736
SM	1	40	1,124	5	120	15,808	2	60	4,320

Table 4.6 Using a grid with 21 equidistant shooting points

Code	$\lambda = 0$			$\lambda = 100$			$\lambda = 10,000$		
	it	NIVP	NFUN	it	NIVP	NFUN	it	NIVP	NFUN
MS	1	140	6,860	6	740	124,502	4	500	70,110
SM	1	80	2,076	4	200	16,880	2	120	5,332

4.6 MATLAB PROGRAMS

In this section, we present a simple implementation of the multiple shooting method in the MATLAB programming language. This implementation can be used to solve the BVPs given in the exercises as well as own problems of the reader. Here, the equidistributed shooting points $a = \tau_0, \ldots, \tau_m = b$ must be given at advance by entering the number m. The numerical solution is presented at the shooting points in form of a table. Moreover, all components of the solution vector are plotted, where the points of the table (red stars) are interpolated by cubic splines. The right-hand side of the ODE and the boundary conditions must be defined in Algorithms 4.3 and 4.4, respectively. Some standard BVPs are pre-formulated (see Algorithm 4.2). Moreover, the problems can depend on an external parameter $\lambda \in \mathbb{R}$. The value of λ must be entered by the user. Setting manually starting values for the vector function $y(x)$ at the given shooting points requires a good understanding of the mathematical background. For the preformulated BVPs we have set the starting values as follows (see Algorithm 4.5):

- BVPs 1 and 3: we use $y(\tau_j) = 0, j = 0, \ldots, m - 1$,
- BVP 2: first, we determine the solution for $\lambda = 0$ and use this result as the starting trajectory for $\lambda > 0$,
- BVPs 4 and 5 (see also Chap. 5): the solutions of both problems are periodic. Considering this behavior of the solutions and the given boundary conditions, we have set $y(\tau_j) = c * \cos(\pi \tau_j)$ for BVP 4 and $y(\tau_j) = c * \sin(\pi \tau_j)$ for BVP 5, $j = 0, \ldots, m - 1$, where c is a free parameter.

Algorithm 4.2 Main program

```
%
% Multiple Shooting Method
%
clear all, %close all
global nr zz la
zz=0;
```

```
nn=[2,5,2,2,2,2];
disp('the following problems are prepared:')
disp(' 1 - Troesch')
disp(' 2 - Scott/Watts')
disp(' 3 - Bratu')
disp(' 4 - Euler Bernoulli rod')
disp(' 5 - la*(y+y^2)')
disp(' 6 - BVP on an infinite interval')
nrs=input('Please enter the number of the problem : ','s');
nr=str2num(nrs);
n=nn(nr);
m=input('Please enter the number of shootingpoints m = ');
las=[];
if nr ~= 6
 las=input('Please enter the value of parameter lambda = ','s');
 la=str2num(las);
end
options=optimset('Display','iter','Jacobian','on','TolFun', ...
         1e-8,'TolX',1e-8,'MaxIter',1000,'MaxFunEvals',500*n*m);
[s0,a,b]=startvalues(n,m,nr);
tic,ua=fsolve(@msnl,s0(:),options,a,b,n,m);toc,
% solution matrix y(n,m)
% each column-vector corresponds to the solution
% at a shooting point
y=reshape(ua,n,m);
options = odeset('RelTol',1e-8,'AbsTol',1e-8);
d=(b-a)/(m);
t=a:d:b;
[~,yend]=ode45(@ode,[t(end-1),t(end)],y(:,m),options);
y=[y,yend(end,:)'];
z=1;
% Table
disp('    ')
disp('Table of solution')
ki=fix(n/3);
es='              ';
form='sprintf(''%#15.8e''';
for ii=0:ki
s=['    i          x(i)                 y(i,',int2str(1+ii*3),')',es];
for i=2+ii*3:min((ii+1)*3,n)
    s=[s,'y(i,',int2str(i),')                '];
end
disp(s)
for i=1:m+1
    s=['disp([','sprintf(''%3i'',i),''     '',',form,',t(i))'];
    for j=1+ii*3:min((ii+1)*3,n)
        js=int2str(j);
        s=[s,',''      '',',form,',y(',js,',i))'];
    end
    s=[s,'])'];
    eval(s)
end
disp('   ')
end
% solving IVPs to determine approximate solutions between
% the shooting points
yk=zeros(m*20,n);
d=d/20;
for i=1:m
```

```
    xx=t(i):d:t(i+1);
    [~,yy]=ode45(@ode,xx,y(:,i),options);
    yk((i-1)*20+1:i*20,:)=yy(1:20,:);
end
yk=[yk;yy(end,:)];
tk=a:d:b;
% plot of y1 to yn
for k=1:n
    figure(k);clf
    hold on
% plot of the shooting points (red points)
    plot(t,y(k,:),'r*')
% plot of the graph between the shooting points (blue curve)
    plot(tk,yk(:,k)')
    hst=' and \lambda = ';
    if nr == 6, hst=[];end
    title(['Solution of problem ',nrs,hst,las])
    ylabel(['y_',int2str(k),'(x)'])
    xlabel('x')
    shg
end
disp('End of program')
```

Algorithm 4.3 Function defining the ODEs

```
function f=ode(x,y)
global nr zz la xi
switch nr
    case 1
        f=[y(2);la*sinh(la*y(1))];
    case 2
        y3my1=y(3)-y(1);
        y3my5=y(3)-y(5);
        f=[la*y(1)*y3my1/y(2);
            -la*y3my1;
            (.9-1e3*y3my5-la*y(3)*y3my1)/y(4);
            la*y3my1;
            1e2*y3my5];
    case 3
        f=[y(2);-la*exp(y(1))];
    case 4
        f=[y(2);-la*sin(y(1))];
    case 5
        f=[y(2);la*(y(1)+y(1)^2)];
    case 6
        f=[y(2);-2*y(1)*y(2)];
end
zz=zz+1;
```

Algorithm 4.4 Function defining the boundary conditions

```
function g=bc(ya,yb)
global nr
switch nr
    case 1
        g=[ya(1);yb(1)-1];
    case 2
```

```
            g=[ya(1)-1;ya(2)-1;ya(3)-1;ya(4)+10;yb(5)-yb(3)];
    case 3
            g=[ya(1);yb(1)];
    case 4
            g=[ya(2);yb(2)];
    case 5
            g=[ya(1);yb(1)];
    case 6
            g=[ya(1);yb(1)-1];
end
```

Algorithm 4.5 Function defining the starting values

```
function [s,a,b]=startvalues(n,m,nr)
global la
a=0;b=1;
switch nr
    case 1
        s=zeros(n*m,1);
    case 2
        s=[1;1;1;-10;.91]*ones(1,m);
        laa=la;la=0;
        options=optimset('Display','iter','Jacobian','on', ...
                        'TolFun',1e-8);
        disp('computing starting value with lambda=0')
        s=fsolve(@msnl,s(:),options,0,1,n,m);
        disp('ready')
        s=s(:);
        la=laa;
        disp(['now lambda=',num2str(la)])
    case 3
        % for the lower branch, uncomment the next row
        s=zeros(n*m,1);
        %
        % for the upper branch with lambda >= 0.2 and m >= 20
        % or with lambda >= 0.025 and m >= 50,
        % uncomment the next row
        %s=zeros(n*m,1)+6;
    case 4
        t=0:1/m:1-1/m;
        % for the upper branch, choose c=2,
        % for the lower branch, choose c=-2
        c=2;
        % for the i-th branch choose d=i
        d=3;
        s=c*[cos(d*pi*t);-d*pi*sin(d*pi*t)];

    case 5
        t=0:1/m:1-1/m;
        % for the lower first branch (la<pi^2),
        % uncomment the next row
        % c=-4;d=1;
        % for the upper first branch (la>pi^2),
        % uncomment the next row
        % c=4;d=1;
        % for the other upper branches, choose c=.5,
        % for the other lower branches, choose c=-.5
        c=.5;
```

```
        % for the i-th (i>2) branch choose d=i
        d=3;
        s=c*[sin(d*pi*t);d*pi*cos(d*pi*t)];
    case 6
        s=zeros(n*m,1);
        b=input('Please enter the right boundary b = ');
end
```

Algorithm 4.6 Function handling the nonlinear algebraic equations

```
function [Fm,Mm]=msnl(s,a,b,n,m)
% msvnl - computes the vector Fm(s) and the Jacobian Mm(s)
%         of multiple shooting
%
% Fm = msvnl(s,ab)
% [Fm,Mm] = msvnl(s,ab)
% s  - nxm-vector, where Fm(s) and Mm have to be computed
% a,b - left and right end of interval [a,b], a>b or b>a
% n  - number of differential equations
% m  - number of (actual) shooting points (t(m+1)=b)
% Fm - nxm-vector Fm of the nonlinear system
% Mm - n*m x n*m Jacobian Mm of Fm in as sparse-matrix
%
s=reshape(s,n,m);
tol=1e-8; hd=tol^.5;
options = odeset('RelTol',tol,'AbsTol',tol);
d=(b-a)/m;t=a:d:b+d/10;
% memory is allocating, Fm is a nxm Matrix for more
% structured programming
u=zeros(n,m); Fm=u;
%  Compute Fm
for j=1:m
    [~,uj]=ode45(@ode,[t(j),t(j+1)],s(:,j),options);
    u(:,j)=uj(end,:)';
    if j ~=m,    % matching conditions
        Fm(:,j+1)=u(:,j)-s(:,j+1);
    else         % boundary condition
        Fm(:,1)=bc(s(:,1),u(:,m));
    end
end
%  Compute Mm by difference approximation
if nargout > 1
% X is a (n x n) x (m+1) (non-sparse) compact matrix
    X=zeros(n,n,m+1);
    for j=1:m
        sh=s(:,j);
        for i=1:n
            h=(1e-3+abs(s(i)))*hd;
            sh(i)=sh(i)+h;
            [~,uh]=ode45(@ode,[t(j),t(j+1)],sh,options);
            uh=uh(end,:)';
            if j==1        % cpmputing of Ba
                Fh=bc(sh,u(:,m));
                X(:,i,m+1)=(Fh(:)-Fm(:,1))/h;
            end
            if j<m        % matrix Xj is computed
                X(:,i,j)=(uh-s(:,j+1)-Fm(:,j+1))/h;
```

```
            else            % Bb*Xm is computed
                Fh=bc(s(:,1),uh);
                X(:,i,m)=(Fh(:)-Fm(:,1))/h;
            end
            sh(i)=s(i,j);
        end
    end
end
Fm=Fm(:); % F is now a column-vector
if m>1      % multiple shooting
    e=ones(n*m,1);
    Mm=spdiags(-e,0,n*m,n*m); % now Mm is a sparse matrix
    Mm(1:n,1:n)=X(:,:,m+1);
    for i=1:m-1
        Mm(i*n+1:n*(i+1),(i-1)*n+1:n*i)=X(:,:,i);
    end
    Mm(1:n,(m-1)*n+1:n*m)=X(:,:,m);
else        % simple shooting
    Mm=X(:,:,1)+X(:,:,2);    % only single shooting (non-sparse)
end
```

4.7   Exercises

Exercise 4.1 Given the parametrized BVP

$$y_1'(x) = y_2(x), \quad y_2'(x) = \lambda\, y_1(x)^2,$$
$$y_1(0) = 4, \quad y_1(1) = 1. \tag{4.80}$$

This BVP is a generalization of a frequently used test problem, where $\lambda = 3/2$ (see, e.g., [57, 117]). For each $\lambda \in [1, 20]$ there are exactly two isolated solutions. For different values of λ determine the corresponding two solutions with the simple shooting method and plot each pair of solutions in a diagram $y_1(x)$ versus x. Moreover, visualize the solution manifold of (4.80) by plotting $y_2(0)$ versus λ, $\lambda \in [1, 20]$.

Exercise 4.2 Given the BVP

$$y_1'(x) = y_2(x), \quad y_2'(x) = 50\, \sin(y_1(x)),$$
$$y_1(0) = 0, \quad y_1(1) = 0. \tag{4.81}$$

This BVP has exactly five isolated solutions. Try to compute all solutions with the simple shooting method and the multiple shooting method.

Exercise 4.3 Consider the BVP (4.79). Set $\lambda = 100$ and determine the solution of this BVP with the multiple shooting method. Generate a starting trajectory by solving (4.79) for the parameter value $\lambda = 0$. Since the ODEs are stiff integrate the associated IVPs with an implicit solver.

Exercise 4.4 Given the periodic BVP (see [103])

$$y''(x) = -0.05\, y'(x) - 0.02\, y(x)^2 \sin(x) + 0.00005\, \sin(x)\cos(x)^2$$
$$- 0.05\, \cos(x) - 0.0025\, \sin(x), \tag{4.82}$$
$$y(0) = y(2\pi), \quad y'(0) = y'(2\pi).$$

It has the exact solution $y(x) = 0.05\cos(x)$. Approximate this solution with the multiple shooting method.

Exercise 4.5 Consider the BVP

$$y''(x) = \exp(y(x)), \quad y(0) = 0, \quad y(1) = -\ln\left(\cos(1/\sqrt{2})^2\right), \tag{4.83}$$

with the exact solution $y(x) = -\ln\left(\cos(x/\sqrt{2})^2\right)$. Approximate this solution with a shooting method of your choice.

Exercise 4.6 Consider the BVP

$$y''(x) = \frac{1}{2}\cos^2(y'(x)), \quad y(0) = 0, \quad y(1) = \arctan(1/2) - \ln(5/4), \tag{4.84}$$

with the exact solution $y(x) = x\arctan((1/2)x) - \ln(1 + (1/4)x^2)$. Approximate this solution with the nonlinear method of complementary functions.

Exercise 4.7 Consider the BVP

$$y''(x) = y(x)\, y'(x),$$
$$y'(0) = -0.5/\cosh^2(1/4), \quad y(1) = -\tanh(1/4). \tag{4.85}$$

The exact solution is $y(x) = \tanh((1 - 2x)/4)$. Approximate this solution with the multiple shooting method.

Exercise 4.8 Given the BVP

$$y''(x) = a\, y(x)\, y'(x). \quad y(0) = \tanh(a/4), \quad y(1) = -\tanh(a/4). \tag{4.86}$$

The exact solution is $y(x) = \tanh(a(1 - 2x)/4)$. Approximate this solution with the multiple shooting method.

Exercise 4.9 Consider the BVP for a system of two second-order ODEs

$$y_1''(x) = y_1(x) + 3\exp(y_2(x)),$$
$$y_2''(x) = y_1(x) - \exp(y_2(x)) + \exp(-x),$$
$$y_1(0) = 0, \quad y_1(1) = \exp(-2) - \exp(-1),$$
$$y_2(0) = 0, \quad y_2(1) = -2, \tag{4.87}$$

with the exact solution

$$y_1(x) = \exp(-2x) - \exp(-x), \quad y_2(x) = -2x.$$

Transform the ODEs into a system of four first-order equations and solve this problem with a shooting method of your choice.

Exercise 4.10 Given the BVP

$$y^{(iv)}(x) = 6\exp(-4y(x)) - \frac{12}{(1+x)^4},$$

$$y(0) = 0, \quad y(1) = \ln(2), \quad y''(0) = -1, \quad y''(1) = -0.25. \tag{4.88}$$

The exact solution is

$$y(x) = \ln(1+x).$$

Approximate this solution with the stabilized march method and the multiple shooting method, and compare the results.

Exercise 4.11 Consider the parametrized BVP

$$y_1''(x) = \lambda^2 y_1(x) + y_2(x) + x + (1 - \lambda^2)\exp(-x),$$
$$y_2''(x) = -y_1(x) + \exp(y_2(x)) + \exp(-\lambda x),$$
$$y_1(0) = 2, \quad y_1(1) = \exp(-\lambda) + \exp(-1), \tag{4.89}$$
$$y_2(0) = 0, \quad y_2(1) = -1.$$

The exact solution is

$$y_1(x) = \exp(-\lambda x) + \exp(-x), \quad y_2(x) = -x.$$

Use the multiple shooting method to approximate this solution for the parameter values $\lambda = 10$ and $\lambda = 20$.

Exercise 4.12 A well-studied nonlinear BVP (see, e.g., [47, 87, 100, 108]) is the so-called Bratu's problem, which is given by

$$y''(x) + \lambda e^{y(x)} = 0,$$
$$y(0) = y(1) = 0, \tag{4.90}$$

where $\lambda > 0$. The analytical solution can be written in the following form

$$y(x) = -2\log\left\{\frac{\cosh(0.5(x - 0.5)\theta)}{\cosh(0.25\,\theta)}\right\},$$

where θ is the solution of the nonlinear algebraic equation

$$f(\theta) \equiv \theta - \sqrt{2\lambda} \cosh(0.25\,\theta) = 0.$$

The Bratu's problem has none, one or two solutions when $\lambda > \lambda^*$, $\lambda = \lambda^*$ and $\lambda < \lambda^*$, respectively. In the next chapter, it will be shown that the critical value $\lambda^* \approx 3.513830719$ is the simple *turning point* of the curve $y'(0)$ versus λ and satisfies the equation

$$1 = 0.25\,\sqrt{2\lambda^*}\,\sinh(0.25\,\theta).$$

For the parameter values $\lambda = 1$ and $\lambda = 2$ approximate and plot the corresponding solutions of Bratu's problem with the simple shooting method.

Exercise 4.13 A modification of the BVP (4.90) is the following problem

$$\begin{aligned}
y''(x) - 2e^{y(x)} &= 0, \\
y(0) = y'(\pi) &= 0.
\end{aligned} \qquad (4.91)$$

Note, in contrast to the standard Bratu problem, the parameter λ is negative. Determine a solution of this BVP with the multiple shooting method using $y(x) \equiv 0$ as starting trajectory and $m = 100$.

In [125], an IVP for the ODE used in formula (4.91) is presented. The corresponding initial conditions are $y(0) = y'(0) = 0$. The exact solution of this IVP is

$$y(x) = -2\ln(\cos(x)). \qquad (4.92)$$

This solution solves the BVP (4.91), too. But the transition from the IVP to the BVP changes the number of solutions. Does the multiple shooting method determine the solution (4.92), or another solution?

Exercise 4.14 Consider the three-point BVP for the third-order nonlinear ODE

$$\begin{aligned}
y'''(x) &= 1 + x\,\sinh(y(x)), \\
y(0) = 0, \quad y(0.25) &= 1, \quad y(1) = 0.
\end{aligned} \qquad (4.93)$$

The exact solution is unknown. Modify the simple shooting method such that the resulting algorithm can be used to determine numerically a solution of the BVP (4.93).

Exercise 4.15 Given the BVP

$$\begin{aligned}
y''(x) &= -y(x) + \frac{2y'(x)^2}{y(x)}, \\
y(-1) = y(1) &= 0.324027137.
\end{aligned} \qquad (4.94)$$

The exact solution is

$$y(x) = \frac{1}{e^x + e^{-x}}.$$

Approximate this solution with a shooting method of your choice.

Exercise 4.16 Consider the two-point BVP for the third-order nonlinear ODE

$$y'''(x) = -\sqrt{1 - y(x)^2},$$
$$y(0) = 0, \quad y'(0) = 1, \quad y(\pi/2) = 1. \tag{4.95}$$

The exact solution is $y(x) = \sin(x)$. Approximate this solution with the nonlinear method of complementary functions.

Exercise 4.17 Given the BVP of Bratu-type (see [125])

$$y''(x) - \pi^2 e^{y(x)} = 0,$$
$$y(0) = y(1) = 0. \tag{4.96}$$

Note, in contrast to the standard Bratu problem (4.90), the parameter λ is negative. The exact solutions are

$$y_{1,2}(x) = -\ln\left(1 \pm \cos\left((0.5 + x)\pi\right)\right).$$

Determine a numerical solution of this BVP with the multiple shooting method using $y(x) \equiv 0$ as starting trajectory and $m = 100$. Which exact solution is approximated by this numerical solution?

Chapter 5
Numerical Treatment of Parametrized Two-Point Boundary Value Problems

5.1 Introduction

One of the simplest and oldest examples to demonstrate the phenomenon of bifurcation is the so-called Euler–Bernoulli rod. Here, a thin and homogeneous rod of length l is gripped at either end. The left end is fixed whereas the right end is allowed to vary along the x-axis when it is subjected to a constant horizontal force P (see Fig. 5.1).

In the unloaded state, the rod coincides with the section $[0, l]$ of the x-axis. If a compressive force P is applied to the right end of the rod, two different states are possible. In the case that the force is very small, the rod is only compressed. However, our experience shows that for larger forces there must be transversal deflections, too.

For the mathematical modelling, let us assume that all deformations of the rod are small and the buckling occurs in the x, y-plane. We are now studying the equilibrium of forces on a piece of the rod (including the left end), see Fig. 5.2.

Let X be the original x-coordinate of a point on the rod. After the buckling this point has moved to the point $(X + u, v)$. Let φ denote the angle between the tangent to the deflected rod and the x-axis. Moreover, s is the arc length measured from the left endpoint. An important assumption for our model is that the rod cannot extend. The Euler–Bernoulli law says that the change in the curvature $(d\varphi/ds)$ is proportional to the moment of force M, i.e., we have the *constitutive relations*

$$s = X \quad \text{and} \quad M = -E I \frac{d\varphi}{ds}, \tag{5.1}$$

where the constants E and I denote the elastic modulus and the second moment of area of the rod's cross-section (see, e.g., [15]), respectively. In addition, we have the *geometric relation*

$$\frac{dv}{ds} = \sin(\varphi), \tag{5.2}$$

© Springer India 2016
M. Hermann and M. Saravi, *Nonlinear Ordinary Differential Equations*,
DOI 10.1007/978-81-322-2812-7_5

Fig. 5.1 Euler–Bernoulli
rod

Fig. 5.2 Equilibrium of
forces of the rod

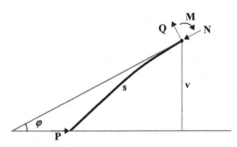

and the *equilibrium relation*

$$M = P\,v. \tag{5.3}$$

These four equations can be reduced to a pair of nonlinear first-order ODEs as
follows. From (5.1), we obtain

$$\frac{M}{E\,I} = -\frac{d\varphi}{ds}.$$

Using (5.3), it follows

$$\frac{P}{E\,I}v = -\frac{d\varphi}{ds}.$$

Defining the dimensionless parameter

$$\lambda \equiv \frac{P}{E\,I}$$

and adding (5.2), we obtain the two equations

$$\varphi'(s) = -\lambda\,v(s), \quad v'(s) = \sin(\varphi(s)). \tag{5.4}$$

The associated boundary conditions are

$$v(0) = v(l) = 0. \tag{5.5}$$

If $\lambda = 0$, i.e., there is no force acting, the first ODE is reduced to $\varphi'(s) = 0$. From
the second ODE, we obtain $v''(s) = \cos(\varphi(s))\,\varphi'(s)$. Substituting $\varphi'(s) = 0$ into
this equation yields $v''(s) = 0$. If we also take into account the boundary conditions
(5.5), we get $v(s) \equiv 0$, which means that no deflection occurs.

The two first-order ODEs (5.4) and the boundary conditions (5.5) can be combined to the following (parametrized) BVP for a single second-order ODE

$$\varphi''(s) + \lambda \sin(\varphi(s)) = 0, \quad 0 < s < l,$$
$$\varphi'(0) = \varphi'(l) = 0. \tag{5.6}$$

Since we have assumed that v and φ are small, it seems reasonable to linearize the ODE. Therefore, we replace $\sin(\varphi)$ by φ and obtain the following linearized BVP

$$\varphi''(s) + \lambda \varphi(s) = 0, \quad 0 < s < l,$$
$$\varphi'(0) = \varphi'(l) = 0. \tag{5.7}$$

Obviously, $\varphi(s) \equiv 0$ is a solution of this BVP for all values of λ. But we are interested in those values of λ for which nontrivial solutions of (5.7) exist. In order to pursue this goal, we proceed as follows. To determine the general solution of the ODE, we use the ansatz $\varphi(s) = e^{ks}$ and obtain the characteristic equation $\rho(k) \equiv k^2 + \lambda = 0$. The corresponding solutions are $k_{1,2} = \pm i\sqrt{\lambda}$. Thus, the general solution of the ODE is

$$\varphi(s) = \hat{c}_1 e^{i\sqrt{\lambda}s} + \hat{c}_2 e^{-i\sqrt{\lambda}s}.$$

Using Euler's formula, we can write

$$\varphi(s) = c_1 \cos(\sqrt{\lambda}s) + c_2 \sin(\sqrt{\lambda}s),$$

where c_1 and c_2 are real constants. It follows

$$\varphi'(s) = -c_1 \sqrt{\lambda} \sin(\sqrt{\lambda}s) + c_2 \sqrt{\lambda} \cos(\sqrt{\lambda}s).$$

Considering the boundary conditions, we get

$$\varphi'(0) = c_2 \sqrt{\lambda} \doteq 0, \quad \text{i.e., } c_2 = 0,$$

and

$$\varphi'(l) = -c_1 \sqrt{\lambda} \sin(\sqrt{\lambda}l) \doteq 0.$$

If $\sin(\sqrt{\lambda}l) = 0$, i.e., $\sqrt{\lambda}l = n\pi$, the above equation is satisfied for all c_1. In analogy to the algebraic eigenvalue problems, we call the numbers

$$\lambda_n = \frac{n^2 \pi^2}{l^2}, \quad n = 1, 2, \ldots \tag{5.8}$$

the *eigenvalues* of the linearized BVP. The corresponding *eigensolutions* are

$$\varphi^{(n)}(s) = c \cos(\sqrt{\lambda_n}s) = c \cos\left(\frac{n\pi}{l}s\right), \quad n = 1, 2, \ldots . \tag{5.9}$$

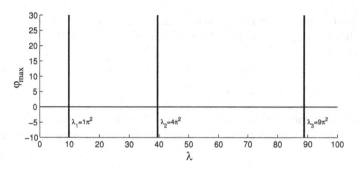

Fig. 5.3 The solution manifold of the linearized BVP (5.7) with $l = 1$

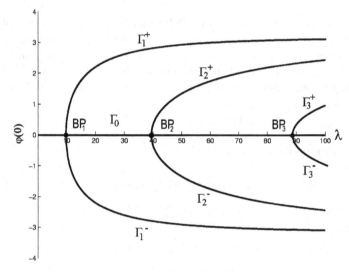

Fig. 5.4 The solution manifold of the nonlinear BVP (5.6); BP_i—primary simple bifurcation points, $\Gamma_i^{\pm}$—solution branches (see the next section)

Now, we want to display graphically φ in dependence of λ. Since φ is a function of s, in Fig. 5.3, we use as ordinate the maximum deflection φ_{max}. As can be seen, there are no deflections before the first critical value $\lambda_1 = (\pi/l)^2$ has been reached. For $\lambda = \lambda_1$, the rod is buckling (the state is $c \cos((\pi/l)s)$), but the magnitude of the deflection is indefinite (since the constant c is arbitrary). If the parameter λ exceeds the critical value λ_1, the rod adopts the undeformed state until the next critical value $\lambda_2 = (4\pi^2)/l^2$ is reached, etc. This shows that in the linearized problem (5.7) the rod is only buckling for the critical values (5.8), i.e., for the loads $P_n \equiv (n^2\pi^2 EI)/l^2$. This result does not correspond to the reality. We can state that only the nonlinear model (5.6) is appropriate to describe the actual buckling behavior of the rod. In Figs. 5.4 and 5.5, the solution manifold ($\varphi(0)$ versus λ) and the corresponding

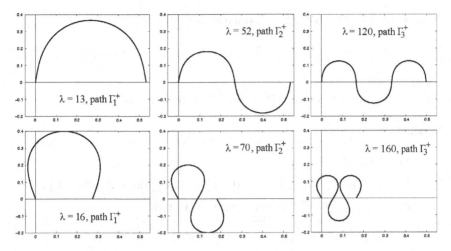

Fig. 5.5 Deformed states of the rod on the branches $\Gamma_i^+, i = 1, 2, 3$

deformed states of the rod, respectively, are shown when the nonlinear BVP is studied with the numerical techniques, which we will present in the next sections.

5.2 Two-Point BVPs and Operator Equations

Let us consider nonlinear parametrized BVPs of the following form

$$
\begin{aligned}
y'(x) &= f(x, y; \lambda), \quad a < x < b, \\
B_a y(a) &+ B_b y(b) = 0,
\end{aligned}
\tag{5.10}
$$

where

$$
y \in C^1([a, b], \mathbb{R}^n), \quad \lambda \in \mathbb{R}, \quad f : D_f \to \mathbb{R}^n, \quad D_f \subset [a, b] \times \mathbb{R}^n \times \mathbb{R},
$$
$$
B_a, B_b \in \mathbb{R}^{n \times n}, \quad \text{rank}[B_a | B_b] = n.
$$

In this notation, the BVP (5.6) reads

$$
\begin{aligned}
y_1'(x) &= y_2(x), \\
y_2'(x) &= -\lambda \sin(y_1(s)),
\end{aligned}
$$

and

$$
\underbrace{\begin{pmatrix} 0 & 1 \\ 0 & 0 \end{pmatrix}}_{B_a} \begin{pmatrix} y_1(0) \\ y_2(0) \end{pmatrix} + \underbrace{\begin{pmatrix} 0 & 0 \\ 0 & 1 \end{pmatrix}}_{B_b} \begin{pmatrix} y_1(l) \\ y_2(l) \end{pmatrix} = \begin{pmatrix} 0 \\ 0 \end{pmatrix}.
$$

Now, we represent the BVP (5.10) as an operator equation

$$T(y, \lambda) = 0, \tag{5.11}$$

where

$$T : Z \equiv Y \times \mathbb{R} \to W, \quad Y, W \text{ Banach spaces,}$$

by defining

$$T(y, \lambda) \equiv y' - f(\cdot, y; \lambda),$$

$$Y = B\mathbb{C}^1([a, b], \mathbb{R}^n) \equiv \left\{ y \in \mathbb{C}^1([a, b], \mathbb{R}^n) : B_a y(a) + B_b y(b) = 0 \right\},$$

$$\|y : B\mathbb{C}^1\| = \|y : \mathbb{C}^1\| \equiv \sup_{x \in [a,b]} \{|y(x)| + |y'(x)|\},$$

$$W \equiv \mathbb{C}([a, b], \mathbb{R}^n), \quad \|w : \mathbb{C}\| \equiv \sup_{x \in [a,b]} \{|w(x)|\},$$

$$|x| \equiv (x^T x)^{1/2} \text{ for } x \in \mathbb{R}^n.$$

To simplify the representation, we often use the abbreviation $z \equiv (y, \lambda) \in Z$ and write (5.11) in the form

$$T(z) = 0. \tag{5.12}$$

For the following considerations, some definitions are required.

Definition 5.1 An element $z_0 \in Z$ is called a *solution* of problem (5.12) if and only if $T(z_0) = 0$. □

Definition 5.2 The operator T is *Fréchet differentiable* at $z_0 \in Z$ if and only if there exists a linear map $L(z_0) : Z \to W$ such that

$$\lim_{h \to 0} \frac{1}{\|h\|} \|T(z_0 + h) - T(z_0) - L(z_0)h\| = 0, \quad h \in Z.$$

If it exists, this $L(z_0)$ is called the *Fréchet derivative* of T at z_0. □

Definition 5.3 The operator $T \equiv T(y, \lambda)$ is *partially Fréchet differentiable* w.r.t. y at $z_0 = (y_0, \lambda_0)$ if and only if there exists a linear map $L(y_0) : Y \to W$ such that

$$\lim_{h \to 0} \frac{1}{\|h\|} \|T(y_0 + h, \lambda_0) - T(y_0, \lambda_0) - L(y_0)h\| = 0, \quad h \in Y.$$

If it exists, this $L(y_0)$ is called the *partial Fréchet derivative* of T w.r.t. y at z_0 and is denoted by T_y^0. □

Definition 5.4 An element $z_0 \in Z$ is an *isolated solution* of the operator equation (5.12) if and only if z_0 is a solution of (5.12) and the Fréchet derivative $T'(z_0)$ is bijective. □

Definition 5.5 A linear map $L \in \mathscr{L}(Z, W)$ is called a *Fredholm operator* if and only if the following three conditions are satisfied:

- $\mathcal{R}(L)$ is closed,
- $\dim \mathcal{N}(L) < \infty$, and
- $\operatorname{codim} \mathcal{R}(L) < \infty$.

Here, the *null space* of L is denoted as $\mathcal{N}(L)$ and the *range* as $\mathcal{R}(L)$. The number

$$\operatorname{ind}(L) \equiv \dim \mathcal{N}(L) - \operatorname{codim} \mathcal{R}(L)$$

is called the *index* of the Fredholm operator. $\qquad \square$

For the BVP (5.10), the following can be shown.

Theorem 5.6 *Let $z_0 \equiv (y_0, \lambda_0) \in Z$ be a solution of the operator equation* (5.11). *Then $T_y^0 \equiv T_y(z_0)$ is a Fredholm operator with index zero.*

Proof See, e.g., [123]. $\qquad \blacksquare$

Remark 5.7 To simplify the representation, we have restricted ourselves to BVPs of the form (5.10). For more general parametrized nonlinear BVPs,

$$\begin{aligned}
y'(x) &= f(x, y; \lambda), \quad a < x < b, \\
r(y(a), y(b); \lambda) &= 0,
\end{aligned} \tag{5.13}$$

the operator T, which is used in the operator equation (5.11), has to be defined as

$$T(y, \lambda) \equiv \begin{pmatrix} y' - f(\cdot, y; \lambda) \\ r(y(a), y(b); \lambda) \end{pmatrix}, \tag{5.14}$$

where the Banach spaces Y, W must be adapted accordingly. If it can be shown that T_y^0 is a Fredholm operator with index zero, the theoretical and numerical techniques presented in the next sections are still valid for the BVP (5.13). $\qquad \square$

Let $E(T)$ be the manifold of all solutions $z \equiv (y, \lambda)$ of the nonlinear operator equation (5.11), which we call the *solution field* of the operator T. The solution field can be represented graphically in form of a *bifurcation diagram* where the values of a linear functional l^* at the solutions y are plotted versus λ. In Fig. 5.6, an example of a bifurcation diagram is given. In this diagram, the solutions are arranged in curves Γ_i, which we call the *solution curves*. These curves may intersect at so-called *bifurcation points* (BP). The bifurcation points are singular solutions of the operator equation as we will see later. There is another type of singular solutions, namely points on a solution curve where this curve turns back. These points are called *limit points* or *turning points* (TP).

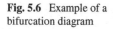

Fig. 5.6 Example of a
bifurcation diagram

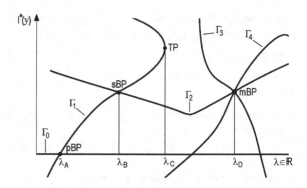

Our bifurcation diagram contains qualitatively different types of singular points. If Γ_0 is a known (primary) solution curve, the point λ_A is called *primary simple* bifurcation point (pBP) since only two curves (the primary curve Γ_0 and the secondary curve Γ_1) are intersecting in λ_A. In general, it can be assumed that $\Gamma_0 = \{(y, \lambda) = (0, \lambda)$ for all $\lambda\}$. The point λ_B is a *secondary simple* bifurcation point (sBP) since only two curves (the secondary curves Γ_1 and Γ_2) are intersecting in λ_B. Consequently, λ_D is a *secondary multiple* bifurcation point (mBP) since more than two curves (the secondary curves Γ_2, Γ_3 and Γ_4) are intersecting in λ_D. Finally, λ_C is called *simple turning* point.

The great variety in the quality of the possible singular points must be taken into account, if a numerical treatment of the nonlinear parametrized operator equation (5.11) is realized.

5.3 Analytical and Numerical Treatment of Limit Points

5.3.1 Simple Solution Curves

In this section, the study of $E(T)$ is restricted to regions, where no branching of solution curves occur. For this, we need the statement of the following theorem.

Theorem 5.8 (IMPLICIT FUNCTION THEOREM)
Suppose that X, Y, Z are Banach spaces, $U \subset X$, $V \subset Y$ are open sets, $T \in \mathbb{C}^p(U \times V, Z)$, $p \geq 1$, $z_0 \equiv (x_0, y_0) \in U \times V$, $T(z_0) = 0$ and $T_x(z_0)$ is a linear homeomorphism from X onto Z. Then, there exist a neighborhood $U_1 \times V_1 \in U \times V$ of z_0 and a function $g \in \mathbb{C}^p(V_1, X)$, $g(y_0) = x_0$ such that $T(x, y) = 0$ for $(x, y) \in U_1 \times V_1$ if and only if $x = g(y)$.

Proof See, e.g., [27].   ∎

Let $z_0 \equiv (x_0, \lambda_0) \in Z$ be a solution of (5.11). Then, y can be parametrized with respect to λ in a neighborhood of z_0, if T_y^0 is a linear homeomorphism.

Definition 5.9 A solution $z_0 \in Z$ of (5.11) is called *simple*, if the Fréchet derivative

$$T'(z_0) = (T_y^0, T_\lambda^0) \in \mathscr{L}(Z, W)$$

satisfies

$$\dim \mathcal{N}(T'(z_0)) = 1, \quad \mathcal{R}(T'(z_0)) = W. \tag{5.15}$$

$\square$

To find the simple solutions of the operator equation (5.11), let us consider the equation

$$T'(z_0)h \equiv (T_y^0, T_\lambda^0) \begin{pmatrix} w \\ \mu \end{pmatrix} = 0, \quad \text{i.e., } T_y^0 w + \mu T_\lambda^0 = 0.$$

Obviously, this equation is homogeneous and we have to distinguish between two cases:

$$\text{case 1:} \quad h_1 = \begin{pmatrix} w_1 \\ 1 \end{pmatrix}, \quad \text{case 2:} \quad h_2 = \begin{pmatrix} w_2 \\ 0 \end{pmatrix}.$$

- *Case 1*: To determine w_1 uniquely, it must be assumed that $\mathcal{N}(T_y^0) = \{0\}$. Thus, w_1 is defined by $T_y^0 w_1 = -T_\lambda^0$ and the corresponding solution z_0 is called *isolated*.
- *Case 2*: To determine w_2 uniquely, it must be assumed that $\mathcal{N}(T_y^0) = \text{span}\{\varphi\}$, $0 \neq \varphi \in Y$, and $T_\lambda^0 \notin \mathcal{R}(T_y^0)$. The corresponding solution z_0 is called *limit point*.

In case of isolated solutions, Theorem 5.8 guarantees the existence of a unique solution curve $\{y(\lambda), \lambda\}$ for $|\lambda - \lambda_0| \leq \varepsilon$.

We now want to derive a local statement for limit points. Let us start with the assumption (see case 2) that

$$\mathcal{N}(T_y^0) = \text{span}\{\varphi_0\}, \quad \|\varphi_0\| = 1. \tag{5.16}$$

From Theorem 5.6 we know, that T_y^0 is a Fredholm operator with index zero. Therefore, the dual operator $T_y^{0*} : W^* \to Y^*$ also has a one-dimensional null space, i.e.,

$$\mathcal{N}(T_y^{0*}) = \text{span}\{\psi_0^*\}, \quad \psi_0^* \in W^*, \quad \|\psi_0^*\| = 1, \tag{5.17}$$

where Y^* and W^* are the dual spaces of Y and W, respectively. For any Banach space X the dual Banach space X^* is the set of bounded linear functionals on X equipped with an appropriate norm.

The theorem on biorthogonal bases (see, e.g., [134]) implies the existence of elements $\varphi_0^* \in Y^*$ and $\psi_0 \in W$ such that

$$\varphi_0^* \varphi_0 = 1 \quad \text{and} \quad \psi_0^* \psi_0 = 1. \tag{5.18}$$

Now we can define the following decompositions of the Banach spaces Y and W:

$$Y = \text{span}\{\varphi_0\} \oplus Y_1, \quad W = \text{span}\{\psi_0\} \oplus W_1, \tag{5.19}$$

where

$$Y_1 \equiv \{y \in Y : \varphi_0^* y = 0\}, \quad W_1 \equiv \{w \in W : \psi_0^* w = 0\}. \tag{5.20}$$

Finally, we have the *Fredholm alternative* (see, e.g., [43])

$$W_1 = \mathcal{R}(T_y^0). \tag{5.21}$$

A well-known analytical technique to study the operator equation (5.11) is the *Liapunov–Schmidt reduction*. Let Q_2 denote the projection of W onto $W_1 = \mathcal{R}(T_y^0)$, i.e.,

$$Q_2 w = w - (\psi_0^* w)\psi_0, \quad w \in W. \tag{5.22}$$

Obviously, for $w \in W$ it holds

$$w = 0 \quad \text{if and only if } Q_2 w = 0 \text{ and } (I - Q_2)w = 0.$$

Thus, the operator equation (5.11) can be expanded to an equivalent pair of equations

$$\begin{aligned} Q_2\, T(z) &= 0, \quad z \equiv (y, \lambda) \in Z, \\ Q_1\, T(z) &\equiv (I - Q_2)\, T(z) = 0. \end{aligned} \tag{5.23}$$

We still need a second pair of projections, namely

$$P_1 : Y \to \mathcal{N}(T_y^0) \quad \text{and} \quad P_2 = I - P_1.$$

Let $z_0 = (y_0, \lambda_0)$ be a fixed solution of (5.11). Because of the decomposition (5.19), we may split any $y \in Y$ in the form

$$y = y_0 + \varepsilon \varphi_0 + w, \quad \varepsilon \in \mathbb{R}, \quad P_2 w = w. \tag{5.24}$$

The condition $P_2 w = w$ says that $w \in Y_1$. In the same way, we split $\lambda \in \mathbb{R}$ as

$$\lambda = \lambda_0 + \xi, \quad \xi \in \mathbb{R}. \tag{5.25}$$

Substituting (5.24) and (5.25) into the first equation of (5.23) yields

$$G(\xi, \varepsilon, w) \equiv Q_2\, T(y_0 + \varepsilon \varphi_0 + P_2 w, \lambda_0 + \xi) = 0.$$

It holds $G(0, 0, 0) = 0$ and $G_w(\xi, \varepsilon, w) = Q_2\, T_y(y_0 + \varepsilon \varphi_0 + P_2 w, \lambda_0 + \xi)P_2$. Let us set $G_w^0 \equiv G_w(0, 0, 0) = Q_2 T_y^0 P_2$.

The operator G_w^0 is an *injective* map of Y_1 onto $\mathcal{R}(T_y^0)$. This can be shown as follows: Given the equation

$$G_w^0 v = 0, \quad \text{i.e.,} \quad Q_2 T_y^0 P_2 v = 0.$$

Since Q_2 is a projection onto the range of T_y^0, we get $T_y^0 P_2 v = 0$. From $P_2 v \notin \mathcal{N}(T_y^0)$ we deduce $v = 0$.

To show that the operator G_w^0 is also a *surjective* map, we need the following lemma:

Lemma 5.10 *If X and Y are Banach spaces and $L \in \mathcal{L}(X, Y)$ is bijective, then the inverse operator L^{-1} is continuous.*

Proof This statement is a direct conclusion from the Open Mapping Theorem (see, e.g., [132]). ∎

For any $h \in \mathcal{R}(T_y^0)$, the equation $T_y^0 y = h$ is solvable. Obviously, the corresponding solution y also satisfies the equation $Q_2 T_y^0 P_2 y = G_w^0 y = h$.

Thus, G_w^0 is *bijective*. With Lemma 5.10, we can conclude that G_w^0 is a homeomorphism of Y_1 onto $\mathcal{R}(T_y^0)$. This result and the Implicit Function Theorem 5.8 imply the claim of the next theorem.

Theorem 5.11 *Assume z_0 is a limit point of the operator T. Then, there exist two positive constants ξ_0, ε_0, and a $\mathbb{C}^p$-map $w : [-\xi_0, \xi_0] \times [-\varepsilon_0, \varepsilon_0] \to Y_1$ such that*

$$G(\xi, \varepsilon, w(\xi, \varepsilon)) = 0, \quad w(0, 0) = 0.$$

Now, we use the information just received in the second equation of (5.23)

$$Q_1 T(y_0 + \varepsilon\varphi_0 + w(\xi, \varepsilon), \lambda_0 + \xi) = 0.$$

This equation is transformed into a scalar equation by applying the functional ψ_0^* on both sides. We obtain

$$
\begin{aligned}
0 &= \psi_0^* Q_1 T(y_0 + \varepsilon\varphi_0 + w(\xi, \varepsilon), \lambda_0 + \xi) \\
&= \psi_0^* (I - Q_2) T(y_0 + \varepsilon\varphi_0 + w(\xi, \varepsilon), \lambda_0 + \xi)
\end{aligned}
$$

Using the definition (5.22) of the projection Q_2, it follows:

$$0 = \psi_0^* \big(T(\cdot) - T(\cdot) + (\psi_0^* T(\cdot)) \, \psi_0 \big).$$

Thus, the solution of the operator equation (5.11) in the neighborhood of z_0 is reduced to the solution of a finite-dimensional algebraic equation in the neighborhood of the origin $(0, 0)$

$$F(\xi, \varepsilon) \equiv \psi_0^* T(y_0 + \varepsilon\varphi_0 + w(\xi, \varepsilon), \lambda_0 + \xi) = 0. \tag{5.26}$$

This important equation is called the *limit point equation*. The next theorem shows that ξ can be parametrized with respect to ε.

Theorem 5.12 *Let $z_0 \in Z$ be a limit point of the operator T. Then, there exist a constant $\varepsilon_0 > 0$ and a $\mathbb{C}^p$-map $\xi : [-\varepsilon_0, \varepsilon_0] \to \mathbb{R}$ such that*

$$F(\xi(\varepsilon), \varepsilon) = 0, \quad |\varepsilon| \le \varepsilon_0, \quad \xi(0) = 0.$$

Proof It is $F(0, 0) = 0$ and

$$\frac{\partial F}{\partial \varepsilon}(\xi, \varepsilon) = \psi_0^* T_y(y_0 + \varepsilon\varphi_0 + w(\xi, \varepsilon), \lambda_0 + \xi)[\varphi_0 + w_\varepsilon(\xi, \varepsilon)].$$

Thus

$$\frac{\partial F}{\partial \varepsilon}(0, 0) = \psi_0^* T_y^0[\varphi_0 + w_\varepsilon(0, 0)] = \psi_0^* T_y^0 w_\varepsilon(0, 0) = 0,$$

since $T_y^0 w_\varepsilon(0, 0) \in \mathcal{R}(T_y^0) = \{x : \psi_0^* x = 0\}$.

For a limit point the condition $T_\lambda^0 \notin \mathcal{R}(T_y^0)$ is satisfied. Let us write this condition in the form

$$c \equiv \psi_0^* T_\lambda^0 \ne 0, \tag{5.27}$$

where c is the *characteristic coefficient*.

Now, we compute

$$\frac{\partial F}{\partial \xi}(\xi, \varepsilon) = \psi_0^* T_\lambda(y_0 + \varepsilon\varphi_0 + w(\xi, \varepsilon), \lambda_0 + \xi)$$

$$+ \psi_0^* T_y(y_0 + \varepsilon\varphi_0 + w(\xi, \varepsilon), \lambda_0 + \xi)w_\xi(\xi, \varepsilon).$$

The second term on the right-hand side vanishes, since $T_y^0(\cdot)w_\xi(\cdot) \in \mathcal{R}(T_y^0)$. Because of the condition (5.27), we obtain

$$\frac{\partial F}{\partial \xi}(0, 0) \ne 0.$$

Using the Implicit Function Theorem 5.8, one obtains the result stated in the theorem. ∎

Corollary 5.13 *Let $z_0 \equiv (y_0, \lambda_0)$ be a limit point of the operator T. Then, in the neighborhood of $z_0 \equiv (y_0, \lambda_0)$ there exist a curve $\{y(\varepsilon), \lambda(\varepsilon)\}$, $|\varepsilon| \le \varepsilon_0$, of solutions of the equation (5.11). Moreover, $\lambda : [-\varepsilon_0, \varepsilon_0] \to \mathbb{R}$ and $y : [-\varepsilon_0, \varepsilon_0] \to Y$ are $\mathbb{C}^p$-functions, which can be represented in the form*

$$\lambda(\varepsilon) = \lambda_0 + \xi(\varepsilon),$$
$$y(\varepsilon) = y_0 + \varepsilon\varphi_0 + w(\xi(\varepsilon), \varepsilon), \quad |\varepsilon| \le \varepsilon_0. \tag{5.28}$$

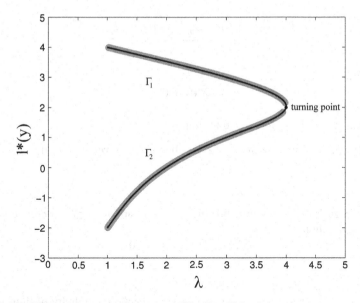

Fig. 5.7 The difference between solution *curve* (*black*) and branch (*gray*)

Definition 5.14

- A continuously differentiable curve of solutions $z(\varepsilon) \equiv \{y(\varepsilon), \lambda(\varepsilon)\} \in Z, \varepsilon \in [\varepsilon_0, \varepsilon_1]$, of the equation (5.11) is called *simple*, if it exclusively consists of simple solutions (isolated solutions, limit points) and $\dot{z}(\varepsilon) \neq 0$ for all $\varepsilon \in [\varepsilon_0, \varepsilon_1]$. Here, the dot indicates the derivative w.r.t. ε.
- A *solution branch* is a continuously differentiable solution curve $\{y(\lambda), \lambda\} \in Z$, $\lambda \in [\lambda_0, \lambda_1]$, which consists exclusively of isolated solutions. □

In Fig. 5.7 it is shown that the solution curve is split into two solution branches Γ_1 and Γ_2 if the limit point is extracted from the curve.

We now want to make a classification of the limit points. Let us start with the limit point equation

$$F(\xi(\varepsilon), \varepsilon) = \psi_0^* T(y_0 + \varepsilon\varphi_0 + w(\xi(\varepsilon), \varepsilon), \lambda_0 + \xi(\varepsilon)).$$

Differentiating this equation gives

$$\frac{dF}{d\varepsilon}(\xi(\varepsilon), \varepsilon) = \psi_0^* T_y(y_0 + \varepsilon\varphi_0 + w(\xi(\varepsilon), \varepsilon), \lambda_0 + \xi(\varepsilon)) \dot{y}(\varepsilon)$$

$$+ \psi_0^* T_\lambda(y_0 + \varepsilon\varphi_0 + w(\xi(\varepsilon), \varepsilon), \lambda_0 + \xi(\varepsilon)) \dot{\lambda}(\varepsilon).$$

In the proof of Theorem 5.12, we have shown that $\dfrac{dF}{d\varepsilon}(0, 0) = 0$. Thus, we obtain

$$0 = \frac{dF}{d\varepsilon}(0,0) = \underbrace{\psi_0^* T_y^0 \, \dot{y}(0)}_{=0} + \underbrace{\psi_0^* T_\lambda^0 \, \dot{\lambda}(0)}_{c \neq 0}.$$

It follows

$$\dot{\lambda}(0) = 0. \tag{5.29}$$

This gives rise to the following definition.

Definition 5.15 A limit point $z_0 \in Z$ is called *simple turning point* if and only if

$$\ddot{\lambda}(0) \neq 0. \tag{5.30}$$

Otherwise, z_0 is called *multiple turning point*. In particular, z_0 is a *double turning point* if and only if

$$\ddot{\lambda}(0) = 0, \quad \dddot{\lambda}(0) \neq 0. \tag{5.31}$$

$\square$

Now, we want to find easy-to-verify criteria for the multiplicity of limit points. The existence of the parametrization (5.28) we have shown. Therefore, we can insert the parametrization into (5.11) and obtain

$$T(y(\varepsilon), \lambda(\varepsilon)) = 0, \quad |\varepsilon| \leq \varepsilon_0.$$

Differentiating this equation w.r.t. ε gives

$$T_y(y(\varepsilon), \lambda(\varepsilon)) \, \dot{y}(\varepsilon) + T_\lambda(y(\varepsilon), \lambda(\varepsilon)) \, \dot{\lambda}(\varepsilon) = 0. \tag{5.32}$$

Setting $\varepsilon = 0$ yields

$$T_y^0 \, \dot{y}(0) + T_\lambda^0 \, \dot{\lambda}(0) = 0.$$

Taking into account that the relation (5.29) is satisfied for a limit point, we get

$$T_y^0 \, \dot{y}(0) = 0.$$

Thus, $\dot{y}(0) \in \mathcal{N}(T_y^0)$, i.e. $\dot{y}(0) = c \, \varphi_0$.
Differentiating the identity (5.32) again with respect to ε, we obtain

$$T_y(y(\varepsilon), \lambda(\varepsilon)) \, \ddot{y}(\varepsilon) + T_{yy}(y(\varepsilon), \lambda(\varepsilon)) \, \dot{y}(\varepsilon)^2 + 2 T_{y\lambda}(y(\varepsilon), \lambda(\varepsilon)) \, \dot{y}(\varepsilon) \dot{\lambda}(\varepsilon)$$
$$+ T_{\lambda\lambda}(y(\varepsilon), \lambda(\varepsilon)) \, \dot{\lambda}(\varepsilon)^2 + T_\lambda(y(\varepsilon), \lambda(\varepsilon)) \, \ddot{\lambda}(\varepsilon) = 0.$$

Setting $\varepsilon = 0$ yields

$$T_y^0 \, \ddot{y}(0) + T_{yy}^0 \, \dot{y}(0)^2 + 2 T_{y\lambda}^0 \, \dot{y}(0) \underbrace{\dot{\lambda}(0)}_{=0} + T_{\lambda\lambda}^0 \, \underbrace{\dot{\lambda}(0)^2}_{=0} + T_\lambda^0 \, \ddot{\lambda}(0) = 0.$$

It follows
$$T_y^0 \, \ddot{y}(0) = -\left(T_{yy}^0 \, \dot{y}(0)^2 + T_\lambda^0 \, \ddot{\lambda}(0)\right).$$

This equation is solvable if the right-hand side is in the range of T_y^0, i.e.,

$$\psi_0^* T_{yy}^0 \, \dot{y}(0)^2 + \ddot{\lambda}(0)\psi_0^* T_\lambda^0 = 0.$$

Thus,

$$\ddot{\lambda}(0) = -\frac{\psi_0^* T_{yy}^0 \, \dot{y}(0)^2}{c},$$

where the characteristic coefficient c is defined in (5.27). Since $\dot{y}(0) \in \mathcal{N}(T_y^0)$ we can write

$$\ddot{\lambda}(0) = -\alpha^2 \frac{\psi_0^* T_{yy}^0 \, \varphi_0^2}{c}, \quad \alpha \in \mathbb{R}, \quad \alpha \neq 0.$$

Corollary 5.16 *The limit point $z_0 \in Z$ is a simple turning point if and only if the second bifurcation coefficient a_2 satisfies*

$$a_2 \equiv \psi_0^* T_{yy}^0 \, \varphi_0^2 \neq 0. \tag{5.33}$$

A similar calculation yields the following result.

Corollary 5.17 *The limit point $z_0 \in Z$ is a double turning point if and only if*

$$a_2 = 0 \quad and \quad a_3 \equiv \psi_0^* \left(6 T_{yy}^0 \, \varphi_0 w_0 + T_{yyy}^0 \, \varphi_0^3\right) \neq 0, \tag{5.34}$$

where $w_0 \in Y_1$ is the unique solution of the equation $T_y^0 \, w_0 = -\dfrac{1}{2} T_{yy}^0 \, \varphi_0^2$.

5.3.2 Extension Techniques for Simple Turning Points

Let $z_0 \equiv (y_0, \lambda_0) \in Z$ be a simple turning point. A direct determination of y_0 from the original problem (5.11) (for the fixed value $\lambda = \lambda_0$) by numerical standard techniques is not possible since in that case y_0 is not an isolated solution. One possibility to overcome this difficulty is to embed (5.11) into an enlarged problem. The new problem is constructed such that some components of an isolated solution of the enlarged problem coincide with the components of z_0. We call this solution strategy *extension technique*.

A first extension technique is based on the enlarged operator

$$\tilde{T} : \begin{cases} \tilde{Z} \equiv Y \times \mathbb{R} \times Y \to \tilde{W} \equiv W \times W \times \mathbb{R} \\ \tilde{z} \equiv (z, \varphi) = (y, \lambda, \varphi) \to \begin{pmatrix} T(y, \lambda) \\ T_y(y, \lambda)\varphi \\ \varphi_0^*\varphi - 1 \end{pmatrix}, \end{cases} \tag{5.35}$$

where $\varphi_0^* \in Y^*$ such that $\varphi_0^*\varphi_0 = 1$. Obviously, $\tilde{z}_0 = (z_0, \varphi_0) \in \tilde{Z}$ satisfies the equation

$$\tilde{T}(\tilde{z}) = 0. \tag{5.36}$$

The question we have to answer now is: do the simple turning points z_0 of problem (5.11) actually correspond to isolated solutions $\tilde{z}_0$ of (5.36)?

Let us study the null space of $\tilde{T}'(\tilde{z}_0)$. It holds

$$\tilde{T}'(\tilde{z}_0) = \begin{pmatrix} T_y^0 & T_\lambda^0 & 0 \\ T_{yy}^0\varphi_0 & T_{y\lambda}^0\varphi_0 & T_y^0 \\ 0 & 0 & \varphi_0^* \end{pmatrix}. \tag{5.37}$$

If $(u, \mu, w) \in \mathcal{N}(\tilde{T}'(\tilde{z}_0))$, with $u, w \in Y$ and $\mu \in \mathbb{R}$, we have

$$\tilde{T}'(\tilde{z}_0) \begin{pmatrix} u \\ \mu \\ w \end{pmatrix} = \begin{pmatrix} T_y^0 u + \mu T_\lambda^0 \\ T_{yy}^0\varphi_0 u + \mu T_{y\lambda}^0\varphi_0 + T_y^0 w \\ \varphi_0^* w \end{pmatrix} = 0. \tag{5.38}$$

The first equation reads

$$T_y^0 u = -\mu T_\lambda^0.$$

The condition $T_\lambda^0 \notin \mathcal{R}(T_y^0)$ implies $\mu = 0$. Therefore, $u = \alpha\varphi_0$, $\alpha \in \mathbb{R}$. Substituting this representation into the second and third equation of (5.38) gives

$$T_y^0 w = -\alpha T_{yy}^0\varphi_0^2, \quad \varphi_0^* w = 0. \tag{5.39}$$

As defined in (5.19), let Y_1 be the complement of $\mathcal{N}(T_y^0)$. The condition $\varphi_0^* w = 0$ indicates that $w \in Y_1$. Problem (5.39) is solvable if

$$0 = \alpha \psi_0^* T_{yy}^0\varphi_0^2 = \alpha a_2. \tag{5.40}$$

Now, two cases must be distinguished:

- $z_0 \equiv (y_0, \lambda_0)$ is a *simple turning point*. Then, equation (5.40) implies $\alpha = 0$ since $a_2 \neq 0$. Substituting this result into (5.39) we have $w \in Y_1 \cap \mathcal{N}(T_y^0)$, i.e., $w = 0$. Furthermore $u = \alpha\varphi_0 = 0$. Thus,

$$\mathcal{N}(\tilde{T}'(\tilde{z}_0)) = \{0\}.$$

- $z_0 \equiv (y_0, \lambda_0)$ is a *multiple turning point*. In that case, we have $a_2 = 0$ and

$$\mathcal{N}(\tilde{T}'(\tilde{z}_0)) = \{\alpha\,\Phi_0 : \alpha \in \mathbb{R}\}, \tag{5.41}$$

where $\Phi_0 \equiv (\varphi_0, 0, w_0) \in Y \times \mathbb{R} \times Y$ and w_0 is the unique solution of the linear problem

$$T_y^0 w_0 = -T_{yy}^0 \varphi_0^2, \quad w_0 \in Y_1.$$

We are now able to formulate the following theorem.

Theorem 5.18 *Let $z_0 \equiv (y_0, \lambda_0)$ be a simple turning point of the operator T. Then, $\tilde{T}'(\tilde{z}_0)$ is a linear homeomorphism from $\tilde{Z}$ onto $\tilde{W}$, i.e.,*

$$\mathcal{N}(\tilde{T}'(\tilde{z}_0)) = \{0\}, \quad \mathcal{R}(\tilde{T}'(\tilde{z}_0)) = \tilde{W}.$$

Let $z_0 \equiv (y_0, \lambda_0)$ be a multiple turning point of the operator T. Then, it holds

$$\dim \mathcal{N}(\tilde{T}'(\tilde{z}_0)) = 1, \quad \mathrm{codim}\,\mathcal{R}(\tilde{T}'(\tilde{z}_0)) = 1.$$

The null space is given in (5.41), whereas the range satisfies

$$\mathcal{R}(\tilde{T}'(\tilde{z}_0)) = \{f \equiv (v, x, \sigma) \in W \times W \times \mathbb{R} : \Psi_0^* f = 0\}, \tag{5.42}$$

where

$$\Psi_0^* \equiv (\xi_0^*, \psi_0^*, 0) \in W^* \times W^* \times \mathbb{R},$$

and ξ_0^* is defined by the equations

$$\begin{aligned}
\xi_0^* T_\lambda^0 &= -\psi_0^* T_{y\lambda}^0 \varphi_0, \\
\xi_0^* T_y^0 &= -\psi_0^* T_{yy}^0 \varphi_0.
\end{aligned} \tag{5.43}$$

Proof The first part of the proof can be seen above. The rest is shown in the monograph [123]. ∎

If z_0 denotes again a simple turning point, then the reversal of the claim of Theorem 5.18 is true.

Theorem 5.19 *Let $\tilde{z}_0 \equiv (y_0, \lambda_0, \varphi_0)$ be an isolated solution of (5.36), i.e., $\tilde{T}'(\tilde{z}_0)$ is a linear homeomorphism from $\tilde{Z}$ onto $\tilde{W}$. Then, $z_0 \equiv (y_0, \lambda_0)$ is a simple turning point.*

Proof

1. Suppose $\varphi_1 \in Y$ satisfies $T_y^0 \varphi_1 = 0$, $\varphi_0^* \varphi_1 = 0$. It can be easily seen that $(0, 0, \varphi_1) \in \mathcal{N}(\tilde{T}'(\tilde{z}_0))$. Since $\tilde{z}_0$ is an isolated solution, we must have $\varphi_1 = 0$, i.e., the null space of T_y^0 is one-dimensional with $\mathcal{N}(T_y^0) = \mathrm{span}\{\varphi_0\}$ and $T_y^0 \varphi_0 = 0$, $\varphi_0^* \varphi_0 = 1$.

2. Suppose $a_2 = 0$ (z_0 is a multiple turning point). Then, $\Phi_0 \equiv (\varphi_0, 0, w_0) \in \mathcal{N}(\tilde{T}'(\tilde{z}_0))$, with $\varphi_0 \neq 0$ and $w_0 \neq 0$. However, since $\mathcal{N}(\tilde{T}'(\tilde{z}_0)) = \{0\}$, we must have $a_2 \neq 0$.
3. Let us assume that $c \equiv \psi_0^* T_\lambda^0 = 0$. Then, there must exist a $\varphi_1 \in Y$, which satisfies $T_y^0 \varphi_1 = -T_\lambda^0$. With a suitably chosen $\alpha \in \mathbb{R}$ and $w \in Y_1 \equiv \{v \in Y : \varphi_0^* v = 0\}$, it holds $(\alpha \varphi_0 + \varphi_1, 1, w) \in \mathcal{N}(\tilde{T}'(\tilde{z}_0))$. This is a contradiction to the assumption $\mathcal{N}(\tilde{T}'(\tilde{z}_0)) = \{0\}$. ∎

Theorems 5.18 and 5.19 say that there is a one-to-one relationship between the simple turning points of (5.11) and the isolated solutions of the enlarged problem (5.36).

Another extension technique is based on the following enlarged operator

$$\hat{T} : \begin{cases} \hat{Z} \equiv Y \times \mathbb{R} \times W^* \to \hat{W} \equiv W \times Y^* \times \mathbb{R} \\ \hat{z} \equiv (z, \psi^*) = (y, \lambda, \psi^*) \to \begin{pmatrix} T(y, \lambda) \\ T_y^*(y, \lambda) \psi^* \\ \psi^* T_\lambda(y, \lambda) - 1 \end{pmatrix} \end{cases} . \tag{5.44}$$

Obviously, $\hat{z}_0 \equiv (z_0, \psi_0^*) \in \hat{Z}$ satisfies the equation

$$\hat{T}(\hat{z}) = 0. \tag{5.45}$$

The next theorem shows that the operator $\hat{T}$ has the same favorable properties as the operator $\tilde{T}$.

Theorem 5.20 *Let $z_0 \equiv (y_0, \lambda_0)$ be a simple turning point of the operator T. Then, $\hat{T}'(\hat{z}_0)$ is a linear homeomorphism from $\hat{Z}$ onto $\hat{W}$, i.e.,*

$$\mathcal{N}(\hat{T}'(\hat{z}_0)) = \{0\}, \quad \mathcal{R}(\hat{T}'(\hat{z}_0)) = \hat{W}.$$

Proof See, e.g., [123]. ∎

We now want to come back to the BVP (5.10) and apply the corresponding extension techniques. For this BVP the enlarged operator equation (5.36) is

$$\begin{aligned} y'(x) &= f(x, y(x); \lambda), & B_a y(a) + B_b y(b) &= 0, \\ \varphi'(x) &= f_y(x, y(x); \lambda)\varphi(x), & B_a \varphi(a) + B_b \varphi(b) &= 0, \\ 0 &= \varphi_0^* \varphi(x) - 1. \end{aligned} \tag{5.46}$$

Here, the question arises how the linear functional φ_0^* can be adequately formulated. Obviously, the following formula defines a functional on the Banach space Y:

$$\varphi_0^* v(x) \equiv \int_a^b \varphi_0(x)^T v(x) dx, \quad v \in Y. \tag{5.47}$$

The condition

$$1 = \varphi_0^* \varphi_0(x) = \int_a^b \varphi_0(x)^T \varphi_0(x) dx$$

can be realized with the function

$$\xi(x) \equiv \int_a^x \varphi_0(s)^T \varphi_0(s) ds$$

as $\xi(b) = 1$. The function $\xi(x)$ satisfies the following IVP

$$\xi'(x) = \varphi_0(x)^T \varphi_0(x), \quad \xi(a) = 0. \tag{5.48}$$

Replacing the last equation in (5.46) by the IVP (5.48) and the condition $\xi(b) = 1$ leads to a BVP where the number of ODEs and the number of boundary conditions do not match. To correct this gap, we use a trick and add the trivial ODE $\lambda' = 0$ (λ is constant) to the resulting system and obtain the following enlarged BVP of dimension $2n + 2$ for the numerical determination of simple turning points

$$
\begin{aligned}
y'(x) &= f(x, y(x); \lambda), & B_a y(a) + B_b y(b) &= 0, \\
\varphi'(x) &= f_y(x, y(x); \lambda)\varphi(x), & B_a \varphi(a) + B_b \varphi(b) &= 0, \\
\xi'(x) &= \varphi(x)^T \varphi(x), & \xi(a) &= 0, \quad \xi(b) = 1, \\
\lambda'(x) &= 0.
\end{aligned}
\tag{5.49}
$$

Example 5.21 In Exercise 4.12, we have already considered Bratu's BVP. The governing equations are

$$y''(x) = -\lambda \exp(y(x)), \quad y(0) = y(1) = 0. \tag{5.50}$$

As can be seen in Fig. 5.8, the corresponding solution curve has a simple turning point $z_0 \equiv (y_0, \lambda_0)$.

To determine numerically z_0, we have used the enlarged BVP (5.49), which reads in that case

$$
\begin{aligned}
\begin{pmatrix} y_1(x) \\ y_2(x) \end{pmatrix}' &= \begin{pmatrix} y_2(x) \\ -\lambda \exp(y_1(x)) \end{pmatrix}, & y_1(0) = y_1(1) &= 0, \\
\begin{pmatrix} \varphi_1(x) \\ \varphi_2(x) \end{pmatrix}' &= \begin{pmatrix} \varphi_2(x) \\ -\lambda \exp(y_1(x))\varphi_1(x) \end{pmatrix}, & \varphi_1(0) = \varphi_1(1) &= 0, \\
\xi'(x) &= \varphi_1(x)^2 + \varphi_2(x)^2, & \xi(0) = 0, \quad \xi(1) &= 1, \\
\lambda'(x) &= 0.
\end{aligned}
$$

Applying the multiple shooting method (see Sect. 4.4) with $m = 10$ and the starting trajectories $y_1(x) = y_2(x) = \varphi_1(x) = \varphi_2(x) = \xi_0(x) = \lambda(x) \equiv 1$, we obtained

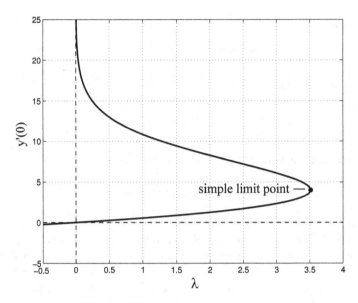

Fig. 5.8 Solution curve of Bratu's BVP

after eight iteration steps the approximations $\tilde{\lambda}_0 = 3.51383071912$ for the critical parameter λ_0 and $\tilde{y}_1'(0) = 4$ for the missing initial value $y_1'(0)$. □

Note, in (5.47) any linear functional φ_0^* can be used. Therefore, it is possible to replace (5.47) by the functional

$$\varphi_0^* v(x) \equiv \left(e^{(k)}\right)^T v(a), \quad v \in Y, \tag{5.51}$$

where $e^{(k)}$ denotes the k-the unit vector in $\mathbb{R}^n$ (see, e.g., [112]). The resulting enlarged BVP of dimension $2n + 1$ is

$$
\begin{aligned}
y'(x) &= f(x, y(x); \lambda), & B_a y(a) + B_b y(b) &= \mathbf{0}, \\
\varphi'(x) &= f_y(x, y(x); \lambda)\varphi(x), & B_a \varphi(a) + B_b \varphi(b) &= \mathbf{0}, \\
\lambda'(x) &= 0, & \left(e^{(k)}\right)^T \varphi(a) &= 1.
\end{aligned}
\tag{5.52}
$$

The advantage of (5.52) is that its dimension is one less than the dimension of (5.51). However, the major drawback of (5.52) is the specification of the index k. It must be specified a priori which component of φ_0 does not vanish. This is only possible if a good approximation is at hand.

Another possibility for the realization of the functional φ_0^* is

$$\varphi_0^* v(x) \equiv \varphi_0(a)^T v(a). \tag{5.53}$$

In this way, we obtain the following enlarged BVP

$$
\begin{aligned}
y'(x) &= f(x, y(x); \lambda), & B_a y(a) + B_b y(b) &= 0, \\
\varphi'(x) &= f_y(x, y(x); \lambda)\varphi(x), & B_a\varphi(a) + B_b\varphi(b) &= 0, & (5.54)\\
\lambda'(x) &= 0, & \varphi(a)^T \varphi(a) &= 1,
\end{aligned}
$$

which is of the same dimension as problem (5.52).

Let us now come to the realization of the other extension technique (5.45). In the same way as before, we define the functional $\psi^* \in W^*$ as

$$
\psi^* w(x) \equiv \int_a^b \psi(x)^T w(x)dx, \quad w \in W, \tag{5.55}
$$

where $\psi(x)$ is the solution of the adjoint BVP

$$
\begin{aligned}
\psi'(x) &= - \left(f_y(x, y(x); \lambda)\right)^T \psi(x), \quad a < x < b, \\
B_a^*\psi(a) + B_b^*\psi(b) &= 0.
\end{aligned} \tag{5.56}
$$

The adjoint matrices $B_a^*, B_b^* \in \mathbb{R}^{n \times n}$ are to be determined such that

$$
B_a \left(B_a^*\right)^T - B_b \left(B_b^*\right)^T = 0, \quad \text{rank} \left(B_a^* | B_b^*\right) = n.
$$

This can be realized by computing the QR factorization of the rectangular matrix $[B_a | B_b]^T$ via Givens transformations, i.e.,

$$
\begin{pmatrix} B_a^T \\ B_b^T \end{pmatrix} = Q \begin{pmatrix} R \\ 0 \end{pmatrix} = \begin{pmatrix} Q_{11} & Q_{12} \\ Q_{21} & Q_{22} \end{pmatrix} \begin{pmatrix} R \\ 0 \end{pmatrix}.
$$

Since Q is an orthogonal quadratic matrix, i.e., $Q^{-1} = Q^T$, we have

$$
Q^T \begin{pmatrix} B_a^T \\ B_b^T \end{pmatrix} = \begin{pmatrix} Q_{11}^T & Q_{21}^T \\ Q_{12}^T & Q_{22}^T \end{pmatrix} \begin{pmatrix} B_a^T \\ B_b^T \end{pmatrix} = \begin{pmatrix} R \\ 0 \end{pmatrix}.
$$

Now, the desired adjoint matrices are $B_a^* = Q_{12}^T$ and $B_b^* = -Q_{22}^T$.

In view of the above considerations, the enlarged problem (5.45) can be written for the given problem (5.10) as the following BVP of dimension $n + 2$

$$
\begin{aligned}
y'(x) &= f(x, y(x), \lambda), & B_a y(a) + B_b y(b) &= 0, \\
\psi'(x) &= - \left(f_y(x, y(x); \lambda)\right)^T \psi(x), & B_a^*\psi(a) + B_b^*\psi(b) &= 0 \\
\xi'(x) &= -\psi(x)^T f_\lambda(x, y(x); \lambda), & \xi(a) &= 0, \quad \xi(b) = 1, & (5.57)\\
\lambda'(x) &= 0.
\end{aligned}
$$

The answer to the question, which of the two enlarged BVPs (5.49) or (5.57) should be used, depends on the function vector $f(x, y; \lambda)$ and the problem that has to be solved afterwards (whether φ_0 or ψ_0 is required).

5.3.3 An Extension Technique for Double Turning Points

Double (multiple) turning points can only be calculated numerically stable as two- (multiple-) parameter problems. If the problem does not already contain two parameters, we must embed the original problem (5.11) into the two-parameter problem

$$\tilde{T}(y, \lambda, \tau) = 0, \quad \tilde{T} : Y \times \mathbb{R} \times \mathbb{R} \to W. \tag{5.58}$$

For a fixed $\tau = \tau^*$, we have our one-parameter standard problem

$$T_{(\tau^*)}(y, \lambda) \equiv \tilde{T}(y, \lambda, \tau^*) = 0, \quad T_{(\tau^*)} : Y \times \mathbb{R} \to W. \tag{5.59}$$

A possible behavior of the solution curve $y_{(\tau^*)}(\lambda) \equiv y(\lambda; \tau^*)$ of (5.59) in dependence of τ^* is displayed in Fig. 5.9. Here, the solution curve has two separated simple turning points for $\tau^* > \tau_0$. But for $\tau^* = \tau_0$ these two limit points collapse into a double turning point, whereas for $\tau^* < \tau_0$ the solution curve $y(\lambda; \tau^*)$ consists of only isolated solutions.

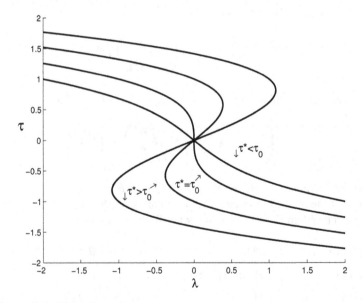

Fig. 5.9 $\tau = 0$: double turning point

Fig. 5.10 $\tau = 0$: double
turning point

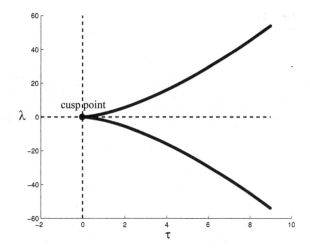

Let $\tilde{z}_0 \equiv (y_0, \lambda_0, \tau_0)$ be a double limit point w.r.t. λ of the equation

$$\tilde{T}(y, \lambda, \tau_0) = 0.$$

If the parameter τ is also varied in (5.58) and the corresponding simple turning points
are determined for each value of τ, then the projection of the limit points onto the
(λ, τ)-plane has the behavior of a cusp catastrophe (see Fig. 5.10). The canonical
equation of the cusp catastrophe is studied in the standard literature of this topic
(see, e.g., [43, 97, 133] and has the form

$$x^3 - \tau x + \lambda = 0, \quad x, \lambda, \tau \in \mathbb{R}.$$

It can be easily seen that:

- if $\lambda = 0$ is fixed, the qualitative behavior of the solution curves x versus τ is that
  of a pitchfork bifurcation point (see Fig. 5.11),
- if $\tau = 0$ is fixed, the corresponding solution curve x versus λ has a double turning
  point $(x, \lambda) = (0, 0)$ (see Fig. 5.9),
- if $\tau > 0$, there are two simple turning points (x_0, λ_0), with $\lambda_0 = \pm 2(\tau/3)^{3/2}$ and
  $x_0 = \pm(\tau/3)^{1/2}$. It is $\lambda_0^2 = (4/27)\tau^3$. This relation describes a cusp curve and the
  corresponding cusp point is $(\lambda, \tau) = (0, 0)$ (see Fig. 5.10).
- if $\tau < 0$, there are no limit points.

This situation is described by the following theorem.

Theorem 5.22 *Assume that*

$$\xi_0^* \tilde{T}_\tau(y_0, \lambda_0, \tau_0) + \psi_0^* \tilde{T}_{y\tau}(y_0, \lambda_0, \tau_0)\varphi_0 \neq 0, \tag{5.60}$$

Fig. 5.11 $\lambda = 0$: pitchfork bifurcation

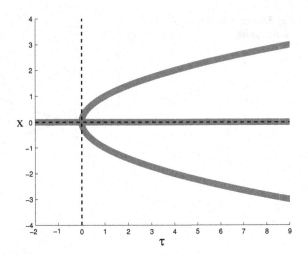

where $\xi_0^* \in W^*$ is defined by (5.43). Then, the double turning point (y_0, λ_0, τ_0) of the 2-parameter problem (5.58) corresponds to a cusp point.

Proof See [116].   ∎

The correlation between a double turning point and the cusp catastrophe is also shown by the following statement.

Theorem 5.23 *Let the condition* (5.60) *be satisfied. Then, there exists a linear transformation*

$$[\hat{\lambda}, \hat{\tau}] = [\lambda, \tau]\, P, \quad P \in \mathbb{R}^{2\times 2},$$

such that the double turning point (y_0, λ_0, τ_0) *of* (5.58) *is a pitchfork bifurcation point with respect to* $\hat{\tau}$.

Proof See [104].   ∎

A suitable extended problem for the numerical determination of double turning points has been proposed in [104] and is based on the enlarged operator

$$\hat{T}: \begin{cases} \hat{Z} \equiv Y \times \mathbb{R} \times \mathbb{R} \times \mathbb{R} \times Y \times W^* \to \hat{W} \equiv W \times W \times Y^* \times \mathbb{R} \times \mathbb{R} \times \mathbb{R} \\[4pt] \hat{z} \equiv (y, \lambda, \tau, \mu, \varphi, \psi^*) \to \begin{pmatrix} \tilde{T}(y, \lambda, \tau) \\ \tilde{T}_y(y, \lambda, \tau)\,\varphi \\ \tilde{T}_y^*(y, \mu, \tau)\psi^* \\ \varphi_0^*\varphi - 1 \\ \psi^*\psi_0 - 1 \\ \psi^*\tilde{T}_{yy}(y, \lambda, \tau)\varphi^2 \end{pmatrix}, \end{cases}$$

$$(5.61)$$

with $\varphi_0^* \in Y^*$, $\psi_0 \in W$ such that

$$\varphi_0^* \varphi_0^* \varphi_0 = \psi_0^* \psi_0 = 1.$$

The equation

$$\hat{T}(\hat{z}) = 0 \tag{5.62}$$

can be solved by standard methods as the following theorem shows.

Theorem 5.24 *Let $\tilde{z}_0 \equiv (y_0, \lambda_0, \tau_0)$ be a double turning point of the equation (5.58) w.r.t. the parameter λ. Suppose that*

$$\xi_0^* \tilde{T}_\tau(y_0, \lambda_0, \tau_0) + \psi_0^* \tilde{T}_{y\tau}(y_0, \lambda_0, \tau_0)\varphi_0 \neq 0$$

and

$$\psi_0^* \tilde{T}_{y\lambda}(y_0, \lambda_0, \tau_0)\varphi_0 \neq 0,$$

where ξ_0^* is defined by (5.43). Then, $\hat{T}'(\hat{z}_0)$ is a linear homeomorphism from $\hat{Z}$ onto $\hat{W}$, i.e.,

$$\mathcal{N}(\hat{T}'(\hat{z}_0)) = \{0\}, \quad \mathcal{R}(\hat{T}'(\hat{z}_0)) = \hat{W}.$$

Proof See, e.g., the paper [104]. ∎

Obviously, the inversion of the statement postulated in Theorem 5.19 is not valid. Only the isolated solutions of (5.62) with $\mu = 0$ are in relationship with the double turning points of the original problem (5.11).

A possible realization of the extended problem (5.62) for the BVP (5.10) is

$$
\begin{aligned}
y'(x) &= \tilde{f}(x, y(x); \lambda, \tau), & B_a y(a) + B_b y(b) &= \mathbf{0}, \\
\varphi'(x) &= \tilde{f}_y(x, y(x); \lambda, \tau)\varphi(x), & B_a \varphi(a) + B_b \varphi(b) &= \mathbf{0}, \\
\psi'(x) &= -\left(\tilde{f}_y(x, y(x); \mu, \tau)\right)^T \psi(x), & B_a^* \psi(a) + B_b^* \psi(b) &= \mathbf{0}, \\
\xi_1'(x) &= \varphi(x)^T \varphi(x), & \xi_1(a) &= 0, \quad \xi_1(b) = 1, \\
\xi_2'(x) &= \psi(x)^T \psi(x), & \xi_2(a) &= 0, \quad \xi_2(b) = 1, \\
\xi_3'(x) &= \psi(x)^T \left(\tilde{f}_{yy}(x, y(x); \lambda, \tau)\varphi(x)^2\right), & \xi_3(a) &= 0, \quad \xi_3(b) = 0, \\
\lambda'(x) &= 0, \quad \mu'(x) = 0, \quad \tau'(x) = 0.
\end{aligned} \tag{5.63}
$$

This enlarged BVP of dimension $3n + 6$ is again in standard form and can be solved with the shooting methods described in Chap. 4.

5.3.4   Determination of Solutions in the Neighborhood of a Simple Turning Point

Let $z \equiv (y, \lambda) \in Z$ be an isolated solution of (5.11). If this solution is in the direct neighborhood of a limit point, then its numerical determination is very complicated since in that case the operator equation (5.11) represents a bad conditioned problem. The idea is now to convert (5.11) into a modified problem, which is well-posed in the neighborhood of the limit point. We call this strategy *transformation technique*. In the following, we restrict our analysis to simple turning points. Multiple turning points can be handled analogously.

According to Corollary 5.13, formula (5.28), in the neighborhood of limit points there exists a unique curve of solutions of (5.11), which can be represented in the form

$$\lambda(\varepsilon) = \lambda_0 + \xi(\varepsilon),$$
$$y(\varepsilon) = y_0 + \varepsilon\varphi_0 + w(\xi(\varepsilon), \varepsilon), \quad |\varepsilon| \leq \varepsilon_0.$$

If $\xi(\varepsilon)$ satisfies $\ddot{\xi}(0) = \ddot{\lambda}(0) \neq 0$, then z_0 is a simple turning point and (5.28) can be specified as

$$\lambda(\varepsilon) = \lambda_0 + \varepsilon^2 \tau(\varepsilon),$$
$$y(\varepsilon) = y_0 + \varepsilon\varphi_0 + \varepsilon^2 u(\varepsilon), \quad |\varepsilon| \leq \varepsilon_0, \tag{5.64}$$

where $\tau(0) \neq 0$ and $u(\varepsilon) \in Y_1$, i.e., $\varphi_0^* u(\varepsilon) = 0$. In the ansatz (5.64), the unknown quantities are $\tau(\varepsilon)$ and $u(\varepsilon)$. We will now show how these quantities can be determined. At first, we expand $T(y, \lambda)$ at the simple turning point $z_0 = (y_0, \lambda_0)$ into a Taylor series and obtain

$$0 = T(y, \lambda) = \overbrace{T(y_0, \lambda_0)}^{=0} + T_y^0(y - y_0) + T_\lambda(\lambda - \lambda_0) + \frac{1}{2} T_{yy}^0 (y - y_0)^2$$
$$+ T_{y\lambda}^0(\lambda - \lambda_0)(y - y_0) + \frac{1}{2} T_{\lambda\lambda}^0(\lambda - \lambda_0)^2 + R(y, \lambda). \tag{5.65}$$

The remainder R consists only of third or higher order terms in $(y - y_0)$ and $(\lambda - \lambda_0)$. Now, we substitute the ansatz (5.64) into (5.65). We get

$$0 = T_y^0\big(\varepsilon\varphi_0 + \varepsilon^2 u(\varepsilon)\big) + \varepsilon^2 T_\lambda^0 \tau(\varepsilon) + \frac{1}{2} T_{yy}^0\big(\varepsilon\varphi_0 + \varepsilon^2 u(\varepsilon)\big)^2$$
$$+ \varepsilon^2 T_{y\lambda}^0 \tau(\varepsilon)\big(\varepsilon\varphi_0 + \varepsilon^2 u(\varepsilon)\big) + \frac{\varepsilon^4}{2} T_{\lambda\lambda}^0 \tau(\varepsilon)^2 + R(y(\varepsilon), \lambda(\varepsilon))$$
$$= \varepsilon^2\big(T_y^0 u(\varepsilon) + T_\lambda^0 \tau(\varepsilon)\big) + \frac{\varepsilon^2}{2} T_{yy}^0\big(\varphi_0^2 + 2\varepsilon\varphi_0 u(\varepsilon) + \varepsilon^2 u(\varepsilon)^2\big)$$
$$+ \varepsilon^3 T_{y\lambda}^0\big(\varphi_0 \tau(\varepsilon) + \varepsilon u(\varepsilon)\tau(\varepsilon)\big) + \frac{\varepsilon^4}{2} T_{\lambda\lambda}^0 \tau(\varepsilon)^2 + R(y(\varepsilon), \lambda(\varepsilon)).$$

It can easily be seen that $R(y(\varepsilon), \lambda(\varepsilon)) = O(\varepsilon^3)$. Dividing both sides by ε^2 yields

$$T_y^0 u(\varepsilon) + T_\lambda^0 \tau(\varepsilon) = -w(u(\varepsilon), \tau(\varepsilon); \varepsilon), \quad \varphi_0^* u(\varepsilon) = 0,$$

where

$$w(u(\varepsilon), \tau(\varepsilon); \varepsilon) \equiv \frac{1}{2} T_{yy}^0 \left(\varphi_0^2 + 2\varepsilon\varphi_0 u(\varepsilon) + \varepsilon^2 u(\varepsilon)^2 \right)$$

$$+ \varepsilon T_{y\lambda}^0 \left(\varphi_0 \tau(\varepsilon) + \varepsilon u(\varepsilon)\tau(\varepsilon) \right) + \frac{\varepsilon^2}{2} T_{\lambda\lambda}^0 \tau(\varepsilon)^2$$

$$+ \frac{1}{\varepsilon^2} R \left(y_0 + \varepsilon\varphi_0 + \varepsilon^2 u(\varepsilon), \lambda_0 + \varepsilon^2 \tau(\varepsilon) \right).$$

These two equations can be written in the form

$$\begin{pmatrix} T_y^0 & T_\lambda^0 \\ \varphi_0^* & 0 \end{pmatrix} \begin{pmatrix} u(\varepsilon) \\ \tau(\varepsilon) \end{pmatrix} = - \begin{pmatrix} w(u(\varepsilon), \tau(\varepsilon); \varepsilon) \\ 0 \end{pmatrix}, \quad |\varepsilon| \le \varepsilon_0. \qquad (5.66)$$

In order to prove the next theorem, the following lemma is required:

Lemma 5.25 *Let X and Y be Banach spaces. Assume that the linear operator* $\mathbb{A}$: $X \times \mathbb{R}^m \to Y \times \mathbb{R}^m$ *has the following block structure*

$$\mathbb{A} \equiv \begin{pmatrix} A & B \\ C^* & D \end{pmatrix},$$

where $A : X \to Y$, $B : \mathbb{R}^m \to Y$, $C^* : X \to \mathbb{R}^m$ *and* $D : \mathbb{R}^m \to \mathbb{R}^m$. *Then, it holds:*

1. *A is bijective* $\Rightarrow$ {$\mathbb{A}$ *is bijective* $\Leftrightarrow$ $\det(D - C^* A^{-1} B) \ne 0$},
2. *A is not bijective and* $\dim \mathcal{N}(A) = \text{codim}\mathcal{R}(A) = m \ge 1$
   $\Rightarrow$ {$\mathbb{A}$ *is bijective* $\Leftrightarrow$ $(c_0) - (c_3)$ *are satisfied*},
   where
(c_0)	$\dim \mathcal{R}(B) = m,$	(c_1)	$\mathcal{R}(B) \cap \mathcal{R}(A) = \{0\},$
(c_2)	$\dim \mathcal{R}(C^*) = m,$	(c_3)	$\mathcal{N}(A) \cap \mathcal{N}(C^*) = \{0\}.$

3. *A is not bijective and* $\dim \mathcal{N}(A) > m$ $\Rightarrow$ $\mathbb{A}$ *is not bijective.*

Proof See, e.g., [31]. ∎

Theorem 5.26 *Let z_0 be a simple turning point. Then, there exists a real constant* $\varepsilon_0 > 0$, *such that for* $|\varepsilon| \le \varepsilon_0$ *the transformed problem (5.66) has a continuous curve of isolated solutions* $\eta(\varepsilon) \equiv (u(\varepsilon), \tau(\varepsilon))^T$. *Substituting these solutions into the ansatz (5.64), a continuous curve of solutions of the original operator equation (5.11) results, which represents a simple solution curve passing through the simple turning point* $z_0 \equiv (y_0, \lambda_0) \in Z$.

Proof For $\varepsilon \to 0$ problem (5.66) turns into

$$A \begin{pmatrix} u(0) \\ \tau(0) \end{pmatrix} \equiv \begin{pmatrix} T_y^0 & T_\lambda^0 \\ \varphi_0^* & 0 \end{pmatrix} \begin{pmatrix} u(0) \\ \tau(0) \end{pmatrix} = \begin{pmatrix} -\dfrac{1}{2} T_{yy}^0 \varphi_0^2 \\ 0 \end{pmatrix}. \tag{5.67}$$

We set

$$A \equiv T_y^0, \quad B \equiv T_\lambda^0, \quad C^* \equiv \varphi_0^*, \quad D \equiv 0, \quad m \equiv 1$$

and apply Lemma 5.25. Since T_y^0 is not bijective, the second case is true. Now, we have to verify (c_0)–(c_3).

(c_0): $\dim \mathcal{R}(T_\lambda^0) = 1 = m$,
(c_1): $\mathcal{R}(T_\lambda^0) \cap \mathcal{R}(T_\lambda^0) = \{0\}$,
        because $\psi_0^* T_y^0 \varphi = 0$ and $\underbrace{\psi_0^* T_\lambda^0}_{\neq 0} \zeta \neq 0$ for $\zeta \neq 0$,

(c_2): $\dim \mathcal{R}(\varphi_0^*) = 1 = m$,
(c_3): $\mathcal{N}(T_y^0) \cap \mathcal{N}(\varphi_0^*) = \{0\}$,
        because in $\mathcal{N}(T_y^0)$ are the elements $c\varphi_0$ and $\varphi_0^*(c\varphi_0) = c \underbrace{\varphi_0^* \varphi_0}_{=1} = c$.

    Thus, it must hold $c = 0$.

The assumptions of Lemma 5.25 are satisfied and we can conclude that the linear operator A is bijective. Therefore, there exists a unique solution $\eta^0 \equiv (u(0), \tau(0))^T$ of (5.67).

    If problem (5.66) is written in the form

$$F(\varepsilon; \eta) = 0, \quad \eta \equiv (u, \tau)^T,$$

where $F : \mathbb{R} \times \mathrm{dom}(A) \to W \times \mathbb{R}$, then we have $F(0; \eta^0) = A\eta^0 - g, g \in W \times \mathbb{R}$. Thus, $F_\eta(0; \eta^0) = A$. The bijectivity of A implies that A is a linear homeomorphism of $\mathrm{dom}(A)$ onto $W \times \mathbb{R}$. The claim of the theorem follows immediately from the Implicit Function Theorem (see Theorem 5.8). ∎

    Thus, we can conclude that for all $\varepsilon : |\varepsilon| \le \varepsilon_0$, the transformed problem (5.66) represents a well-posed problem, which can be solved by numerical standard techniques.

    Let us now formulate the transformed problem for the BVP (5.11). Here, we use again the functional $\varphi_0^* \in Y^*$ defined in (5.47), i.e.,

$$\varphi_0^* v(x) \equiv \int_a^b \varphi(x)^T v(x) dx, \quad v \in Y.$$

Then, the condition $\varphi_0^* u(x) = 0$ can be written as

$$\xi'(x) = \varphi(x)^T u(x), \quad \xi(a) = \xi(b) = 0.$$

The fact that the parameter τ does not depend on x can be written in ODE formulation as $\tau' = 0$. Thus, we obtain the following transformed BVP of dimension $n + 2$

$$u'(x; \varepsilon) = f^0_y u(x; \varepsilon) + f^0_\lambda \tau(\varepsilon) + \tilde{w}(u(x; \varepsilon), \tau(\varepsilon); \varepsilon), \quad |\varepsilon| \le \varepsilon_0,$$

$$\xi'(x; \varepsilon) = \varphi_0(x)^T u(x; \varepsilon), \quad \tau'(x) = 0, \tag{5.68}$$

$$B_a u(a; \varepsilon) + B_b u(b; \varepsilon) = \mathbf{0}, \quad \xi(a) = 0, \quad \xi(b) = 0,$$

where

$$\tilde{w}(u(x; \varepsilon), \tau(\varepsilon); \varepsilon) \equiv \frac{1}{2} f^0_{yy} \left(\varphi^2_0 + 2\varepsilon \varphi_0 u(x; \varepsilon) + \varepsilon^2 u(x; \varepsilon)^2 \right)$$

$$+ \varepsilon f^0_{y\lambda} \left(\tau(\varepsilon) \varphi_0 + \varepsilon \tau(\varepsilon) u(x, \varepsilon) \right) + \frac{\varepsilon^2}{2} f^0_{\lambda\lambda} \tau(\varepsilon)^2$$

$$+ \frac{1}{\varepsilon^2} \tilde{R}(y_0 + \varepsilon \varphi_0 + \varepsilon^2 u(x; \varepsilon), \lambda_0 + \varepsilon^2 \tau(\varepsilon)),$$

and

$$\tilde{R}(y, \lambda) \equiv f(y, \lambda) - \left\{ f^0 + f^0_y (y - y_0) + f^0_\lambda (\lambda - \lambda_0) + \frac{1}{2} f^0_{yy} (y - y_0)^2 \right.$$

$$\left. + f^0_{y\lambda} (\lambda - \lambda_0)(y - y_0) + \frac{1}{2} f^0_{\lambda\lambda} (\lambda - \lambda_0)^2 \right\}.$$

Remark 5.27 Obviously, the right-hand side of the transformed problem (5.68) is based on the solution $(y_0, \lambda_0, \varphi_0)$ of the enlarged problem (5.49). If a shooting method is used to solve (5.68) both BVPs must be computed on the same grid. Since in a shooting technique IVP-solvers with automatic step-size control are commonly used it is appropriate to combine (5.68) and (5.49) into one BVP.   □

5.4 Analytical and Numerical Treatment of Primary Simple Bifurcation Points

5.4.1 Bifurcation Points, Primary and Secondary Bifurcation Phenomena

In this section, we will study solutions of the operator equation (5.11), which do not lie on a simple solution curve. These are the so-called *bifurcation points* or *branching points*.

Definition 5.28 A solution $z_0 \in Z$ of the equation (5.11) is called bifurcation point (see Fig. 5.12), if

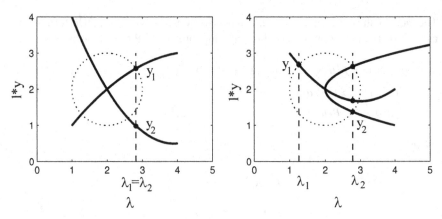

Fig. 5.12 Illustration of Definition 5.28, part 1 and 2

1. in any open neighborhood of z_0 there exist two distinct solutions $z_1 \equiv (y_1, \lambda_1)$ and $z_2 \equiv (y_2, \lambda_2)$, with $\lambda_1 = \lambda_2$;
2. any open neighborhood of z_0 contains at least one solution of (5.11) for $\lambda > \lambda_0$ as well as for $\lambda < \lambda_0$.

$\square$

A necessary condition that a solution of (5.11) is a bifurcation point is given in the following theorem.

Theorem 5.29 *Let* $T : U(z_0) \subseteq Y \times \mathbb{R} \to W$ *be a* $\mathbb{C}^1$*-map in a neighborhood of* $z_0 \in Z$. *If* z_0 *is a bifurcation point of (5.11), then* $T_y(z_0)$ *is not a homeomorphism.*

Proof If $T_y(z_0)$ is a homeomorphism, then the Implicit Function Theorem (see Theorem 5.8) implies that in a neighborhood of z_0 all solutions of (5.11) lie on a curve $\{y(\lambda), \lambda\}, |\lambda - \lambda_0| \le \varepsilon$. This is in contradiction to Definition 5.28. ■

From a numerical point of view, it is useful to distinguish between primary and secondary bifurcation. In practice, $\{0, \lambda\}, \lambda \in \mathbb{R}$, is often a solution curve of (5.11). This curve is called the *trivial solution curve*. We speak of *primary bifurcation* if nontrivial solution curves are branching off from the trivial solution curve. The branching curves are called *primary solution curves*. The situation where at least two nontrivial solution curves are intersecting is called *secondary bifurcation* and the corresponding curves are the *secondary solution curves* (see also Fig. 5.6).

Here, we always assume that the secondary solution curves are not explicitly known. Otherwise, if for a curve a parametrization $\{y(\lambda), \lambda\}, \lambda \in [\lambda_0, \lambda_1]$, is available, the given problem can be reduced to the case of primary bifurcation by considering the operator equation

$$\tilde{T}(w, \lambda) \equiv T(y(\lambda) + w, \lambda) = 0.$$

5.4.2 Analysis of Primary Simple Bifurcation Points

In this section, we assume that the operator T in (5.11) satisfies

$$T(0, \lambda) = 0 \quad \text{for all } \lambda \in \mathbb{R}. \tag{5.69}$$

Thus, there is the possibility that in the solution field of (5.11) primary bifurcation occurs.

Definition 5.30 The parameter $\lambda_0 \in \mathbb{R}$ is called *geometrically simple eigenvalue* of $T_y(0, \lambda)$ if

$$\mathcal{N}(T_y(z_0)) = \text{span}\{\varphi_0\}, \quad z_0 \equiv (0, \lambda_0) \in Z, \quad \varphi_0 \in Y, \quad \|\varphi_0\| = 1. \tag{5.70}$$

□

Looking at Theorem 5.6, we can conclude that $T_y(0, \lambda)$ is a Fredholm operator with index zero. Thus, the Riesz–Schauder theory presented in Sect. 5.3.1 is valid here as well. However, for the characteristic coefficient (see formula (5.27)), we now have

$$c \equiv \psi_0^* T_\lambda^0 = 0, \tag{5.71}$$

since $T_\lambda^0 = T_{\lambda\lambda}^0 = T_{\lambda\lambda\lambda}^0 \cdots = 0$.

The decompositions (5.19) of the Banach spaces Y and W imply that $y \in Y$ can be uniquely represented in the form

$$y = \alpha\varphi_0 + w, \quad \alpha \in \mathbb{R}, \quad P_2 w = w. \tag{5.72}$$

We set

$$\lambda = \lambda_0 + \xi, \tag{5.73}$$

and write the operator equation (5.11) in the paired form (see formula (5.23))

$$\begin{aligned} Q_2 T(z) &= 0, \quad z \equiv (0, \lambda) \in Z, \\ Q_1 T(z) &\equiv (I - Q_2) T(z) = 0. \end{aligned} \tag{5.74}$$

Substituting (5.72) and (5.73) into the first equation of (5.74), we obtain

$$G(\xi, \alpha, w) \equiv Q_2 T(\alpha\varphi_0 + P_2 w, \lambda_0 + \xi) = 0, \tag{5.75}$$

with the $\mathbb{C}^p$-map $G : \mathbb{R} \times \mathbb{R} \times Y_1 \to \mathcal{R}(T_y^0)$. Here, the same arguments as in the case of limit points (see Theorem 5.11) lead to the result that there exist two positive constants ξ_0, α_0 and a unique $\mathbb{C}^p$-map

$$w : [-\xi_0, \xi_0] \times [-\alpha_0, \alpha_0] \to Y_1$$

such that

$$G(\xi, \alpha, w(\xi, \alpha)) = 0, \quad w(0, 0) = 0.$$

If the second equation in (5.74) is also taken into account, we see that the treatment of (5.11) in the neighborhood of $z_0 \equiv (0, \lambda_0)$ requires the solution of the so-called *bifurcation point equation*

$$F(\xi, \alpha) \equiv \psi_0^* T(\alpha \varphi_0 + w(\xi, \alpha), \lambda_0 + \xi) = 0 \qquad (5.76)$$

in the neighborhood of the origin $(0, 0)$.

Differentiating (5.76) gives

$$F(0, 0) = \frac{\partial F}{\partial \xi}(0, 0) = \frac{\partial F}{\partial \alpha}(0, 0) = 0,$$

$$\frac{\partial^2 F}{\partial \xi^2}(0, 0) = 0, \quad \frac{\partial^2 F}{\partial \xi \partial \alpha} = a_1, \quad \frac{\partial^2 F}{\partial \alpha^2}(0, 0) = a_2,$$

where

$$a_1 \equiv \psi_0^* T_{y\lambda}^0 \varphi_0, \quad a_2 \equiv \psi_0^* T_{yy}^0 \varphi_0^2 \qquad (5.77)$$

are the first and the second bifurcation coefficient, respectively.

Now, let us suppose that the solution $z_0 \equiv (0, \lambda_0) \in Z$ of the equation (5.11) is a *simple bifurcation point*.

Definition 5.31 A solution of the operator equation (5.11) is called simple bifurcation point, if λ_0 is a geometrically simple eigenvalue of T_y^0 und the first bifurcation coefficient a_1 satisfies

$$a_1 \equiv \psi_0^* T_{y\lambda}^0 \varphi_0 \neq 0. \qquad (5.78)$$

□

Now, we will present the often used *Morse Lemma*. Assume that the equation

$$F(y) = 0$$

is given, where $F \in C^p(\Omega, \mathbb{R}), 0 \in \Omega \subset \mathbb{R}^d, \Omega$ open region.

Lemma 5.32 *Suppose that*

$$F(0) = 0, \quad F_y(0) = 0, \quad \text{and} \quad F_{yy}(0) \text{ is nonsingular.}$$

Then, there exist a C^{p-2}-function $w(y) \in \mathbb{R}^d$, which is defined in the neighborhood of the origin and satisfies

$$w(0) = 0, \quad w_y(0) = I,$$

and $F(y) = \dfrac{1}{2} F_{yy}(0) w(y)^2$ in the neighborhood of the origin.

Proof See, e.g., [92]   ■

The following corollary can easily be proved.

Corollary 5.33 *Let the assumptions of Lemma 5.32 be satisfied. Moreover, suppose that $d = 2$ and the quadratic form $F_{yy}(0)w^2$ is indefinite. Then, the solution field $E(F)$ in the neighborhood of the origin consists of exactly two $\mathbb{C}^{p-2}$ curves, which intersect (for $p > 2$ transversally) at the origin.*

Let us apply Corollary 5.33 to the bifurcation equation (5.76). Here, $d = 2$ and $y = (\xi, \alpha)^T$. It holds

$$F(0,0) = 0, \quad F_y(0,0) = \begin{pmatrix} F_\xi(0,0) \\ F_\alpha(0,0) \end{pmatrix} = \begin{pmatrix} 0 \\ 0 \end{pmatrix},$$

and

$$F_{yy}(0,0) = \begin{pmatrix} F_{\xi\xi} & F_{\xi\alpha} \\ F_{\alpha\xi} & F_{\alpha\alpha} \end{pmatrix} (0,0) = \begin{pmatrix} 0 & a_1 \\ a_1 & a_2 \end{pmatrix}.$$

Since $\det(F_{yy}(0,0)) = -a_1^2 < 0$, the quadratic form is indefinite. Thus, we can conclude that in the neighborhood of the origin the solutions of the equation (5.76) lie on two curves of the class $\mathbb{C}^{p-2}$, which intersect transversally at $(0, 0)$.

For the original problem (5.11) this result implies that the statement of the following theorem is valid.

Theorem 5.34 *Let $z_0 \equiv (0, \lambda_0) \in Z$ be a simple bifurcation point of the operator equation (5.11). Then, in the neighborhood of z_0 the solutions of (5.11) belong to two curves of the class $\mathbb{C}^{p-2}$, which intersect transversally at z_0.*

Under the assumption (5.69), we have primary bifurcation and one of the solution curves corresponds to the trivial solution ($y \equiv 0, \ \lambda = \lambda_0 + \xi$). Let us parameterize the other (nontrivial) solution curve in dependence of an artificial parameter ε as follows

$$\lambda(\varepsilon) = \lambda_0 + \xi(\varepsilon), \quad y(\varepsilon) = \alpha(\varepsilon)\varphi_0 + w(\xi(\varepsilon), \alpha(\varepsilon)),$$
$$\xi(\varepsilon) = \varepsilon\sigma(\varepsilon), \quad \alpha(\varepsilon) = \varepsilon\delta(\varepsilon), \tag{5.79}$$

with $w \in Y_1$ satisfying $w(0, 0) = 0$ (see formula (5.75)).

We define the function

$$\boldsymbol{H} : (\varepsilon, \sigma, \delta) \in [-1, 1] \times [-\xi_0, \xi_0] \times [-\alpha_0, \alpha_0] \to \mathbb{R}^2,$$

$$\boldsymbol{H}(\varepsilon, \sigma, \delta) \equiv \left[\frac{F(\varepsilon\sigma, \varepsilon\delta)}{\varepsilon^2}, \sigma^2 + \delta^2 - 1 \right]^T,$$

and consider the equation

$$\boldsymbol{H}(\varepsilon, \sigma, \delta) = \boldsymbol{0}. \tag{5.80}$$

The second equation in (5.80) represents only a certain normalization of the solution. We can now state the following theorem.

Theorem 5.35 *Let $z_0 \equiv (0, \lambda_0) \in Z$ be a simple bifurcation point of (5.11). Then, there exist a positive constant ε_0 and two $\mathbb{C}^{p-2}$-functions*

$$\sigma : [-\varepsilon_0, \varepsilon_0] \to \mathbb{R}, \quad \delta : [-\varepsilon_0, \varepsilon_0] \to \mathbb{R},$$

with

$$\boldsymbol{H}(\varepsilon, \sigma(\varepsilon), \delta(\varepsilon)) = \boldsymbol{0}.$$

Proof Since $F \in \mathbb{C}^p$, $p \geq 2$, and $(0, 0)$ is a critical point of F, it holds

$$F(\varepsilon\sigma, \varepsilon\delta) = \frac{\varepsilon^2}{2} \left(a_2\delta^2 + 2a_1\delta\sigma\right) + O(\varepsilon^3), \quad \varepsilon \to 0.$$

If we substitute this expression into (5.80), we obtain

$$\boldsymbol{H}(0, \sigma, \delta) = \begin{pmatrix} \frac{1}{2}\left[a_2\delta^2 + 2a_1\delta\sigma\right] \\ \sigma^2 + \delta^2 - 1 \end{pmatrix}.$$

The assumption $a_1 \neq 0$ implies that $a_2\delta^2 + 2a_1\delta\sigma = 0$ is the equation of two straight lines in the (σ, δ)-plane, which intersect at the origin. These two lines are

- $\{\delta_1^0 = 0, \ \sigma_1^0 \text{ arbitrary}\}$, and
- $\{\sigma_2^0 = -\dfrac{a_2\delta_2^0}{2a_1}, \ \delta_2^0 \neq 0\}$.

The first line is the trivial solution curve of (5.11). For the second line we have

$$J \equiv \left.\frac{\partial \boldsymbol{H}}{\partial(\sigma, \delta)}\right|_{(0,\sigma_2^0,\delta_2^0)} = \begin{pmatrix} a_1\delta_2^0 & a_2\delta_2^0 + a_1\sigma_2^0 \\ 2\sigma_2^0 & 2\delta_2^0 \end{pmatrix}.$$

Thus,

$$\det(J) = 2\left[a_1\left(\delta_2^0\right)^2 - a_1\left(\sigma_2^0\right)^2 - a_2\delta_2^0\sigma_2^0\right]$$

$$= 2\left(\delta_2^0\right)^2\left[a_1 + \frac{a_2^2}{4a_1}\right] \neq 0.$$

Now, the application of the Implicit Function Theorem (see Theorem 5.8) to $\boldsymbol{H}$ at $(0, \sigma_2^0, \delta_2^0)$ leads to the following result. There exist real $\mathbb{C}^{p-2}$-functions $\sigma_2(\varepsilon)$ and $\delta_2(\varepsilon)$ such that the relations

$$\boldsymbol{H}(\varepsilon, \sigma_2(\varepsilon), \delta_2(\varepsilon)) = \boldsymbol{0}, \quad \sigma_2(0) = \sigma_2^0, \quad \delta_2(0) = \delta_2^0$$

are satisfied.   ∎

Let $\{(\xi(\varepsilon) = \varepsilon\sigma(\varepsilon), \alpha(\varepsilon) = \varepsilon\delta(\varepsilon)); \ |\varepsilon| \leq \varepsilon_0\}$ be the primary solution curve of (5.76), which intersects the trivial solution curve $\{(\xi, \alpha = 0); \ \xi \text{ arbitrary}\}$ at the origin. Then, $\{(y(\varepsilon), \lambda(\varepsilon)); \ |\varepsilon| \leq \varepsilon_0\}$ and $\{(y = 0, \lambda); \ \lambda \text{ arbitrary}\}$ are the corresponding solution curves of (5.11).

The currently used parametrization of the primary solution curve in the neighborhood of a simple bifurcation point is based on an artificial real parameter ε. However, sometimes it is possible to parameterize y directly by the control parameter λ. For this purpose, we will discuss two different types of bifurcations.

- *asymmetric bifurcation point*: $a_2 \neq 0$
  It holds $\xi(\varepsilon) = \varepsilon\sigma(\varepsilon)$. Thus
  $$\frac{d\xi}{d\varepsilon}(0) = \sigma(0).$$

Using the formula
$$\sigma_2^0 = -\frac{a_2\, \delta_2^0}{2a_1} \tag{5.81}$$

presented in the proof of Theorem 5.35, we obtain $\sigma(0) \neq 0$. Now, the Implicit Function Theorem implies that ε can be represented as a $\mathbb{C}^{p-2}$-function of ξ. If this function is substituted into the ansatz (5.79), we get

$$\lambda = \lambda_0 + \xi, \quad y = y(\xi) = \alpha(\xi)\varphi_0 + v(\xi, \alpha(\xi)), \tag{5.82}$$

where α is a $\mathbb{C}^{p-2}$-function, which is defined in $[-\xi_0, \xi_0]$, and ξ_0 is a sufficiently small number.

To develop $\alpha(\xi)$ into a Taylor series, we compute

$$\varepsilon = \frac{\xi}{\sigma(\xi)}, \ \text{i.e.,} \ \alpha(\xi) = \xi\,\frac{\delta(\xi)}{\sigma(\xi)},$$

$$\alpha(0) = 0, \quad \frac{d\alpha}{d\xi}(0) = \frac{\delta(0)}{\sigma(0)}.$$

Using (5.81), we obtain $\dfrac{d\alpha}{d\xi}(0) = -\dfrac{2a_1}{a_2} \neq 0$, and $\alpha(\xi) = -\dfrac{2a_1}{a_2}\xi + O(\xi^2)$.

Thus, the ansatz (5.82) can be written in more detail as

$$y = y(\xi) = -\frac{2a_1}{a_2}\,\varphi_0\,\xi + O(\xi^2), \quad \xi \equiv \lambda - \lambda_0. \tag{5.83}$$

- *symmetric bifurcation point*: $a_2 = 0, a_3 \neq 0$
  Since $a_2 = 0$, we get now from (5.81) that $\sigma(0) = 0$. Thus, differentiating the relation $\xi(\varepsilon) = \varepsilon\sigma(\varepsilon)$, we see that

$$\frac{d\xi}{d\varepsilon}(0) = \sigma(0) = 0,$$

and ε cannot be represented as a function of ξ. However, from

$$\alpha(\varepsilon) = \varepsilon\delta(\varepsilon)$$

we obtain

$$\frac{d\alpha}{d\varepsilon}(0) = \delta(0).$$

Using the equation

$$\sigma(0)^2 + \delta(0)^2 = 1,$$

we obtain

$$\frac{d\alpha}{d\varepsilon}(0) = \delta(0) = 1.$$

Thus, ε can be written as a $\mathbb{C}^{p-2}$-function of α and the primary solution curve takes on the form

$$\lambda = \lambda(\alpha) = \lambda_0 + \xi(\alpha), \quad y = y(\alpha) = \alpha\,\varphi_0 + v(\xi(\alpha), \alpha), \tag{5.84}$$

where $\xi(\alpha)$ is a $\mathbb{C}^{p-2}$-function, which is defined in $[-\alpha_0, \alpha_0]$, α_0 sufficiently small.

In [24, 29] it is shown that the above representation can be written in more detail as

$$\lambda = \lambda(\alpha) = \lambda_0 - \frac{a_3}{6\,a_1}\,\alpha^2 + O(\alpha^3),$$
$$y = y(\alpha) = \alpha\varphi_0 + O(\alpha^3). \tag{5.85}$$

Under the assumptions $a_3 \neq 0$ and

$$C_2 \equiv -\frac{6a_1}{a_3} > 0, \tag{5.86}$$

the first equation in (5.85) can be written in the form $\alpha = \alpha(\lambda)$. Substituting this relation into the second equation, we obtain the following representation of y in fractional powers of $\lambda - \lambda_0$:

$$y = y(\eta) = \sqrt{C_2}\,\varphi_0\,\eta + O(\eta^2), \quad \eta^2 \equiv \lambda - \lambda_0. \tag{5.87}$$

In Fig. 5.13, both types of a primary simple bifurcation point are graphically illustrated.

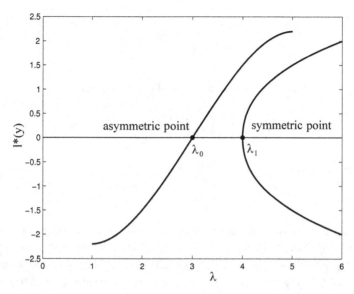

Fig. 5.13 Two different types of primary simple bifurcation points

5.4.3 An Extension Technique for Primary Simple Bifurcation Points

Let $z_0 \equiv (0, \lambda_0)$ be a simple bifurcation point. An often used extension technique for the determination of λ_0 is based on the enlarged operator

$$\tilde{T} : \begin{cases} \tilde{Z} \equiv \mathbb{R} \times Y \rightarrow \tilde{W} \equiv W \times \mathbb{R} \\ \tilde{z} \equiv (\lambda, \varphi) \rightarrow \begin{pmatrix} T_y(0, \lambda)\, \varphi \\ \varphi_0^* \varphi - 1 \end{pmatrix} \end{cases}, \qquad (5.88)$$

where $\varphi_0^* \in Y^*$ such that $\varphi_0^* \varphi_0 = 1$. Since λ_0 is a geometrically simple eigenvalue, the point $\tilde{z}_0 = (\lambda_0, \varphi_0) \in \tilde{Z}$ satisfies the equation

$$\tilde{T}(\tilde{z}) = 0. \qquad (5.89)$$

For the extended problem (5.89) we have the following result.

Theorem 5.36 *Let $z_0 \equiv (0, \lambda_0) \in Z$ be a simple bifurcation point of T. Then, the extended operator $\tilde{T}'(\tilde{z}_0)$ is a linear homeomorphism from $\tilde{Z}$ onto $\tilde{W}$, i.e.,*

$$\mathcal{N}(\tilde{T}'(\tilde{z}_0)) = \{0\}, \quad \mathcal{R}(\tilde{T}'(\tilde{z}_0)) = \tilde{W}.$$

Proof It is

$$\tilde{T}'(\tilde{z}_0) = \begin{pmatrix} T_{y\lambda}^0 \varphi_0 & T_y^0 \\ 0 & \varphi_0^* \end{pmatrix}.$$

1. If $(\mu, w)^T \in \mathcal{N}(\tilde{T}'(\tilde{z}_0))$, where $w \in Y$ and $\mu \in \mathbb{R}$, we have

$$\tilde{T}'(\tilde{z}_0) \begin{pmatrix} \mu \\ w \end{pmatrix} = \begin{pmatrix} \mu T_{y\lambda}^0 \varphi_0 + T_y^0 w \\ \varphi_0^* w \end{pmatrix} = 0. \tag{5.90}$$

The first (block) equation of (5.90) reads

$$T_y^0 w = -\mu T_{y\lambda}^0 \varphi_0. \tag{5.91}$$

This equation is solvable, if the right-hand side is in the range of the linear operator T_y^0. The Fredholm alternative says, that this is the case if the condition $\mu \psi_0^* T_{y\lambda}^0 \varphi_0 = 0$ is satisfied (see Eqs. (5.20) and (5.21)). Now, the assumption $a_1 \equiv \psi_0^* T_{y\lambda}^0 \varphi_0 \neq 0$ implies $\mu = 0$. Inserting this value into (5.91), we get $w = c \varphi_0, c \in \mathbb{R}$. The second (block) equation of (5.90) is

$$\varphi_0^* w = 0.$$

Now, we have $\varphi_0^* w = c \varphi_0^* \varphi_0 = 0$. The condition $\varphi_0^* \varphi_0 = 1$ implies $c = 0$, i.e., $w = 0$.

2. Let $(u, \sigma)^T \in \mathcal{R}(\tilde{T}'(\tilde{z}_0))$ be satisfied. Now, instead of the homogeneous equation (5.90) we consider the inhomogeneous equation

$$\tilde{T}'(\tilde{z}_0) \begin{pmatrix} \mu \\ w \end{pmatrix} = \begin{pmatrix} \mu T_{y\lambda}^0 \varphi_0 + T_y^0 w \\ \varphi_0^* w \end{pmatrix} = \begin{pmatrix} u \\ \sigma \end{pmatrix}. \tag{5.92}$$

The first (block) equation of (5.92) is

$$T_y^0 w = u - \mu T_{y\lambda}^0 \varphi_0. \tag{5.93}$$

There exists a solution if $\psi_0^* (u - \mu T_{y\lambda}^0 \varphi_0) = \psi_0^* u - \mu \psi_0^* T_{y\lambda}^0 \varphi_0 = 0$. Thus, μ is uniquely determined as

$$\mu = \frac{\psi_0^* u}{a_1}.$$

The solution of (5.93) can be written in the form

$$w = \alpha \varphi_0 + \hat{w}, \quad \hat{w} \in Y_1,$$

where α is an arbitrary real constant. Inserting this representation of w into the second (block) equation of (5.92),

$$\varphi_0^* w = \sigma,$$

we see that the constant α is uniquely determined as

$$\alpha = -\varphi_0^* \hat{w} + \sigma,$$

and thus, w is uniquely determined, too. ∎

If z_0 denotes a primary simple bifurcation point, then the reversal of the statement of Theorem 5.36 is also true.

Theorem 5.37 *Let $\tilde{z}_0 = (\lambda_0, \varphi_0)$ be an isolated solution of the extended problem (5.89). Then, $z_0 = (0, \lambda_0)$ is a primary simple bifurcation point of the original problem (5.11).*

Proof We show the claim of the theorem in two steps.

1. Assume that there exists a $0 \neq \varphi_1 \in Y$ such that $T_y^0 \varphi_1 = 0$ and $\varphi_0^* \varphi_1 = 0$. Then, $(0, \varphi_1) \in \mathcal{N}(\tilde{T}'(\tilde{z}_0))$. Due to the assumption of our theorem it must hold $\varphi_1 = 0$.
2. Assume that $a_1 \equiv \psi_0^* T_{y\lambda}^0 \varphi_0 = 0$. Let w_0 denote the unique solution of the problem

$$T_y^0 w_0 = -T_{y\lambda}^0 \varphi_0, \quad w_0 \in Y_1,$$

and $\Phi_0 \equiv \alpha \begin{pmatrix} 1 \\ w_0, \end{pmatrix}$, $\alpha \neq 0$. Then, it can be easily seen that $\Phi_0 \in \mathcal{N}(\tilde{T}'(\tilde{z}_0))$. This is in contradiction to our assumption $\mathcal{N}(\tilde{T}'(\tilde{z}_0)) = \{0\}$. Thus, it must hold that $a_1 \neq 0$. ∎

Theorems 5.36 and 5.37 show that the extended problem (5.89) can be solved with numerical standard techniques. A possible realization of the extended problem (5.89) for the BVP (5.10) is

$$\begin{aligned} \varphi'(x) &= f_y(x, \mathbf{0}; \lambda)\varphi(x), & B_a\varphi(a) + B_b\varphi(b) &= \mathbf{0}, \\ \xi'(x) &= \varphi(x)^T \varphi(x), & \xi(a) &= 0, \ \xi(b) = 1, \\ \lambda'(x) &= 0. \end{aligned} \tag{5.94}$$

Obviously, the dimension of the extended BVP is $n + 2$.

Example 5.38 In Sect. 5.1 we have shown that the primary simple bifurcation points of the BVP (5.6),

$$\varphi''(s) = -\lambda \sin(\varphi(s)),$$
$$\varphi'(0) = \varphi'(1) = 0,$$

are $z_0^{(k)} = (0, \lambda_0^{(k)})$, with $\lambda_0^{(k)} = (k\pi)^2$, $k = 1, 2, \ldots$.

The vector-matrix formulation of this BVP is

$$y'(x) = \begin{pmatrix} y_1'(x) \\ y_2'(x) \end{pmatrix} = f(x, y(x); \lambda) = \begin{pmatrix} y_2(x) \\ -\lambda \sin(y_1(x)) \end{pmatrix},$$

$$B_a y(0) + B_b y(1) = \begin{pmatrix} 0 & 1 \\ 0 & 0 \end{pmatrix} y(0) + \begin{pmatrix} 0 & 0 \\ 0 & 1 \end{pmatrix} y(1) = 0.$$

We have

$$f_y(x, y(x); \lambda) = \begin{pmatrix} 0 & 1 \\ -\lambda \cos(y_1(x)) & 0 \end{pmatrix}, \quad \text{and thus}$$

$$f_y(x, 0; \lambda) = \begin{pmatrix} 0 & 1 \\ -\lambda & 0 \end{pmatrix}.$$

Now, it can be easily seen that the associated extended problem (5.94) is

$$\begin{aligned}
\varphi_1'(x) &= \varphi_2(x), & \varphi_2(0) &= \varphi_2(1) = 0, \\
\varphi_2'(x) &= -\lambda(x)\,\varphi_1(x) & & \\
\xi'(x) &= \varphi_1(x)^2 + \varphi_2(x)^2, & \xi(0) &= 0, \ \ \xi(1) = 1, \\
\lambda'(x) &= 0. & &
\end{aligned} \qquad (5.95)$$

We have used the multiple shooting code RWPM to determine the solutions of the extended problem (5.95). Here we present the results for the first bifurcation point. Our computations are based on ten shooting points $\tau_j = j \cdot 0.1$, $j = 0, 1, \ldots, 9$, the IVP-solver ODE45 from MATLAB and a tolerance of 10^{-6}. For the starting values it is important to impress the qualitative behavior of the eigensolution $\varphi_1(x)$ upon the starting trajectory. Therefore, we have used the following starting values

$$\varphi_1^{(0)}(\tau_j) = 1 - \tau_j, \quad \varphi_2^{(0)}(\tau_j) = -1, \quad \varphi_3^{(0)}(\tau_j) = \tau_j, \quad \varphi_4^{(0)}(\tau_j) = 5; \quad j = 0, \ldots, 9.$$

After six iterations, we obtained the approximation $\tilde{\lambda}_0^{(1)} = 9.869040$ for the first critical value $\lambda_0^{(1)} = \pi^2$. The corresponding approximation of the eigensolution $\varphi_1(x)$ is shown in Fig. 5.14. $\square$

5.4.4   Determination of Solutions in the Neighborhood of a Primary Simple Bifurcation Point

Let $z \equiv (y, \lambda) \in Z$ be an isolated solution of the operator equation (5.11), which is located in the neighborhood of a primary simple bifurcation point $z_0 \equiv (0, \lambda_0)$.

An analytical technique to determine z has been proposed by Levi-Cività, see [74]. It is based on the following representation of the nontrivial solution branch

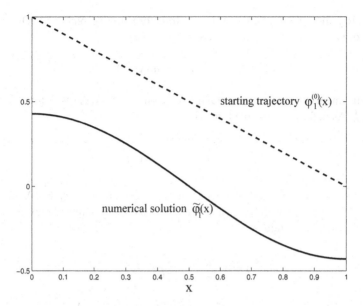

Fig. 5.14 Numerically determined eigensolution $\tilde{\varphi}_1(x) \approx \varphi_1(x)$ of problem (5.95)

$$\lambda(\varepsilon) = \lambda_0 + \sum_{i=1}^{\infty} s_i \varepsilon^i,$$

$$y(\varepsilon) = \varepsilon \varphi_0 + \sum_{i=2}^{\infty} u_i \varepsilon^i, \quad \varphi_0^* u_i = 0. \tag{5.96}$$

Here, ε is a sufficiently small parameter, say $|\varepsilon| \leq \varepsilon_0$. The unknown u_k and s_k are determined by

- expanding the operator T at z_0 into a Taylor series and substituting this series into the operator equation (5.11),
- substituting the ansatz (5.96) into the resulting equation,
- comparing the terms, which have the same power of ε, and using the orthogonality relation.

Example 5.39 For the governing equations (5.6) of the Euler-Bernoulli rod, we will demonstrate the method of Levi-Cività. In Sect. 5.1 we have shown that the eigenvalues and the corresponding (normalized) eigensolutions of the linearized problem are

$$\lambda_0^{(k)} = (k\pi)^2, \quad \varphi_0^{(k)}(x) = \sqrt{2}\cos(k\pi x), \quad k = 1, 2, \ldots.$$

Here, we set $k = 1$. Let us consider the first eigenvalue and the corresponding eigensolution. Therefore, we set

$$\lambda_0 \equiv \lambda_0^{(1)} = \pi^2, \quad \varphi_0 \equiv \varphi_0^{(1)}(x) = \sqrt{2}\cos(\pi x).$$

To simplify the computation, we do not transform the second-order ODE into a system of two first-order equations, but instead, we set

$$T(y, \lambda) \equiv y''(x) + \lambda \sin(y(x)) = 0. \tag{5.97}$$

Since our linearized second-order ODE (see formula (5.7)) is self-adjoint, we have $\psi_0^* = \varphi_0^*$. As before, we use the functional

$$\varphi_0^* y(x) \equiv \int_0^1 \varphi_0(x) y(x) dx.$$

The first step of the method of Levi-Cività is to expand T at $z_0 \equiv (0, \lambda_0)$ into a Taylor series. In our case, this means that the sinus function must be expanded into a Taylor series. We obtain

$$\sin(y) = y - \frac{y^3}{3!} + \frac{y^5}{5!} - \frac{y^7}{7!} \pm \cdots,$$

$$T(y, \lambda) = y'' + \lambda \left(y - \frac{y^3}{3!} + \frac{y^5}{5!} - \frac{y^7}{7!} \pm \cdots \right) = 0. \tag{5.98}$$

In the second step, the ansatz (5.96) has to be inserted into (5.98). It results

$$\varepsilon \varphi_0'' + \sum_{i=2}^{\infty} u_k i' \varepsilon^i + \left(\lambda_0 + \sum_{i=1}^{\infty} s_i \varepsilon^i \right) \left[\left(\varepsilon \varphi_0 + \sum_{i=2}^{\infty} u_i \varepsilon^i \right) \right.$$
$$\left. - \frac{1}{6} \left(\varepsilon \varphi_0 + \sum_{i=2}^{\infty} u_i \varepsilon^i \right)^3 + \frac{1}{120} \left(\varepsilon \varphi_0 + \sum_{i=2}^{\infty} u_i \varepsilon^i \right)^5 + O(\varepsilon^7) \right] = 0. \tag{5.99}$$

In the last step, we have to compare the terms which have the same power of the parameter ε. We obtain:

• Terms in ε

$$\varphi_0'' + \lambda_0 \varphi_0 = 0.$$

• Terms in ε^2

$$u_2'' + \lambda_0 u_2 = -s_1 \varphi_0, \quad u_2 \in Y_1.$$

This equation is solvable, if $s_1 \psi_0^* \varphi_0 = 0$. It follows $s_1 = 0$. Thus, we have to solve the BVP

$$u_2'' + \lambda_0 u_2 = 0, \quad u_2'(0) = u_2'(1) = 0.$$

Due to the orthogonality condition $\varphi_0^* u_2 = 0$, this homogeneous BVP has the unique solution $u_2(x) \equiv 0$.

- Terms in ε^3

$$u_3'' + \lambda_0 u_3 = -s_2\varphi_0 + \frac{1}{6}\lambda_0\varphi_0^3, \quad u_3 \in Y_1.$$

This equation is solvable, if

$$-s_2\psi_0^*\varphi_0 + \frac{1}{6}\lambda_0\psi_0^*\varphi_0^3 = 0, \quad \text{i.e.,}$$

$$s_2 = \frac{1}{6}\lambda_0\psi_0^*\varphi_0^3 = \frac{1}{6}\lambda_0\int_0^1 \left(\sqrt{2}\cos(\pi x)\right)^4 dx = \frac{1}{6}\lambda_0\frac{3}{2} = \frac{1}{4}\lambda_0.$$

Thus, we have to solve the BVP

$$u_3'' + \lambda_0 u_3 = -\frac{1}{4}\lambda_0\varphi_0 + \frac{1}{6}\lambda_0\varphi_0^3, \quad u_3(0) = u_3(1) = 0. \tag{5.100}$$

It is

$$\begin{aligned}
u_3'' + \lambda_0 u_3 &= -\frac{1}{4}\lambda_0\varphi_0 + \frac{1}{6}\lambda_0\varphi_0^3 \\
&= -\frac{1}{4}\pi^2\sqrt{2}\cos(\pi x) + \frac{1}{6}\pi^2 2\sqrt{2}\cos^3(\pi x) \\
&= -\frac{\sqrt{2}}{4}\pi^2\cos(\pi x) + \frac{\sqrt{2}}{3}\pi^2\left(\frac{3}{4}\cos(\pi x) + \frac{1}{4}\cos(3\pi x)\right) \\
&= \left(-\frac{\sqrt{2}}{4}\pi^2 + \frac{\sqrt{2}}{4}\pi^2\right)\cos(\pi x) + \frac{\sqrt{2}}{12}\pi^2\cos(3\pi x) \\
&= \frac{\sqrt{2}}{12}\pi^2\cos(3\pi x).
\end{aligned}$$

To determine a particular solution of this ODE, we use the ansatz

$$u_3(x) = C\cos(3\pi x),$$

which satisfies the boundary conditions, too. We have

$$\begin{aligned}
&u_3(x) = C\cos(3\pi x), \quad u_3'(x) = -3\pi C\sin(3\pi x), \\
&u_3''(x) = -9\pi^2 C\cos(3\pi x).
\end{aligned}$$

Substituting $u_3(x)$ and $u_3''(x)$ into the ODE, gives

$$-9\pi^2 C \cos(3\pi x) + \pi^2 C \cos(3\pi x) = \frac{\sqrt{2}}{12}\pi^2 \cos(3\pi x).$$

A comparison of the coefficients of the cosine function yields $C = -\dfrac{\sqrt{2}}{96}$. Thus, the solution $u_3 \in Y_1$ of the BVP (5.100) is

$$u_3(x) = -\frac{\sqrt{2}}{96}\cos(3\pi x).$$

- Terms in ε^4

$$u_4'' + \lambda_0 u_4 = -s_1 u_3 - s_2 u_2 - s_3 \varphi_0 + \frac{1}{6}\left(s_1 \varphi_0^3 + \lambda_0 \varphi_0^2 u_2\right)$$

$$= -s_3 \varphi_0, \quad u_4 \in Y_1.$$

This equation is solvable if $s_3 \psi_0^* \varphi_0 = s_3 \varphi_0^* \varphi_0 = 0$. It follows $s_3 = 0$. Thus, we have to solve the BVP

$$u_4'' + \lambda_0 u_4 = 0, \quad u_4(0) = u_4(1) = 0.$$

Due to the condition $\varphi_0^* u_4 = 0$, the unique solution is $u_4(x) \equiv 0$.

- Terms in ε^5

$$u_5'' + \lambda_0 u_5 = \frac{\lambda_0}{2}\varphi_0^2 u_3 - \frac{1}{120}\lambda_0\varphi_0^5 - s_2 u_3 + \frac{1}{6}s_2\varphi_0^3 - s_4\varphi_0 \equiv R,$$

$$u_5 \in Y_1.$$

This equation is solvable if $\psi_0^* R = \varphi_0^* R = 0$, i.e.,

$$s_4 = \frac{\lambda_0}{2}\varphi_0^*\varphi_0^2 u_3 - \frac{1}{120}\lambda_0\varphi_0^*\varphi_0^5 - s_2\varphi_0^* u_3 + \frac{1}{6}s_2\varphi_0^*\varphi_0^3$$

$$= -\frac{1}{48}\pi^2 \int_0^1 \cos^3(\pi x)\cos(3\pi x)dx - \frac{1}{15}\pi^2 \int_0^1 \cos^6(\pi x)dx$$

$$+ \frac{1}{192}\pi^2 \int_0^1 \cos(3\pi x)\cos(\pi x)dx + \frac{1}{6}\pi^2 \int_0^1 \cos^4(\pi x)dx$$

$$= -\frac{1}{48}\pi^2\frac{1}{8} - \frac{1}{15}\pi^2\frac{5}{16} + \frac{1}{6}\pi^2\frac{3}{8}$$

$$= \frac{5}{128}\pi^2.$$

Substituting this value into the right-hand side of the ODE, we obtain

$$
\begin{aligned}
R &= -\frac{\sqrt{2}}{96}\pi^2 \left(\frac{1}{4}\cos(\pi x) + \frac{1}{2}\cos(3\pi x) + \frac{1}{4}\cos(5\pi x) \right) \\
&\quad -\frac{\sqrt{2}}{30}\pi^2 \left(\frac{5}{8}\cos(\pi x) + \frac{5}{16}\cos(3\pi x) + \frac{1}{16}\cos(5\pi x) \right) \\
&\quad +\frac{\sqrt{2}}{384}\pi^2 \cos(3\pi x) + \frac{\sqrt{2}}{12}\pi^2 \left(\frac{3}{4}\cos(\pi x) + \frac{1}{4}\cos(3\pi x) \right) \\
&\quad -\frac{5\sqrt{2}}{128}\pi^2 \cos(\pi x) \\
&= \pi^2 \cos(\pi x) \cdot 0 + \pi^2 \cos(3\pi x) \cdot \frac{\sqrt{2}}{128} + \pi^2 \cos(5\pi x) \cdot \left(-\frac{3\sqrt{2}}{640} \right).
\end{aligned}
$$

Now, we have to solve the BVP

$$
u_5'' + \pi^2 u_5 = R, \quad u_5(0) = u_5(1) = 0. \tag{5.101}
$$

To determine a particular solution of the ODE, we use the ansatz

$$
u_5 = C_3 \cos(3\pi x) + C_5 \cos(5\pi x).
$$

It is

$$
\begin{aligned}
u_5' &= -3\pi C_3 \sin(3\pi x) - 5\pi C_5 \sin(5\pi x), \\
u_5'' &= -9\pi^2 C_3 \cos(3\pi x) - 25\pi^2 C_5 \cos(5\pi x).
\end{aligned}
$$

Therefore

$$
u_5'' + \pi^2 u_5 = -8\pi^2 C_3 \cos(3\pi x) - 24\pi^2 C_5 \cos(5\pi x).
$$

Equating the right-hand side of this ODE with the above calculated expression for R, we obtain

$$
C_3 = -\frac{\sqrt{2}}{1024} \quad \text{and} \quad C_5 = \frac{\sqrt{2}}{5120}.
$$

Thus, the solution $u_5 \in Y_1$ of the BVP (5.101) is

$$
u_5(x) = -\frac{\sqrt{2}}{1024}\cos(3\pi x) + \frac{\sqrt{2}}{5120}\cos(5\pi x).
$$

If we stop the algorithm at this point, we have obtained the following representation of the nontrivial solution curve branching at $z_0 = (0, \lambda_0) = (0, \pi^2)$:

$$y(x; \varepsilon) = \varepsilon \sqrt{2} \cos(\pi x) - \varepsilon^3 \frac{\sqrt{2}}{96} \cos(3\pi x)$$

$$+ \varepsilon^5 \left(-\frac{\sqrt{2}}{1024} \cos(3\pi x) + \frac{\sqrt{2}}{5120} \cos(5\pi x) \right) + O(\varepsilon^7), \qquad (5.102)$$

$$\lambda(\varepsilon) = \pi^2 \left(1 + \frac{1}{4}\varepsilon^2 + \frac{5}{128}\varepsilon^4 + O(\varepsilon^6) \right), \quad |\varepsilon| \le \varepsilon_0.$$

More generally, the nontrivial solution curves branching at $z_0^{(k)} \equiv (0, (k\pi)^2)$, $k = 1, 2, \ldots$, can be parametrized in the form

$$y(x; \varepsilon) = \varepsilon \sqrt{2} \cos(k\pi x) - \varepsilon^3 \frac{\sqrt{2}}{96} \cos(3k\pi x)$$

$$+ \varepsilon^5 \left(-\frac{\sqrt{2}}{1024} \cos(3k\pi x) + \frac{\sqrt{2}}{5120} \cos(5k\pi x) \right) + O(\varepsilon^7), \qquad (5.103)$$

$$\lambda(\varepsilon) = (k\pi)^2 \left(1 + \frac{1}{4}\varepsilon^2 + \frac{5}{128}\varepsilon^4 + O(\varepsilon^6) \right), \quad |\varepsilon| \le \varepsilon_0.$$

In Fig. 5.15 it is shown that the graph $(y(x; \varepsilon), \lambda(\varepsilon))$ comes closer to the exact solution graph $(y(x), \lambda)$ if the number of terms in the series (5.102) is increased. $\qquad\square$

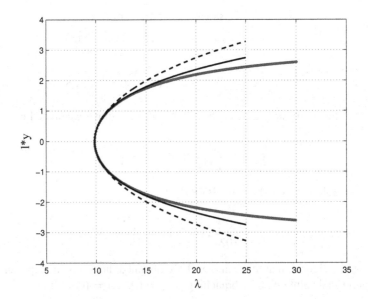

Fig. 5.15 *Dashed line* three terms of the series (5.102) are used, *solid line* four terms of the series (5.102) are used, *gray line* exact solution curve. Here, we have set $l^* y \equiv y(0)$

Before we describe how a nontrivial solution curve branching at a primary simple bifurcation point can be determined numerically, let us demonstrate the Levi-Cività technique for a second example.

Example 5.40   Consider the following BVP

$$y''(x) = \lambda(y(x) + y(x)^2),$$
$$y(0) = y(1) = 0.$$
                                                                                   (5.104)

The primary simple bifurcation points are

$$z_0^{(k)} = (0, \lambda_0^{(k)}), \quad \lambda_0^{(k)} = -(k\pi)^2, \quad k = 1, 2, \dots.$$

The corresponding (normalized) eigenfunctions are

$$\varphi_0^{(k)}(x) = \sqrt{2} \sin(k\pi x), \quad k = 1, 2, \dots.$$

In order to simplify the notation, let us set

$$\lambda_0 \equiv \lambda_0^{(k)} \quad \text{and} \quad \varphi_0(x) \equiv \sqrt{2} \sin(k\pi x).$$

Substituting the ansatz (5.96) into the ODE yields

$$\varphi_0'' \varepsilon + \sum_{i=2}^{\infty} u_i'' \varepsilon^i =$$

$$\left(\lambda_0 + \sum_{i=1}^{\infty} s_i \varepsilon^i \right) \left(\varphi_0 \varepsilon + \sum_{i=2}^{\infty} u_i \varepsilon^i + \left(\varphi_0 \varepsilon + \sum_{i=2}^{\infty} u_i \varepsilon^i \right)^2 \right).$$ (5.105)

Now, we compare the terms, which have the same power of the parameter ε, and obtain:

- Terms in ε
$$\varphi_0'' - \lambda_0 \varphi_0 = 0.$$

- Terms in ε^2
$$u_2'' - \lambda_0 u_2 = s_1 \varphi_0 + \lambda_0 \varphi_0^2, \quad u_2 \in Y_1.$$

This equation is solvable if $\varphi_0^* \{s_1 \varphi_0 + \lambda_0 \varphi_0^2\} = 0$, where we use again the fact that $\psi_0^* = \varphi_0^*$ (see Example 5.39). Thus,

$$
\begin{aligned}
s_1 &= -\lambda_0 \varphi_0^* \varphi_0^2 = -\lambda_0 2\sqrt{2} \int_0^1 \sin^3(k\pi x)dx \\
&= 2\sqrt{2}(k\pi) \left\{ \frac{1}{3} \cos^3(k\pi x) - \cos(k\pi) \right\} \Big|_0^1 \\
&= \begin{cases} 0, & k \text{ even} \\ \dfrac{8\sqrt{2}}{3} k\pi, & k \text{ odd.} \end{cases}
\end{aligned}
$$

Let us substitute s_1 into the ODE and distinguish between even and odd values of k. If k is even, we have to solve the BVP

$$
\begin{aligned}
& u_2'' + (k\pi)^2 u_2 = -2(k\pi)^2 \sin^2(k\pi x) = -(k\pi)^2(1 - \cos(2k\pi x)), \\
& u_2(0) = u_2(1) = 0.
\end{aligned}
\tag{5.106}
$$

To determine a particular solution of this equation, we use the ansatz

$$
\begin{aligned}
u_2 &= C_1 \cos(k\pi x) + C_2 \sin(2k\pi x) + C_3 \cos(2k\pi x) \\
&\quad + C_4 x \cos(k\pi x) + C_5.
\end{aligned}
\tag{5.107}
$$

It follows

$$
\begin{aligned}
u_2' = & - C_1(k\pi) \sin(k\pi x) + 2C_2(k\pi) \cos(2k\pi x) \\
& - 2C_3(k\pi) \sin(2k\pi x) + C_4 [\cos(k\pi x) - x(k\pi) \sin(k\pi x)],
\end{aligned}
$$

and

$$
\begin{aligned}
u_2'' = & - (k\pi)^2 [C_1 \cos(k\pi x) + 4C_2 \sin(2k\pi x) + 4C_3 \cos(2k\pi x)] \\
& - (k\pi)C_4 [2 \sin(k\pi x) + (k\pi)x \cos(k\pi x)].
\end{aligned}
$$

Thus,

$$
\begin{aligned}
u_2'' + (k\pi)^2 u_2 = & - 3C_2(k\pi)^2 \sin(2k\pi x) - 3C_3(k\pi)^2 \cos(2k\pi x) \\
& - 2C_4(k\pi) \sin(k\pi x) + (k\pi)^2 C_5.
\end{aligned}
\tag{5.108}
$$

Comparing the coefficients in (5.106) and (5.108), we obtain

$$
C_2 = 0, \quad C_3 = -\frac{1}{3}, \quad C_4 = 0, \quad C_5 = -1.
$$

However, the coefficient C_1 is still undetermined. Substituting the coefficients into the ansatz (5.107) and using the first boundary condition, we have

$$u_2(0) = C_1 - \frac{1}{3} - 1 \doteq 0, \quad \text{i.e., } C_1 = \frac{4}{3}.$$

Obviously, for this value of C_1 the second boundary condition is automatically satisfied. Therefore, the particular solution of (5.106) is

$$u_2(x) = \frac{1}{3}(4\cos(k\pi x) - \cos(2k\pi x)) - 1. \tag{5.109}$$

Let us now assume that k is odd. Instead of (5.106), we have to solve the BVP

$$u_2'' + (k\pi)^2 u_2 = \frac{16}{3}(k\pi)\sin(k\pi x) - (k\pi)^2(1 - \cos(2k\pi x)),$$
$$u_2(0) = u_2(1) = 0. \tag{5.110}$$

Comparing the coefficients in (5.106) and (5.110), we obtain

$$C_2 = 0, \quad C_3 = -\frac{1}{3}, \quad C_4 = -\frac{8}{3}, \quad C_5 = -1.$$

As before, the coefficient C_1 is still undetermined. Substituting the coefficients into the ansatz (5.107) and using the first boundary condition, we have

$$u_2(0) = C_1 - \frac{1}{3} - 1 \doteq 0, \quad \text{i.e., } C_1 = \frac{4}{3}.$$

For this value of C_1 the second boundary condition is automatically satisfied. Therefore, the particular solution of (5.110) is

$$u_2(x) = \frac{1}{3}(4\cos(k\pi x) - \cos(2k\pi x) - 8x\cos(k\pi x)) - 1. \tag{5.111}$$

- Terms in ε^3

$$u_3'' - \lambda_0 u_3 = s_2\varphi_0 + 2\lambda_0\varphi_0 u_2 + s_1 u_2 + s_1\varphi_0^2, \quad u_3 \in Y_1.$$

This equation is solvable if $\varphi_0^*\{s_2\varphi_0 + 2\lambda_0\varphi_0 u_2 + s_1 u_2 + s_1\varphi_0^2\} = 0$. First, let us consider the case that k is even. It must hold (note that $s_1 = 0$)

$$0 \doteq \varphi_0^*\{s_2\varphi_0 + 2\lambda_0\varphi_0 u_2\}$$
$$= \int_0^1 \sqrt{2}\sin(k\pi x)\{s_2\sqrt{2}\sin(k\pi x) - 2(k\pi)^2\sqrt{2}\sin(k\pi x)u_2\}dx.$$

Substituting the expression (5.109) for u_2 into this equation, we get

$$0 = 2s_2 \int_0^1 \sin^2(k\pi x)dx$$

$$- 4(k\pi)^2 \int_0^1 \sin^2(k\pi x)\left[\frac{4}{3}\cos(k\pi x) - \frac{1}{3}\cos(2k\pi x) - 1\right]dx$$

$$= 2s_2\frac{1}{2} + (k\pi)^2\left[-\frac{1}{3} + 2\right].$$

Thus,

$$s_2 = -\frac{5}{3}(k\pi)^2. \tag{5.112}$$

Now, we assume that k is odd. It must hold

$$0 \doteq \varphi_0^*\{s_2\varphi_0 + 2\lambda_0\varphi_0 u_2 + s_1 u_2 + s_1\varphi_0^2\}$$

$$= \int_0^1 \sqrt{2}\sin(k\pi x)\Bigg[s_2\sqrt{2}\sin(k\pi x) - 2\sqrt{2}(k\pi)^2\Big(\frac{2}{3}\sin(2k\pi x)$$

$$- \frac{1}{3}\sin(k\pi x)\cos(2k\pi x) - \frac{4}{3}x\sin(2k\pi x) - \sin(k\pi x)\Big)$$

$$+ s_1\Big(\frac{4}{3}\cos(k\pi x) - \frac{1}{3}\cos(2k\pi x) - \frac{8}{3}x\cos(k\pi x)$$

$$- 1 + 2\sin^2(k\pi x)\Big)\Bigg]dt$$

$$= 2s_2 \cdot \frac{1}{2} - 4(k\pi)^2\Big(\frac{2}{3} \cdot 0 + \frac{1}{3} \cdot \frac{1}{4} + \frac{4}{3} \cdot \frac{8}{9(k\pi)^2} - \frac{1}{2}\Big)$$

$$\sqrt{2}s_1\Big(\frac{4}{3} \cdot 0 + \frac{1}{3} \cdot \frac{2}{3(k\pi)} + \frac{8}{3} \cdot \frac{1}{4(k\pi)} - 1 \cdot \frac{1}{(k\pi)} + 2 \cdot \frac{4}{3(k\pi)}\Big)$$

$$= s_2 + \frac{5}{3}(k\pi)^2 + \frac{32}{9}.$$

Thus,

$$s_2 = -\frac{32}{9} - \frac{5}{3}(k\pi)^2. \tag{5.113}$$

Summarizing the above computations, we have the following asymptotic expansion of the branching solutions:

- k is odd

$$y(x; \varepsilon) = \left(\sqrt{2} \sin(k\pi x) \right) \varepsilon$$

$$\left(\frac{4}{3} \cos(k\pi x) - \frac{1}{3} \cos(2k\pi x) - \frac{8}{3} x \cos(k\pi x) - 1 \right) \varepsilon^2 + O(\varepsilon^3),$$

$$\lambda(\varepsilon) = -(k\pi)^2 + \left(\frac{8\sqrt{2}}{3} (k\pi) \right) \varepsilon - \left(\frac{32}{9} + \frac{5}{3} (k\pi)^2 \right) \varepsilon^2 + O(\varepsilon^3).$$

(5.114)

- k is even

$$y(x; \varepsilon) = \left(\sqrt{2} \sin(k\pi x) \right) \varepsilon$$

$$+ \left(\frac{4}{3} \cos(k\pi x) - \frac{1}{3} \cos(2k\pi x) - 1 \right) \varepsilon^2 + O(\varepsilon^3),$$ (5.115)

$$\lambda(\varepsilon) = -(k\pi)^2 - \left(\frac{5}{3} (k\pi)^2 \right) \varepsilon^2 + O(\varepsilon^4).$$

Note, these expansions depend on a small parameter ε: $|\varepsilon| \le \varepsilon_0$, which has been artificially introduced into the problem.

In Fig. 5.16, the exact solution graph $(y(x), \lambda)$ and the graph $(y(x; \varepsilon), \lambda(\varepsilon))$ are represented.   □

These two examples demonstrate impressively that the representation of the branching solutions as infinite sums of the form (5.96) is extremely work-intensive

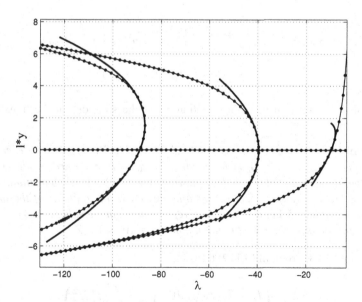

Fig. 5.16 *Dotted/dashed line* exact solution curve, *solid line* the asymptotic expansion (5.114) of the branching solutions. Here, we have set $l^* y \equiv y'(0)$

and very susceptible to arithmetical errors. Therefore, we will now describe a numerical approach (see [129, 130]), which is based on the idea that the infinite sums are expressed as functions, which are the isolated solutions of a standard BVP.

Instead of (5.96), we use the following representation of an isolated solutions $z \equiv (y, \lambda) \in Z$ of the operator equation (5.11), which is located in the neighborhood of a primary simple bifurcation point $z_0 \equiv (0, \lambda_0)$:

$$
\begin{aligned}
y(\varepsilon) &= \varepsilon\varphi_0 + \varepsilon^2 u(\varepsilon), & u(\varepsilon) &\in Y_1, & |\varepsilon| &\leq \varepsilon_0, \\
\lambda(\varepsilon) &= \lambda_0 + \varepsilon\tau(\varepsilon), & \tau(\varepsilon) &\in \mathbb{R}.
\end{aligned}
\tag{5.116}
$$

The assumption (5.69) implies that

$$
T_\lambda^0 = T_{\lambda\lambda}^0 = \cdots = 0.
\tag{5.117}
$$

Therefore, the formula (5.65) simplifies to

$$
0 = T(y, \lambda) = T_y^0 y + \frac{1}{2}T_{yy}^0 y^2 + T_{y\lambda}^0 \lambda y + R(y, \lambda).
\tag{5.118}
$$

The remainder R consists only of third or higher order terms in y and λ. Now, we substitute the ansatz (5.116) into (5.118), and obtain

$$
\begin{pmatrix} T_y^0 & T_{y\lambda}^0\varphi_0 \\ \varphi_0^* & 0 \end{pmatrix} \begin{pmatrix} u \\ \tau \end{pmatrix} = -\begin{pmatrix} w(u, \tau; \varepsilon) \\ 0 \end{pmatrix}, \quad |\varepsilon| \leq \varepsilon_0,
\tag{5.119}
$$

where

$$
\begin{aligned}
w(u, \tau; \varepsilon) &\equiv \frac{1}{2}T_{yy}^0\varphi_0^2 + \varepsilon\frac{1}{2}T_{yy}^0(2\varphi_0 u + \varepsilon u^2) + \varepsilon T_{y\lambda}^0\tau u \\
&\quad + \frac{1}{\varepsilon^2}R(\varepsilon\varphi_0 + \varepsilon^2 u, \lambda_0 + \varepsilon\tau).
\end{aligned}
$$

The following theorem gives information about the solutions of the transformed problem (5.119).

Theorem 5.41 *Let the conditions (5.69), (5.70) and (5.78) be satisfied. Then, there exists a real constant $\varepsilon_0 > 0$, such that for $|\varepsilon| \leq \varepsilon_0$ the transformed problem (5.119) has a continuous curve of isolated solutions $\eta(\varepsilon) \equiv (u(\varepsilon), \tau(\varepsilon))^T$. Substituting these solutions into the ansatz (5.116), a continuous curve of solutions of the original operator equation (5.11) results, which represents a simple solution curve passing through the simple primary bifurcation point $z_0 \equiv (0, \lambda_0) \in Z$.*

Proof For $\varepsilon \to 0$ problem (5.119) turns into

$$
\mathbb{A}\begin{pmatrix} u(0) \\ \tau(0) \end{pmatrix} \equiv \begin{pmatrix} T_y^0 & T_{y\lambda}^0\varphi_0 \\ \varphi_0^* & 0 \end{pmatrix} \begin{pmatrix} u(0) \\ \tau(0) \end{pmatrix} = -\begin{pmatrix} \frac{1}{2}T_{yy}^0\varphi_0^2 \\ 0 \end{pmatrix}.
\tag{5.120}
$$

The assumptions and Lemma 5.25, part 2, imply that

- A is bijective, and
- there exists a unique solution $\eta^0 \equiv (u^0, \tau^0)^T$ of (5.120), with

$$\tau^0 = -\frac{1}{2} \left(\psi_0^* T_{y\lambda}^0 \varphi_0 \right)^{-1} \psi_0^* T_{yy}^0 \varphi_0^2,$$

$$u^0 = -\left(T_y^0 \right)^+ \left\{ T_{y\lambda}^0 \varphi_0 \tau^0 + \frac{1}{2} T_{yy}^0 \varphi_0^2 \right\},$$

where $\left(T_y^0 \right)^+$ denotes the generalized inverse of T_y^0, which maps $\mathcal{R}(T_y^0)$ one-to-one onto Y_1.

If problem (5.119) is written in the form

$$F(\varepsilon; \eta) = 0, \quad \eta \equiv (u, \tau)^T,$$

then the claim follows, as in the proof of Theorem 5.26, from the Implicit Function Theorem (see Theorem 5.8). ∎

Applying, as shown in Sect. 5.3.4, the transformed problem (5.119) to the BVP (5.10) and using the functional (5.47), the following transformed BVP of dimension $n + 2$ results

$$u'(x; \varepsilon) = f_y^0 u(x; \varepsilon) + f_{y\lambda}^0 \varphi_0(x) \tau(\varepsilon) + \tilde{w}(u(x; \varepsilon), \tau(\varepsilon); \varepsilon), \quad |\varepsilon| \le \varepsilon_0,$$

$$\xi'(x; \varepsilon) = \varphi_0(x)^T u(x; \varepsilon), \quad \tau'(x) = 0,$$

$$B_a u(a; \varepsilon) + B_b u(b; \varepsilon) = \mathbf{0}, \quad \xi(a) = 0, \quad \xi(b) = 0,$$

$$(5.121)$$

where

$$\tilde{w}(u, \tau; \varepsilon) \equiv \frac{1}{2} f_{yy}^0 \varphi_0^2 + \frac{\varepsilon}{2} f_{yy}^0 \left(2\varphi_0 u + \varepsilon u^2 \right)$$

$$+ \varepsilon f_{y\lambda}^0 \tau u + \frac{1}{\varepsilon^2} \tilde{R}(\varepsilon \varphi_0 + \varepsilon^2 u, \lambda_0 + \varepsilon \tau),$$

and

$$\tilde{R}(y, \lambda) \equiv f(y, \lambda) - \left\{ f_y^0 y + \frac{1}{2} f_{yy}^0 y^2 + f_{y\lambda}^0 (\lambda - \lambda_0) y \right\}.$$

For all ε with $|\varepsilon| \le \varepsilon_0$, the BVP (5.121) can be solved by the shooting methods presented in Chap. 4.

Remark 5.42 The quantities λ_0 and φ_0, which are required in the ansatz (5.116), can be determined from the extended problem (5.89). Usually, the two problems (5.121) and (5.89) are combined into a single (larger) problem. In the case of parametrized

BVPs, the advantage of this strategy is that both problems are defined on the same grid. Thus, IVP-solver with an automatic step-size control can be used in the shooting methods.   □

The transformation techniques discussed so far, are based on the ansatz (5.96) or (5.116), which depends on an additional (artificial) parameter ε. However, many engineering and physical models require the computation of the solution y for prescribed values of the (control) parameter λ, which is already contained in the given problem. In the neighborhood of primary simple bifurcation points, the approaches (5.83) and (5.87) enable the determination of nontrivial solutions of the operator equation (5.11) in direct dependence of λ. But the disadvantage of this strategy is that the exact type of the bifurcation phenomenon must be known a priori and has to be taken into account.

First, we want to consider the case of a *symmetric bifurcation point*, i.e., let us assume that

$$a_2 \equiv \psi_0^* T_{yy}^0 \varphi_0^2 = 0 \text{ and } a_3 \equiv \psi_0^*(6T_{yy}^0 \varphi_0 v_2 + T_{yyy}^0 \varphi_0^3) \neq 0, \qquad (5.122)$$

where $v_2 \in Y_1$ is the unique solution of the linear operator equation

$$T_y^0 v_2 = -\frac{1}{2} T_{yy}^0 \varphi_0^2. \qquad (5.123)$$

In analogy to the analytical approach of Levi-Cività, the following ansatz for the nontrivial branching solutions has been proposed by Nekrassov, see [90, 91]:

$$y(x; \xi) = \frac{2}{\sqrt{\lambda_0}} \varphi_0(x) \xi + \sum_{i=2}^{\infty} (s_i \varphi_0(x) + u_i(x)) \xi^i, \qquad (5.124)$$
$$\varphi_0^* u_i = 0, \quad \xi^2 \equiv \lambda - \lambda_0.$$

Let us illustrate the determination of the free parameters u_i and s_i for the Euler–Bernoulli problem, which we have studied in Example 5.39.

Example 5.43 Consider the governing equations (5.6) of the Euler–Bernoulli problem. Here, we have
$$\lambda_0 = \pi^2 \text{ and } \varphi_0(x) = \sqrt{2} \cos(\pi x).$$

If we define
$$T(y, \lambda) \equiv y'' + \lambda \sin(y),$$

the associated bifurcation coefficients are

$$a_1 = \varphi_0^* \varphi_0 = \int_0^1 \varphi_0(x)^2 dx = 2 \int_0^1 \cos(\pi x)^2 dx = 1,$$

$$a_2 = \varphi_0^* \cdot 0 \cdot \varphi_0^2 = 0,$$

$$v_2 = 0,$$

$$a_3 = \varphi_0^*(-\lambda_0)\varphi_0^3 = -\lambda_0 \int_0^1 \varphi_0(x)^4 dx = -4\lambda_0 \int_0^1 \cos(\pi x)^4 dx = -\frac{3}{2}\pi^2,$$

$$C_2 = -\frac{6a_1}{a_3} = \frac{4}{\pi^2} > 0.$$

Before continuing, compare the ansatz (5.124) of Nekrassov with the representation (5.87), which we have obtained in Sect. 5.4.2.

Substituting the ansatz (5.124) into the equation (5.98) yields

$$\frac{2}{\pi}\varphi_0''\xi + \sum_{i=2}^{\infty}(s_i\varphi_0'' + u_i'')\xi^i + (\lambda_0 + \xi^2)\left[\frac{2}{\pi}\varphi_0\xi + \sum_{i=2}^{\infty}(s_i\varphi_0 + u_i)\xi^i \right.$$
$$\left. -\frac{1}{6}\left(\frac{2}{\pi}\varphi_0\xi + \sum_{i=2}^{\infty}(s_i\varphi_0 + u_i)\xi^i\right)^3 + O(\xi^5)\right] = 0. \qquad (5.125)$$

Now, we look for the terms, which have the same power of ξ. We obtain:

- Terms in ξ

$$\frac{2}{\sqrt{\pi}}\underbrace{(\varphi_0'' + \lambda_0\varphi_0)}_{=0} = 0.$$

- Terms in ξ^2

$$u_2'' + \lambda_0 u_2 + s_2\underbrace{(\varphi_0'' + \lambda_0\varphi_0)}_{=0} = 0.$$

The BVP

$$u_2'' + \lambda_0 u_2 = 0, \quad u_2'(0) = u_2'(1) = 0, \quad u_2 \in Y_1, \qquad (5.126)$$

has the unique solution $u_2(x) \equiv 0$.

- Terms in ξ^3

$$u_3'' + \lambda_0 u_3 + \frac{2}{\pi}\varphi_0 - \frac{4}{3\pi}\varphi_0^3 = 0.$$

The BVP

$$u_3'' + \lambda_0 u_3 = -\frac{2}{\pi}\varphi_0 + \frac{4}{3\pi}\varphi_0^3 \equiv w_3,$$
$$u_3'(0) = u_3'(1) = 0, \quad u_3 \in Y_1, \qquad (5.127)$$

has a unique solution if $\varphi_0^* w_3 = 0$. We obtain

$$\varphi_0^* w_3 = \varphi_0^* \left(-\frac{2}{\pi} \varphi_0 + \frac{4}{3\pi} \varphi_0^3 \right) = -\frac{4}{\pi} \int_0^1 \cos(\pi x)^2 dx + \frac{16}{3\pi} \int_0^1 \cos(\pi x)^4 dx$$

$$= -\frac{4}{\pi} \frac{1}{2} + \frac{16}{3\pi} \frac{3}{8} = 0.$$

The unique solution of problem the (5.127) is

$$u_3(x) = -\frac{\sqrt{2}}{12\pi^3} \cos(3\pi x).$$

- Terms in ξ^4

$$u_4'' + \lambda u_4 + s_4 (\underbrace{\varphi_0'' + \lambda_0 \varphi_0}_{=0}) - \frac{2\lambda_0}{\pi^2} s_2 \varphi_0^3 + s_2 \varphi_0 = 0.$$

The problem

$$u_4'' + \lambda u_4 = s_2 \left(2\varphi_0^3 - \varphi_0 \right) \equiv w_4,$$

$$u_4'(0) = u_4'(1) = 0, \quad u_4 \in Y_1, \tag{5.128}$$

has a unique solution if $\varphi_0^* w_4 = 0$. We obtain

$$\varphi_0^* w_4 = s_2 \varphi_0^* \left(2\varphi_0^3 - \varphi_0 \right) = s_2 \int_0^1 \varphi_0 \left(2\varphi_0^3 - \varphi_0 \right) dx$$

$$= s_2 \int_0^1 \left(8\cos(\pi x)^4 - 2\cos(\pi x)^2 \right) dx = 2s_2.$$

Thus, $s_2 = 0$ and $u_4(x) \equiv 0$.
- Terms in ξ^5

$$u_5'' + \lambda_0 u_5 - \frac{4}{3\pi^3} \varphi_0^3 - 2s_3 \varphi_0^3 - 2u_3 \varphi_0^2 + s_3 \varphi_0 + u_3 + \frac{4}{15\pi^3} \varphi_0^5 = 0.$$

The problem

$$u_5'' + \lambda_0 u_5 = \frac{4}{3\pi^3} \varphi_0^3 + 2s_3 \varphi_0^3 + 2u_3 \varphi_0^2 - s_3 \varphi_0 - u_3 - \frac{4}{15\pi^3} \varphi_0^5 \equiv w_5,$$

$$u_5'(0) = u_5'(1) = 0, \quad u_5 \in Y_1, \tag{5.129}$$

has a unique solution if $\varphi_0^* w_5 = 0$. Since the next computations by hand are very time-consuming, we have used the symbolic tool of the MATLAB. The equation $\varphi_0^* w_5 = 0$ determines the unknown coefficient s_3 as

$$s_3 = -\frac{5}{8\pi^3}.$$

The unique solution of problem (5.129) is

$$u_5(x) = \frac{3\sqrt{2}}{64\pi^5} \cos(3\pi x) + \frac{\sqrt{2}}{160\pi^5} \cos(5\pi x).$$

- Terms in ξ^6
  Following the strategy used before, we obtain $s_4 = 0$ and $u_6(x) \equiv 0$.
- Terms in ξ^7
  The solvability condition of the resulting BVP for u_7 yields

$$s_5 = \frac{123}{256\pi^5}.$$

Now, the curve of nontrivial solutions (5.124) branching at the symmetric primary simple bifurcation point $z_0 = (0, \lambda_0)$ can be represented in the neighborhood of z_0 in the form

$$y(x; \xi) = \frac{1}{\pi} 2\sqrt{2} \cos(\pi x)\xi + \frac{1}{\pi^3}\left(-\frac{5\sqrt{2}}{8}\cos(\pi x) - \frac{\sqrt{2}}{12}\cos(3\pi x)\right)\xi^3$$

$$+ \frac{1}{\pi^5}\left(\frac{123\sqrt{2}}{256}\cos(\pi x) + \frac{3\sqrt{2}}{64}\cos(3\pi x) + \frac{\sqrt{2}}{160}\cos(5\pi x)\right)\xi^5 + O(\xi^7),$$

(5.130)

where $\xi^2 \equiv \lambda - \lambda_0$. Note, the function $y(x)$ is now determined as a sum of broken powers of λ.

In Fig. 5.17 the exact solution of problem (5.6) and the approximate solution, which has been determined by the series (5.130), are presented. □

The above example shows that the analytical approach is practicable only for relatively simple problems. Therefore, we present now a numerical approach that can be applied to more complex problems. At first we expand the operator T into a Taylor series at $z_0 = (0, \lambda_0)$:

$$T(y, \lambda) = T_y^0 y + T_{y\lambda}^0(\lambda - \lambda_0)y + \frac{1}{2}T_{yy}^0 y^2 + \frac{1}{2}T_{y\lambda\lambda}^0(\lambda - \lambda_0)^2 y$$

$$+ \frac{1}{2}T_{yy\lambda}^0(\lambda - \lambda_0)y^2 + \frac{1}{6}T_{yyy}^0 y^3 + \frac{1}{6}T_{y\lambda\lambda\lambda}^0(\lambda - \lambda_0)^3 y \qquad (5.131)$$

$$+ \frac{1}{4}T_{yy\lambda\lambda}^0(\lambda - \lambda_0)^2 y^2 + \frac{1}{6}T_{yyy\lambda}^0(\lambda - \lambda_0)y^3$$

$$+ \frac{1}{24}T_{yyyy}^0 y^4 + R(y, \lambda).$$

The remainder R contains only terms of fifth and higher order in y and $(\lambda - \lambda_0)$.

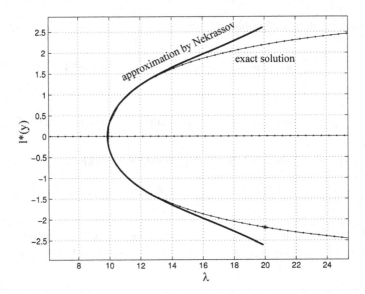

Fig. 5.17 Exact solution and approximate solution of the Euler-Bernoulli rod problem

Assume as before that C_2 satisfies the relation

$$C_2 \equiv -\frac{6a_1}{a_3} > 0.$$

Furthermore, let $v_3 \in Y_1$ be the (unique) solution of the linear problem

$$T_y^0 v_3 = -\left\{ \sqrt{C_2} T_{y\lambda}^0 \varphi_0 + C_2\sqrt{C_2} \left(T_{yy}^0 \varphi_0 v_2 + \frac{1}{6} T_{yyy}^0 \varphi_0^3 \right) \right\}, \qquad (5.132)$$

where v_2 is defined by (5.123). For the curve of nontrivial solutions of (5.11) branching at $z_0 \equiv (0, \lambda_0) \in Z$ we use the following ansatz (have a look at formula (5.87))

$$\begin{aligned}
\lambda(\xi) &= \lambda_0 + \xi^2, \\
y(\xi) &= \sqrt{C_2}\varphi_0\xi + (C_2 v_2 + \tau(\xi)\varphi_0)\xi^2 \\
&\quad + \left(v_3 + 2\sqrt{C_2}\tau(\xi)v_2\right)\xi^3 + \left(u(\xi) + \tau(\xi)^2 v_2\right)\xi^4, \\
\tau(\xi) &\in \mathbb{R}, \quad u(\xi) \in Y_1.
\end{aligned} \qquad (5.133)$$

Substituting (5.131) and (5.133) into (5.11), we get

$$\begin{pmatrix} T_y^0 & T_{y\lambda}^0 \varphi_0 + 3C_2 T_{yy}^0 \varphi_0 v_2 + \frac{1}{2} C_2 T_{yyy}^0 \varphi_0^3 \\ \varphi_0^* & 0 \end{pmatrix} \begin{pmatrix} u(\xi) \\ \tau(\xi) \end{pmatrix} = -\begin{pmatrix} w(u, \tau; \xi) \\ 0 \end{pmatrix}, \qquad (5.134)$$

where

$$
\begin{aligned}
w(u, \tau; \xi) &\equiv T^0_{y\lambda} w_1(u, \tau; \xi) \\
&+ T^0_{yy}\{\sqrt{C_2}\varphi_0(v_3 + [u + \tau^2 v_2])\xi + \tau\varphi_0(w_1(u, \tau; \xi) - C_2 v_2) \\
&\quad + \frac{1}{2} w_1(u, \tau; \xi)^2\} \\
&+ \frac{1}{2} T^0_{y\lambda\lambda} w_2(u, \tau; \xi)\xi + \frac{1}{2} T^0_{yy\lambda} w_2(u, \tau; \xi)^2 \\
&+ \frac{1}{6} T^0_{yyy}\{3C_2\varphi_0^2 w_1(u, \tau; \xi) + 3\sqrt{C_2}\varphi_0(w_1(u, \tau; \xi) + \tau\varphi_0)^2\xi \\
&\quad + (w_1(u, \tau; \xi) + \tau\varphi_0)^3\xi^2\} \\
&+ \frac{1}{6} T^0_{y\lambda\lambda\lambda} w_2(u, \tau; \xi)\xi^3 + \frac{1}{4} T^0_{yy\lambda\lambda} w_2(u, \tau; \xi)^2\xi^2 \\
&+ \frac{1}{6} T^0_{yyy\lambda} w_2(u, \tau; \xi)^3\xi + \frac{1}{24} T^0_{yyyy} w_2(u, \tau; \xi)^4 \\
&+ \frac{1}{\xi^4} R(y(\xi), \lambda(\xi)),
\end{aligned}
$$

$$
w_1(u, \tau; \xi) \equiv C_2 v_2 + \left[v_3 + 2\sqrt{C_2}\tau v_2\right]\xi + \left[u + \tau^2 v_2\right]\xi^2,
$$

$$
w_2(u, \tau; \xi) \equiv \sqrt{C_2}\varphi_0 + [\tau\varphi_0 + w_1(u, \tau; \xi)]\xi.
$$

A statement about the solvability of the transformed problem (5.134) is given in the next theorem.

Theorem 5.44 *Let $z_0 \equiv (0, \lambda_0) \in Z$ be a primary simple bifurcation point. Assume that $a_2 = 0$ and $a_3 \neq 0$. Then, there exists a real number $\xi_0 > 0$ such that for $|\xi| \leq \xi_0$ problem (5.134) has a continuous curve of isolated solutions $\eta(\xi) \equiv (u(\xi), \tau(\xi))^T$. By inserting these solutions into the ansatz (5.133) it is possible to construct a continuous curve of nontrivial solutions of the operator equation (5.11). This curve branches off from the trivial solution curve $\{(0, \lambda), \lambda \in \mathbb{R}\}$ at the simple bifurcation point z_0.*

Proof For $\xi \to 0$, problem (5.134) turns into

$$
\mathbb{A}\begin{pmatrix} u(0) \\ \tau(0) \end{pmatrix} = -\begin{pmatrix} w_0 \\ 0 \end{pmatrix}, \tag{5.135}
$$

with

$$
\mathbb{A} \equiv \begin{pmatrix} T^0_y & T^0_{y\lambda}\varphi_0 + 3C_2 T^0_{yy}\varphi_0 v_2 + \frac{1}{2} C_2 T^0_{yyy}\varphi_0^3 \\ \varphi_0^* & 0 \end{pmatrix}
$$

and

$$w_0 \equiv C_2 T^0_{y\lambda} v_2 + \sqrt{C_2} T^0_{yy} \varphi_0 v_3 + \frac{1}{2} C_2^2 T^0_{yy} v_2^2 + \frac{1}{2} C_2 T^0_{yy\lambda} \varphi_0^2$$
$$+ \frac{1}{2} C_2^2 T^0_{yyy} \varphi_0^2 v_2 + \frac{1}{24} C_2^2 T^0_{yyyy} \varphi_0^4.$$

We set

$$A \equiv T^0_y, \quad B \equiv T^0_{y\lambda} \varphi_0 + 3 C_2 T^0_{yy} \varphi_0 v_2 + \frac{1}{2} C_2 T^0_{yyy} \varphi_0^3,$$
$$C^* \equiv \varphi_0^*, \quad D \equiv 0, \quad m = 1,$$

and apply Lemma 5.25. Since T^0_y is not bijective, the second case is true. Now, we have to verify (c_0)–(c_3):

(c_0): $\dim \mathcal{R}(B) = 1 = m$,
(c_1): $\mathcal{R}(B) \cap \mathcal{R}(A) = \{0\}$,

   because $\psi_0^* B = a_1 + \frac{1}{2} C_2 a_3 = a_1 - 3 \frac{a_1}{a_3} a_3 = -2 a_1 \neq 0$, and

   $\psi_0^* T^0_y \cdot = 0$,
(c_2): $\dim \mathcal{R}(\varphi_0^*) = 1$,
(c_3): $\mathcal{N}(T^0_y) \cap \mathcal{N}(\varphi_0^*) = \{0\}$.

The assumptions of Lemma 5.25 are satisfied and we can conclude that the linear operator $\mathbb{A}$ is bijective. Thus, there exists a unique solution $\eta^0 \equiv (u(0), \tau(0))^T$ of (5.135). Writing (5.134) in the form

$$F(\xi; \eta) = 0, \quad \eta(\xi) \equiv (u(\xi), \tau(\xi))^T,$$

the claim follows as in the proof of Theorem 5.26 from the Implicit Function Theorem (see Theorem 5.8). ∎

Next, we want to deal with the case of an *asymmetric bifurcation point*, i.e., let us now assume that

$$a_2 \equiv \psi_0^* T^0_{yy} \varphi_0^2 \neq 0.$$

Since the analytical approach of Nekrassov is too expensive, we will focus only on numerical techniques. The first step is to expand the operator T into a Taylor series at $z_0 \equiv (0, \lambda_0)$:

$$T(y, \lambda) = T^0_y y + T^0_{y\lambda} (\lambda - \lambda_0) y + \frac{1}{2} T^0_{yy} y^2 + \frac{1}{2} T^0_{y\lambda\lambda} (\lambda - \lambda_0)^2 y$$
$$+ v \frac{1}{2} T^0_{yy\lambda} (\lambda - \lambda_0) y^2 + \frac{1}{6} T^0_{yyy} y^3 + R(y, \lambda). \tag{5.136}$$

The remainder R consists only of fourth or higher order terms in y and $(\lambda - \lambda_0)$.

Let $C_1 \in \mathbb{R}$ be defined as

$$C_1 \equiv -\frac{2a_1}{a_2}.$$ (5.137)

Furthermore, let $v_2 \in Y_1$ be the (unique) solution of the following linear problem

$$T_y^0 v_2 = -C_1 \left\{ T_{y\lambda}^0 \varphi_0 + \frac{C_1}{2} T_{yy}^0 \varphi_0^2 \right\}, \quad \varphi_0^* v_2 = 0.$$ (5.138)

For the curve of nontrivial solutions of (5.11) branching at $z_0 \equiv (0, \lambda_0) \in Z$ we use the following ansatz (have a look at formula (5.83))

$$\begin{aligned}
\lambda(\xi) &= \lambda_0 + \xi, \\
y(\xi) &= C_1 \varphi_0 \xi + \left(v_2 + \tau(\xi)\varphi_0 \right) \xi^2 + u(\xi)\xi^3, \\
\tau(\xi) &\in \mathbb{R}, \quad u(\xi) \in Y_1.
\end{aligned}$$ (5.139)

Substituting (5.136) and (5.139) into (5.11), we get

$$\begin{pmatrix} T_y^0 & T_{y\lambda}^0 \varphi_0 + C_1 T_{yy}^0 \varphi_0^2 \\ \varphi_0^* & 0 \end{pmatrix} \begin{pmatrix} u(\xi) \\ \tau(\xi) \end{pmatrix} = -\begin{pmatrix} w(u, \tau; \xi) \\ 0 \end{pmatrix},$$ (5.140)

where

$$\begin{aligned}
w(u, \tau; \xi) = {}& T_{y\lambda}^0 \{v_2 + u\xi\} \\
& + \frac{1}{2} T_{yy}^0 \{2C_1 \varphi_0 (v_2 + u\xi) + \left(v_2 + \tau\varphi_0 + u\xi \right)^2 \xi\} \\
& + \frac{1}{2} T_{y\lambda\lambda}^0 w_1(u, \tau; \xi) + \frac{1}{2} T_{yy\lambda}^0 w_1(u, \tau; \xi)^2 \\
& + \frac{1}{6} T_{yyy}^0 w_1(u, \tau; \xi)^3 + \frac{1}{\xi^3} R(y(\xi), \lambda_0 + \xi), \\
w_1(u, \tau; \xi) = {}& C_1 \varphi_0 + (v_2 + \tau\varphi_0)\xi + u\xi^2.
\end{aligned}$$

A statement about the solvability of the transformed problem (5.140) is given in the next theorem.

Theorem 5.45 *Let $z_0 \equiv (0, \lambda_0) \in Z$ be a primary simple bifurcation point. Assume that $a_2 \neq 0$. Then, there exists a real number $\xi_0 > 0$ such that for $|\xi| \leq \xi_0$ the problem (5.140) has a continuous curve of isolated solutions $\eta(\xi) \equiv (u(\xi), \tau(\xi))^T$. By inserting these solutions into the ansatz (5.139) it is possible to construct a continuous curve of nontrivial solutions of the operator equation (5.11). This curve branches off from the trivial solution curve $\{(0, \lambda), \lambda \in \mathbb{R}\}$ at the simple bifurcation point z_0.*

Proof For $\xi \to 0$, the problem (5.140) turns into

$$\mathbb{A}\,\eta(0) = -\begin{pmatrix} T^0_{y\lambda} v_2 + C_1 T^0_{yy}\varphi_0 v_2 + \dfrac{C_1}{2} T^0_{y\lambda\lambda}\varphi_0 + \dfrac{C_1^2}{2} T^0_{yy\lambda}\varphi_0^2 + \dfrac{C_1^3}{6} T^0_{yyy}\varphi_0^3 \\ 0 \end{pmatrix},$$

(5.141)

with

$$\mathbb{A} \equiv \begin{pmatrix} T^0_y & T^0_{y\lambda}\varphi_0 + C_1 T^0_{yy}\varphi_0^2 \\ \varphi_0^* & 0 \end{pmatrix}.$$

We set

$$A \equiv T^0_y, \quad B \equiv T^0_{y\lambda}\varphi_0 + C_1 T^0_{yy}\varphi_0^2,$$
$$C^* \equiv \varphi_0^*, \quad D \equiv 0, \quad m = 1,$$

and apply Lemma 5.25. Since T^0_y is not bijective, the second case is true. Now, we have to verify (c_0)–(c_3):

(c_0): $\dim \mathcal{R}(B) = 1 = m$,
(c_1): $\mathcal{R}(B) \cap \mathcal{R}(A) = \{0\}$,

   because $\psi_0^* B = a_1 + C_1 a_2 = a_1 - \dfrac{2a_1}{a_2}a_2 = -a_1 \neq 0$, and

   $\psi_0^* T^0_y \cdot = 0$,
(c_2): $\dim \mathcal{R}(\varphi_0^*) = 1$,
(c_3): $\mathcal{N}(T^0_y) \cap \mathcal{N}(\varphi_0^*) = \{0\}$.

Thus, the linear operator $\mathbb{A}$ is bijective, and there exists a unique solution $\eta^0 \equiv (u^0, \tau^0)^T$ of the equation (5.141). The claim follows from the Implicit Function Theorem (see the proofs of the previous theorems). ∎

To complete the description of some transformation techniques for primary simple bifurcation points, we want to briefly explain how these methods can be applied to BVPs of the form (5.10). As already described in the previous sections (see, e.g., the formulas (5.47), (5.48)), the functional condition $\varphi_0^* u = 0$, $\varphi_0^* \in Y^*$, is written in form of a first-order ODE with two boundary conditions. Moreover, the trivial ODE $\tau' = 0$ is added to the transformed problem. New in the transformation techniques for symmetric and asymmetric bifurcation points is the use of the bifurcation coefficients a_1, a_2 and a_3. Since we have assumed that T^0_y is a Fredholm operator with index zero, the adjoint operator $(T^0_y)^*$ has an one-dimensional null space, too. Now, the strategy is to determine from the adjoint BVP (see (5.56))

$$\psi'(x) = -(f^0_y)^T \psi(x),$$
$$B_a^* \psi(a) + B_b^* \psi(b) = \mathbf{0},$$

a solution $\psi_0(x)$ with $\|\psi_0(x)\| = 1$. Using this solution, the bifurcation coefficients can be written in the form

$$a_1 = -\int_a^b \psi_0(x)^T f_{y\lambda}^0 \varphi_0(x)dx,$$

$$a_2 = -\int_a^b \psi_0(x)^T f_{yy}^0 \varphi_0(x)^2 dx,$$

$$a_3 = -\int_a^b \psi_0(x)^T \left\{ f_{yyy}^0 \varphi_0(x)^3 + 6f_{yy}^0 \varphi_0(x)v_2(x) \right\} dx.$$

When it is difficult to evaluate analytically these integrals, the bifurcation coefficients can be computed easily by the numerical integration of the following IVPs:

$$\alpha_1'(x) = -\psi_0(x)^T f_{y\lambda}^0 \varphi_0(x), \quad \alpha_1(a) = 0,$$
$$a_1 = \alpha_1(b);$$
$$\alpha_2'(x) = -\psi_0(x)^T f_{yy}^0 \varphi_0(x)^2, \quad \alpha_2(a) = 0,$$
$$a_2 = \alpha_2(b);$$
$$\alpha_3'(x) = -\psi_0(x)^T \left\{ f_{yyy}^0 \varphi_0(x)^3 + 6f_{yy}^0 \varphi_0(x)v_2(x) \right\}, \quad \alpha_3(a) = 0,$$
$$a_3 = \alpha_3(b).$$

Remark 5.46 Let us have a look on the above integrands and the right-hand sides of the corresponding ODEs. It is well known that there are only two multiplicative operations for vectors $v, w \in \mathbb{R}^n$, namely: (i) the *inner* product $a = v^T w$, and (ii) the *outer* product $A = vw^T$, where a is a scalar and A is a matrix. Therefore, the derivatives of the vector function f have to be understood as multilinear operators, which are successively applied on a series of vectors. For example, some authors write $f_{yy}^0[v, w]$ instead of $f_{yy}^0 vw$. $\qquad\square$

In the next example, we will show how these expressions can be determined.

Example 5.47 Given the vectors $y = (y_1, y_2)^T$, $\varphi_0 = (\varphi_{0,1}, \varphi_{0,2})^T$ and the following vector function

$$f(y) = \begin{pmatrix} y_1^2 + \sin(y_2) + 5 \\ \cos(y_1) + y_1 y_2 + y_1 \end{pmatrix}.$$

The aim is to evaluate the expression (vector) $f_{yy}^0 \varphi_0^2$ at $y_0 = (0, \pi)^T$. The first step is to determine the Jacobian of f, i.e.,

$$f_y(y) = \begin{pmatrix} 2y_1 & \cos(y_2) \\ -\sin(y_1) + y_2 + 1 & y_1 \end{pmatrix}.$$

Then, using an arbitrary vector $v \in \mathbb{R}^2$, we form the vector

$$f_y(y)v = \begin{pmatrix} 2y_1 & \cos(y_2) \\ -\sin(y_1) + y_2 + 1 & y_1 \end{pmatrix} \begin{pmatrix} v_1 \\ v_2 \end{pmatrix}$$

$$= \begin{pmatrix} 2y_1 v_1 + \cos(y_2)v_2 \\ -\sin(y_1)v_1 + y_2 v_1 + v_1 + y_1 v_2 \end{pmatrix}.$$

The second step is to determine the Jacobian of the vector, i.e.,

$$f_{yy}(y)v = \begin{pmatrix} 2v_1 & -\sin(y_2)v_2 \\ -\cos(y_1)v_1 + v_2 & v_1 \end{pmatrix}.$$

Now, using an arbitrary vector $w \in \mathbb{R}^2$ we form the vector

$$f_{yy}(y)vw = \begin{pmatrix} 2v_1 & -\sin(y_2)v_2 \\ -\cos(y_1)v_1 + v_2 & v_1 \end{pmatrix} \begin{pmatrix} w_1 \\ w_2 \end{pmatrix}$$

$$= \begin{pmatrix} 2v_1 w_1 - \sin(y_2)v_2 w_2 \\ -\cos(y_1)v_1 w_1 + v_2 w_1 + v_1 w_2 \end{pmatrix}.$$

The third step is to insert the vector y_0. We obtain

$$f_{yy}^0 vw = \begin{pmatrix} 2v_1 w_1 \\ -v_1 w_1 + v_2 w_1 + v_1 w_2 \end{pmatrix}.$$

And finally, in the last step we set $v \equiv \varphi_0$ and $w \equiv \varphi_0$. The result is

$$f_{yy}^0 \varphi_0^2 = \begin{pmatrix} 2\varphi_{0,1}^2 \\ -\varphi_{0,1}^2 + 2\varphi_{0,1}\varphi_{0,2} \end{pmatrix}.$$

□

In the next example, we show how the transformation technique can be applied to the BVP (5.104).

Example 5.48 Let us consider once more the BVP (5.104). We should remember that the primary simple bifurcation points are

$$z_0^{(k)} = (0, \lambda_0^{(k)}), \quad \lambda_0^{(k)} = -(k\pi)^2, \quad k = 1, 2, \ldots,$$

and the corresponding (normalized) eigenfunctions are

$$\varphi_0^{(k)}(x) = \sqrt{2}\sin(k\pi x), \quad k = 1, 2, \ldots.$$

Here, we consider only the case $k = 1$. If we define

$$T(y, \lambda) \equiv y'' - \lambda(y + y^2),$$ (5.142)

it follows:

$$
\begin{array}{ll}
T_y = (\cdot)'' - \lambda(1 + 2y), & T_y^0 = (\cdot)'' - \lambda_0 \\
T_{y\lambda} = -(1 + 2y), & T_{y\lambda}^0 = -1 \\
T_{yy} = -2\lambda, & T_{yy}^0 = -2\lambda_0, \\
T_{y\lambda\lambda} = 0, & T_{y\lambda\lambda}^0 = 0, \\
T_{yy\lambda} = -2, & T_{yy\lambda}^0 = -2, \\
T_{yyy} = 0, & T_{yyy}^0 = 0.
\end{array}
$$

The corresponding bifurcation coefficients are

$$a_1 = -\int_0^1 \varphi_0(x)^2 dx = -2 \int_0^1 \sin(\pi x)^2 dx = -2\frac{1}{2} = -1,$$

$$a_2 = 2\pi^2 \int_0^1 \varphi_0(x)^3 dx = 2\pi^2 2\sqrt{2}\frac{4}{2\pi} = \frac{16\sqrt{2}\pi}{3} \neq 0.$$

Thus, $z_0^{(1)} = (0, -\pi^2)$ is an asymmetric bifurcation point. For the coefficient C_1, we obtain

$$C_1 = -\frac{2a_1}{a_2} = \frac{3}{8\pi\sqrt{2}} = 0.084404654639729.$$ (5.143)

Now, we have to determine the solution v_2 of the linear problem (5.138). The first equation reads

$$
\begin{aligned}
v_2'' + \pi^2 v_2 &= -C_1\left\{-\varphi_0 + C_1\pi^2\varphi_0^2\right\} \\
v_2(0) &= 0, \quad v_2(1) = 0.
\end{aligned}
$$ (5.144)

If we write the condition $\varphi_0^* v_2 = 0$ as before in the ODE formulation

$$\xi_1' = \varphi_0 v_2, \quad \xi(0) = \xi(1) = 0,$$ (5.145)

and add this problem to the second-order BVP (5.144), which can be written as two first-order ODEs, we obtain a BVP consisting of three first-order ODEs and four boundary conditions. The difficulty is that the number of the ODEs does not match the number of the boundary conditions. But we can apply a simple trick. We do not insert the value (5.143) into the right-hand side of (5.144), but consider C_1 as a free parameter, write $C_1' = 0$ and add this ODE to the BVP. Now, we have to solve the following BVP of dimension $n = 4$:

$$y_1' = y_2,$$
$$y_2' = -\pi^2 y_1 - y_4 \left\{ -\varphi_0 + y_4 \pi^2 \varphi_0^2 \right\},$$
$$y_3' = \varphi_0 y_1, \tag{5.146}$$
$$y_4' = 0,$$
$$y_1(0) = 0, \quad y_1(1) = 0, \quad y_3(0) = 0, \quad y_3(1) = 0,$$

where we have set

$$y_1 = v_2, \quad y_2 = v_2', \quad y_3 = \xi_1, \quad y_4 = C_1. \tag{5.147}$$

At this point, a further difficulty arises. The BVP (5.146) has two solutions. The first one is the trivial solution $y_i(x) \equiv 0$, $i = 1, \ldots, 4$. But we are interested in the nontrivial solution, which satisfies $y_4 = C_1 = \dfrac{3}{8\pi\sqrt{2}}$.

Our numerical experiments have led to the following result. If we choose positive starting values for y_4 and solve (5.146) with the multiple shooting code presented in Sect. 4.6, then the nontrivial solution v_2 being sought can be determined with just a few iteration steps.

Note: the BVP (5.146) is implemented as *case 6* in our multiple shooting code.

Now, let us formulate the transformed problem (5.140) for our given BVP (5.104). We obtain

$$u'' + \pi^2 u + \left[-\varphi_0 + 2C_1 \pi^2 \varphi_0^2 \right] \tau$$
$$= \Big[-(v_2 + u\xi)$$
$$\quad + \pi^2 \left\{ 2C_1 \varphi_0 (v_2 + u\xi) + (v_2 + \tau\varphi_0 + u\xi)^2 \xi \right\}$$
$$\quad - \left\{ C_1 \varphi_0 + (v_2 + \tau\varphi_0)\xi + u\xi^2 \right\}^2 + \frac{1}{\xi^3} R \Big], \tag{5.148}$$
$$\xi_2' = \varphi_0 u, \quad \tau' = 0,$$
$$u(0) = 0, \quad u(1) = 0, \quad \xi_2(0) = \xi_2(1) = 0,$$

where

$$R \equiv \left[C_1 \varphi_0 \xi + (v_2 + \tau\varphi_0)\xi^2 + u\xi^3 \right]^2 \xi.$$

Setting

$$y_5 = u, \quad y_6 = u', \quad y_7 = \xi_2, \quad y_8 = \tau,$$

and using the notation (5.147), problem (5.148) can be written in the form

$$y_5' = y_6,$$

$$y_6' = -\pi^2 y_5 - \left[-\varphi_0 + 2 y_4 \pi^2 \varphi_0^2 \right] y_8$$

$$+ \left[-(y_1 + y_5 \xi) \right.$$

$$+ \pi^2 \left\{ 2 y_4 \varphi_0 (y_1 + y_5 \xi) + (y_1 + y_8 \varphi_0 + y_5 \xi)^2 \xi \right\}$$

$$- \left\{ y_4 \varphi_0 + (y_1 + y_8 \varphi_0) \xi + y_1 \xi^2 \right\}^2$$

$$+ \frac{1}{\xi^2} \left\{ y_4 \varphi_0 \xi + (y_1 + y_8 \varphi_0) \xi^2 + y_5 \xi^3 \right\}^2 \right],$$

$$y_7' = \varphi_0 y_5, \quad y_8' = 0,$$

$$y_5(0) = 0, \quad y_5(1) = 0, \quad y_7(0) = y_7(1) = 0,$$

(5.149)

To be able to use an IVP-solver with integrated step-size control in the multiple shooting method, the problems (5.146) and (5.149) must be defined on the same grid. Therefore, it is appropriate to merge the two four-dimensional BVPs into a eight-dimensional BVP. Looking at our notation (indexing of the solution components), it can be seen that we have already taken account of this fact.

If the exact eigenfunction φ_0 is not explicitly known, then it is also possible to add the extended problem (5.94) to this large BVP. However, the crucial problem is to find good starting values for all functions $y_i(x)$.

In Fig. 5.18, we present the numerically determined solutions for the control parameter $\lambda = -\pi^2 + \xi$, with $\xi = -2, -1, -0.5., 0.5, 1, 2$. These solutions have been

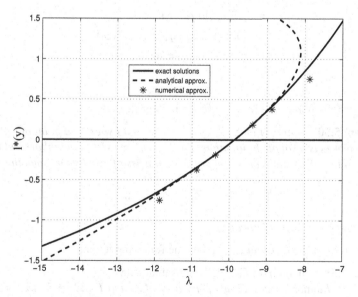

Fig. 5.18 *Solid line* exact solution curve, *dashed line* asymptotic expansion (5.114) of the branching solution curve, *stars* numerical approximations. Here, we have set $l^* y \equiv y'(0)$

computed with the combined BVP (5.146), (5.149) by our multiple shooting code given in Sect. 4.6. In addition, we have added the analytical approximations obtained by the Levi-Città technique for $\varepsilon \in [-0.29, 0.48]$.   □

5.5 Analytical and Numerical Treatment of Secondary Simple Bifurcation Points

5.5.1 Analysis of Secondary Simple Bifurcation Points

In this section, we assume that neither of the two solution curves intersecting at a simple bifurcation point is explicitly known. Thus, the condition $T(0, \lambda) = 0$ for all $\lambda \in \mathbb{R}$, is void.

Let $z_0 \equiv (y_0, \lambda_0) \in Z$ be a *hyperbolic point* as defined in the next definition.

Definition 5.49 The point $z_0 \in Z$ is called a hyperbolic point of the operator T when the following conditions are fulfilled:

- $T(z_0) = 0$,   (solution)
- $\dim \mathcal{N}(T'(z_0)) = 2$, $\operatorname{codim} \mathcal{R}(T'(z_0)) = 1$,   (singular point)
- $\tau \equiv \alpha\gamma - \beta^2 < 0$,   (hyperbolic point)

where

$$\alpha \equiv \psi_0^* T''(z_0) pp, \quad \beta \equiv \psi_0^* T''(z_0) pq,$$
$$\gamma \equiv \psi_0^* T''(z_0) qq, \quad \text{with}$$
$$p, q \in Y : \operatorname{span}\{p, q\} = \mathcal{N}(T'(z_0)),$$
$$\psi_0^* \in W^* : \mathcal{N}(\psi_0^*) = \mathcal{R}(T'(z_0)).$$

□

For hyperbolic points, we have the following result.

Theorem 5.50 *Suppose that $z_0 \equiv (y_0, \lambda_0) \in Z$ is a hyperbolic point of the operator T. Then z_0 is a simple bifurcation point, i.e., there exists a neighborhood U of z_0 in Z such that the solution manifold of $T|_U$ consists of exactly two smooth curves*

$$z_i : [-\varepsilon_i, \varepsilon_i] \to Z, \quad z_i(0) = z_0 \ (i = 1, 2),$$

which intersect transversally at z_0.

The proof of the theorem is based on the following lemma.

Lemma 5.51 (Theorem of Crandall and Rabinowitz)
Let Z, Y be Banach spaces, Ω an open subset of Z, and $T : \Omega \to Y$ a $\mathbb{C}^2$-mapping. Furthermore, $w : [-1, 1] \to \Omega$ is a continuously differentiable curve in Ω, with $T(w(\varepsilon)) = 0, |\varepsilon| \leq 1$. Suppose that the following conditions are satisfied:

- $w'(0) \neq 0$,
- $\dim \mathcal{N}(T'(w(0))) = 2$, $\operatorname{codim} \mathcal{R}(T'(w(0))) = 1$,
- $\mathcal{N}(T'(w(0))) = \operatorname{span}\{w'(0), v\}$,
- $T''(w(0))w'(0)v \notin \mathcal{R}(T'(w(0)))$.

Then, $w(0)$ is a bifurcation point of $T(w) = 0$ with respect to

$$C \equiv \{w(\varepsilon) : \varepsilon \in [-1, 1]\}.$$

In the neighborhood of $w(0)$, the solution manifold of $T(w) = 0$ consists of exactly two continuous curves, which intersect only in $w(0)$.

Proof See the paper of Crandall and Rabinowitz [28]. ∎

We are now in a position to prove Theorem 5.50.

Proof If $\alpha = \gamma = 0$, the following transformation must not be performed.

Without loss of generality, let us assume that $\alpha \neq 0$. Otherwise, only p and q have to be exchanged. Now, we transform (p, q) into (p', q') according to the formula

$$p' = \frac{-\beta + \sqrt{-\tau}}{\alpha} p + q, \quad q' = \frac{-\beta - \sqrt{-\tau}}{\alpha} p + q.$$

It holds

$$
\begin{aligned}
\alpha' &\equiv \psi_0^* T''(z_0) p' p' \\
&= \left(\frac{-\beta + \sqrt{-\tau}}{\alpha}\right)^2 \psi_0^* T''(z_0) pp + 2\frac{-\beta + \sqrt{-\tau}}{\alpha} \psi_0^* T''(z_0) pq \\
&\quad + \psi_0^* T''(z_0) qq \\
&= \frac{\beta^2 - 2\beta\sqrt{-\tau} - \tau}{\alpha^2} \alpha + 2\frac{-\beta + \sqrt{-\tau}}{\alpha} \beta + \gamma \\
&= \frac{\beta^2 - 2\beta\sqrt{-\tau} - \tau}{\alpha} + 2\frac{-\beta + \sqrt{-\tau}}{\alpha} \beta + \frac{\tau + \beta^2}{\alpha} = 0,
\end{aligned}
$$

$$
\begin{aligned}
\beta' &\equiv \psi_0^* T''(z_0) p' q' \\
&= \left(\frac{-\beta + \sqrt{-\tau}}{\alpha}\right)\left(\frac{-\beta - \sqrt{-\tau}}{\alpha}\right) \psi_0^* T''(z_0) pp \\
&\quad + \left(\frac{-\beta + \sqrt{-\tau}}{\alpha} + \frac{-\beta - \sqrt{-\tau}}{\alpha}\right) \psi_0^* T''(z_0) pq + \psi_0^* T''(z_0) qq \\
&= \frac{\beta^2 + \beta\sqrt{-\tau} - \beta\sqrt{-\tau} + \tau}{\alpha^2} \alpha + \frac{-\beta + \sqrt{-\tau} - \beta - \sqrt{-\tau}}{\alpha} \beta + \gamma \\
&= \frac{\beta^2 + \tau - 2\beta^2 + \tau + \beta^2}{\alpha} = \frac{2\tau}{\alpha} \neq 0,
\end{aligned}
$$

$$\gamma' \equiv \psi_0^* T''(z_0) q' q'$$

$$= \left(\frac{-\beta - \sqrt{-\tau}}{\alpha} \right)^2 \psi_0^* T''(z_0) pp + 2 \left(\frac{-\beta - \sqrt{-\tau}}{\alpha} \right) \psi_0^* T''(z_0) pq$$

$$+ \psi_0^* T''(z_0) qq$$

$$= \frac{\beta^2 + 2\beta\sqrt{-\tau} - \tau}{\alpha^2} \alpha + \frac{-2\beta - 2\sqrt{-\tau}}{\alpha} \beta + \gamma$$

$$= \frac{\beta^2 + 2\beta\sqrt{-\tau} - \tau - 2\beta^2 - 2\beta\sqrt{-\tau} + \tau + \beta^2}{\alpha} = 0,$$

and

$$\tau' \equiv \alpha' \gamma' - (\beta')^2 = - \left(\frac{2\tau}{\alpha} \right)^2 \neq 0.$$

From now on, the quantities marked with an apostrophe are denoted again by α, β, γ, τ, p and q. However, we assume that $\alpha = \gamma = 0$.

Let $\bar{Z}$ be the complement of $\mathcal{N}(T'(z_0))$ in Z. In the neighborhood of the secondary bifurcation point $z_0 \in Z$, the solutions of (5.11) can be represented in dependence of an additional parameter ε, $|\varepsilon| \leq \varepsilon_0$, as

$$z(\varepsilon) = z_0 + \varepsilon \{a(\varepsilon) p + b(\varepsilon) q\} + \varepsilon^2 v(\varepsilon), \tag{5.150}$$
$$a(\varepsilon)^2 + b(\varepsilon)^2 - 1 = 0, \quad v(\varepsilon) \in \bar{Z}.$$

At z_0, we expand T into the Taylor series

$$T(z) = T'(z_0)(z - z_0) + \frac{1}{2} T''(z_0)(z - z_0)^2 + R(z), \tag{5.151}$$

where $R(z(\varepsilon)) = O(\varepsilon^3)$. Substituting (5.150) and (5.151) into the operator equation (5.11), we obtain

$$T'(z_0) v(\varepsilon) + \frac{1}{2} T''(z_0) \{a(\varepsilon) p + b(\varepsilon) q + \varepsilon v(\varepsilon)\}^2 + \frac{1}{\varepsilon^2} R(z(\varepsilon)) = 0, \tag{5.152}$$
$$a(\varepsilon)^2 + b(\varepsilon)^2 - 1 = 0, \quad v(\varepsilon) \in \bar{Z}.$$

Let us write (5.152) in compact form as

$$F(x, \varepsilon) = 0, \quad F : X \times \mathbb{R} \to U, \tag{5.153}$$

with

$$x = x(\varepsilon) \equiv (v(\varepsilon), a(\varepsilon), b(\varepsilon)),$$
$$X \equiv \bar{Z} \times \mathbb{R} \times \mathbb{R}, \quad U \equiv W \times \mathbb{R}.$$

Using appropriate norms, X and U are Banach spaces.

Consider (5.153) for $\varepsilon \to 0$. We get

$$T'(z_0)v(0) + \frac{1}{2}T''(z_0)\{a(0)p + b(0)q\}^2 = 0,$$
$$a(0)^2 + b(0)^2 - 1 = 0, \quad v(0) \in \bar{Z}. \tag{5.154}$$

The first equation is solvable if $\psi_0^* T''(z_0)\{a(0)p + b(0)q\}^2 = 0$. Thus,

$$0 = a(0)^2 \psi_0^* T''(z_0)pp + 2a(0)b(0)\psi_0^* T''(z_0)pq + b(0)^2 \psi_0^* T''(z_0)qq$$
$$= a(0)^2\alpha + 2a(0)b(0)\beta + b(0)^2\gamma.$$

The assumption $\alpha = \gamma = 0$ implies

$$a(0)b(0) = 0, \quad a(0)^2 + b(0)^2 - 1 = 0.$$

Obviously, these two nonlinear algebraic equations for the two unknown $a(0)$ and $b(0)$ have four solutions:

$$\xi_1 = (a_1, b_1)^T = (1, 0)^T,$$
$$\xi_2 = (a_2, b_2)^T = (0, 1)^T,$$
$$\xi_3 = -\xi_1,$$
$$\xi_4 = -\xi_2.$$

In the following we consider only the pair of linearly independent solutions (ξ_1, ξ_2). Note, for given $a_i(0) = a_i$ and $b_i(0) = b_i$ there exists a unique solution $v_i(0)$. Let us apply the Implicit Function Theorem (see Theorem 5.8) to the problem (5.153) at

$$x^{(i)} \equiv (v_i(0), a_i(0), b_i(0)), \quad i = 1, 2, \quad \varepsilon = 0.$$

We have to show that the derivative $F_x(x^{(i)}, 0) \in \mathcal{L}(X, U)$ is a linear homeomorphism between X and U. For this we consider the equation

$$0 = F_x(x^{(i)}, 0) h, \quad h \equiv (\tilde{v}, \tilde{a}, \tilde{b}) \in X,$$

or, in more detail

$$0 = T'(z_0)\tilde{v} + \tilde{a}a_i T''(z_0)pp + \tilde{a}b_i T''(z_0)pq + \tilde{b}a_i T''(z_0)pq$$
$$+ \tilde{b}b_i T''(z_0)qq, \tag{5.155}$$
$$0 = 2(\tilde{a}a_i + \tilde{b}b_i), \quad \tilde{v} \in \bar{Z}.$$

There exists a solution of problem (5.155), if the following conditions are fulfilled:

$$\tilde{a}b_i\beta + \tilde{b}a_i\beta = 0, \quad \tilde{a}a_i + \tilde{b}b_i = 0.$$

We substitute the solutions (a_i, b_i), $i = 1, 2$, into the equations (5.155), and obtain $\tilde{a} = \tilde{b} = 0$. Thus, (5.155) is reduced to

$$T'(z_0)\tilde{v} = 0, \quad \tilde{v} \in \bar{Z}.$$

The unique solution of this equation is $\tilde{v} = 0$, i.e., the operator $F_x(x^{(i)}, 0)$ is injective.

Now, we show the subjectivity of $F_x(x^{(i)}, 0)$ for the case $i = 1$. For $g \in W$, $g_1 \in \mathbb{R}$ and $\xi_1 = (1, 0)^T$, we consider the inhomogeneous problem

$$T'(z_0)\tilde{v} + \tilde{a}T''(z_0)pp + \tilde{b}T''(z_0)pq = g, \quad \tilde{v} \in \bar{Z},$$
$$2\tilde{a} = g_1. \tag{5.156}$$

The second equation yields $\tilde{a} = \frac{1}{2}g_1$. We insert this value into the first equation and obtain

$$T'(z_0)\tilde{v} = g - \frac{1}{2}g_1 T''(z_0)pp - \tilde{b}T''(z_0)pq \equiv \hat{g}, \quad \tilde{v} \in \bar{Z}.$$

There exists a solution of that equation if $\hat{g} \in \mathcal{R}(T'(z_0))$, i.e.,

$$\psi_0^* g - \frac{1}{2}g_1 \psi_0^* T''(z_0)pp - \tilde{b}\psi_0^* T''(z_0)pq = \psi_0^* g - \frac{1}{2}g_1\alpha - \tilde{b}\beta = 0.$$

Since we have assumed that $\alpha = 0$ and $\beta \neq 0$, it follows $\tilde{b} = \dfrac{\psi_0^* g}{\beta}$. Now it can be concluded, that the solution $\tilde{v}$ is also uniquely determined.

Similar arguments are used to prove the case $i = 2$.

Using the Open Mapping Theorem (see, e.g., [132]), we can conclude from the bijectivity of $F_x(x^{(i)}, 0) \in \mathcal{L}(X, U)$ that this operator is a linear homeomorphism from X onto U. Then, for $|\varepsilon| \leq \varepsilon_i$ $(i = 1, 2)$, the Implicit Function Theorem implies the existence of two continuous curves of isolated solutions $(v_i(\varepsilon), a_i(\varepsilon), b_i(\varepsilon))$ of (5.152). Inserting these solutions into the ansatz (5.150), we obtain two continuous solution curves of (5.11), which intersect at z_0. To show that in the neighborhood of z_0 the solution manifold of (5.11) consists of exactly these two curves, let us consider one of the two curves, for instance

$$z(\varepsilon) = z_0 + \varepsilon\{a_1(\varepsilon)p + b_1(\varepsilon)q\} + \varepsilon^2 v_1(\varepsilon), \quad v_1 \in \bar{Z}.$$

It holds

$$\left.\frac{dz(\varepsilon)}{d\varepsilon}\right|_{\varepsilon=0} = a_1(0)p + b_1(0)q = p.$$

Moreover, we obtain

$$\beta = \psi_0^* T''(z_0) \underbrace{\frac{dz(0)}{d\varepsilon}}_{p} q \neq 0.$$

Now, the claim follows from the Theorem of Crandall and Rabinowitz (see Lemma 5.51). ∎

5.5.2 Extension Techniques for Secondary Simple Bifurcation Points

Let $z_0 \equiv (y_0, \lambda_0) \in Z$ be a hyperbolic point of the operator T. Assume that at least one of the two solution curves, which intersect at z_0, enables a smooth parametrization with respect to λ, i.e.,

$$T_\lambda(z_0) \in \mathcal{R}(T_y(z_0)). \tag{5.157}$$

As stated in the previous section, we have the following situation:

$$\mathcal{N}(T'(z_0)) = \text{span}\{p, q\}, \quad p, q \in Z,$$

with $p = (p_1, p_2)^T$, $q = (q_1, q_2)^T$, $p_1, q_1 \in Y$ and $p_2, q_2 \in \mathbb{R}$.

Suppose that $\varphi_0^* \in Y^*$ is a suitably chosen functional. Except for the sign, the elements p, q can be uniquely written as

$$p = (\varphi_0, 0)^T, \quad q = (s_0, 1)^T, \quad \varphi_0, s_0 \in Y,$$

where φ_0 and s_0 satisfy $\varphi_0^* \varphi_0 = 1$, $\varphi_0^* s_0 = 0$. Furthermore, we assume that an element $0 \neq \psi_0 \in W$ is known, with

$$\psi_0^* \psi_0 \neq 0. \tag{5.158}$$

An appropriate extension technique for the numerical determination of secondary simple bifurcation points is based on the following operator (see [131])

$$\tilde{T} : \begin{cases} \tilde{Z} \equiv Y \times \mathbb{R} \times Y \times Y \times \mathbb{R} \to \tilde{W} \equiv W \times W \times W \times \mathbb{R} \times \mathbb{R} \\ \tilde{z} \equiv (z, \varphi, s, \mu) = (y, \lambda, \varphi, s, \mu) \to \begin{pmatrix} T(y, \lambda) + \mu \psi_0 \\ T_y(y, \lambda) \varphi \\ T_y(y, \lambda)s + T_\lambda(y, \lambda) \\ \varphi_0^* \varphi - 1 \\ \varphi_0^* s \end{pmatrix} \end{cases}. \tag{5.159}$$

Now we consider the extended operator equation

$$\tilde{T}(\tilde{z}) = 0. \tag{5.160}$$

Obviously, $\tilde{z}_0 \equiv (y_0, \lambda_0, \varphi_0, s_0, 0) \in \tilde{Z}$ is a solution of (5.160). More detailed information about the extended operator is given in the next theorem.

Theorem 5.52 *Let $z_0 \equiv (y_0, \lambda_0) \in Z$ be a hyperbolic point of T. Assume that the conditions (5.157) and (5.158) are satisfied. Then, $\tilde{T}'(\tilde{z}_0)$ is a linear homeomorphism from $\tilde{Z}$ onto $\tilde{W}$, i.e.,*

$$\mathcal{N}(\tilde{T}'(\tilde{z}_0)) = \{0\}, \quad \mathcal{R}(\tilde{T}'(\tilde{z}_0)) = \tilde{W}.$$

Proof The derivative $\tilde{T}'(\tilde{z}_0)$ has the form

$$\tilde{T}'(\tilde{z}_0) = \begin{pmatrix} T_y^0 & T_\lambda^0 & 0 & 0 & \psi_0 \\ T_{yy}^0\varphi_0 & T_{y\lambda}^0\varphi_0 & T_y^0 & 0 & 0 \\ T_{yy}^0 s_0 + T_{y\lambda}^0 & T_{y\lambda}^0 s_0 + T_{\lambda\lambda}^0 & 0 & T_y^0 & 0 \\ 0 & 0 & \varphi_0^* & 0 & 0 \\ 0 & 0 & 0 & \varphi_0^* & 0 \end{pmatrix}. \tag{5.161}$$

(i) Assume that $(\tilde{y}, \tilde{\lambda}, \tilde{\varphi}, \tilde{s}, \tilde{\mu}) \in \mathcal{N}(\tilde{T}'(\tilde{z}_0))$, where $\tilde{y}, \tilde{\varphi}, \tilde{s} \in Y$ and $\tilde{\lambda}, \tilde{\mu} \in \mathbb{R}$. Then,

$$\tilde{T}'(\tilde{z}_0) \begin{pmatrix} \tilde{y} \\ \tilde{\lambda} \\ \tilde{\varphi} \\ \tilde{s} \\ \tilde{\mu} \end{pmatrix} = \begin{pmatrix} T_y^0\tilde{y} + T_\lambda^0\tilde{\lambda} + \tilde{\mu}\psi_0 \\ T_{yy}^0\varphi_0\tilde{y} + T_{y\lambda}^0\varphi_0\tilde{\lambda} + T_y^0\tilde{\varphi} \\ T_{yy}^0 s_0\tilde{y} + T_{y\lambda}^0 s_0\tilde{\lambda} + T_y^0\tilde{s} + T_{y\lambda}^0\tilde{y} + T_{\lambda\lambda}^0\tilde{\lambda} \\ \varphi_0^*\tilde{\varphi} \\ \varphi_0^*\tilde{s} \end{pmatrix} = \begin{pmatrix} 0 \\ 0 \\ 0 \\ 0 \\ 0 \end{pmatrix}. \tag{5.162}$$

There exists a solution of the first equation

$$T_y^0\tilde{y} = -T_\lambda^0\tilde{\lambda} - \tilde{\mu}\psi_0,$$

if

$$\psi_0^* T_\lambda^0\tilde{\lambda} + \tilde{\mu}\psi_0^*\psi_0 = 0.$$

The assumptions (5.157) and (5.158) imply $\tilde{\mu} = 0$. The general solution of the resulting equation

$$T_y^0\tilde{y} = -T_\lambda^0\tilde{\lambda}$$

is $(\tilde{y}, \tilde{\lambda}) = (\delta\varphi_0 + \kappa s_0, \kappa)$. Now, this solution is substituted into the second and third equation of (5.162). We obtain

$$T_y^0 \tilde{\varphi} = -(\delta T_{yy}^0 \varphi_0^2 + \kappa T_{yy}^0 \varphi_0 S_0 + \kappa T_{y\lambda}^0 \varphi_0)$$
$$T_y^0 \tilde{s} = -(\delta T_{yy}^0 s_0 \varphi_0 + \kappa T_{yy}^0 s_0^2 + \kappa T_{y\lambda}^0 s_0 + \delta T_{y\lambda}^0 \varphi_0 + \kappa T_{y\lambda}^0 s_0 + \kappa T_{\lambda\lambda}^0). \qquad (5.163)$$

For $p = (\varphi_0, 0)$ and $q = (s_0, 1)$, the quantities α, β and γ (see Definition 5.49) take the form

$$\alpha = \psi_0^* T_{yy}^0 \varphi_0^2, \quad \beta = \psi_0^* (T_{yy}^0 \varphi_0 S_0 + T_{y\lambda}^0 \varphi_0),$$
$$\gamma = \psi_0^* (T_{yy}^0 s_0^2 + 2 T_{y\lambda}^0 s_0 + T_{\lambda\lambda}^0).$$

The conditions for the solvability of (5.163) are

$$0 = \delta \psi_0^* T_{yy}^0 \varphi_0^2 + \kappa \psi_0^* (T_{yy}^0 \varphi_0 s_0 + T_{y\lambda}^0 \varphi_0),$$
$$0 = \delta \psi_0^* (T_{yy}^0 s_0 \varphi_0 + T_{y\lambda}^0 \varphi_0) + \kappa \psi_0^* (T_{yy}^0 s_0^2 + 2 T_{y\lambda}^0 s_0 + T_{\lambda\lambda}^0),$$

or

$$\begin{pmatrix} \alpha & \beta \\ \beta & \gamma \end{pmatrix} \begin{pmatrix} \delta \\ \kappa \end{pmatrix} = \begin{pmatrix} 0 \\ 0 \end{pmatrix}.$$

The assumption $\tau \equiv \alpha\gamma - \beta^2 < 0$ implies $\delta = 0$ and $\kappa = 0$. Thus $(\tilde{y}, \tilde{\lambda}) = (0, 0)$. Now, the first equation in (5.163) is reduced to

$$T_y^0 \tilde{\varphi} = 0, \quad \text{i.e.,} \quad \tilde{\varphi} = c\varphi_0, \ c \in \mathbb{R}.$$

The fourth equation in (5.162) implies $c = 0$, i.e., $\tilde{\varphi} = 0$. The second equation in (5.163) is reduced to

$$T_y^0 \tilde{s} = 0, \quad \text{i.e.,} \quad \tilde{s} = d\varphi_0, \ d \in \mathbb{R}.$$

The fifth equation in (5.162) implies $d = 0$, i.e., $\tilde{s} = 0$.
Summarizing the above steps, we have the following result

$$\mathcal{N}(\tilde{T}'(\tilde{z}_0)) = \{0\}.$$

(ii) Now, let us consider the inhomogeneous equations

$$T_y^0 \tilde{y} + T_\lambda^0 \tilde{\lambda} + \tilde{\mu}\psi_0 = c_1 \in W,$$
$$T_{yy}^0 \varphi_0 \tilde{y} + T_{y\lambda}^0 \varphi_0 \tilde{\lambda} + T_y^0 \tilde{\varphi} = c_2 \in W,$$
$$T_{yy}^0 s_0 \tilde{y} + T_{y\lambda}^0 s_0 \tilde{\lambda} + T_y^0 \tilde{s} + T_{y\lambda}^0 \tilde{y} + T_{\lambda\lambda}^0 \tilde{\lambda} = c_3 \in W, \qquad (5.164)$$
$$\varphi_0^* \tilde{\varphi} = \mu_1 \in \mathbb{R},$$
$$\varphi_0^* \tilde{s} = \mu_2 \in \mathbb{R},$$

where $(c_1, c_2, c_3, \mu_1, \mu_2) \in \mathcal{R}(\tilde{T}'(\tilde{z}_0))$.

From the condition for the solvability of the first equation, we obtain

$$\tilde{\mu} = \frac{\psi_0^* c_1}{\psi_0^* \psi_0}.$$

Now, we can write

$$(\tilde{y}, \tilde{\lambda}) = (y_1 + \delta \varphi_0 + \kappa s_0, \kappa),$$
$$y_1 \in Y_1 \equiv \{y \in Y : \varphi_0^* y = 0\},$$

where δ and κ are real numbers that are to be determined. If we substitute $\tilde{y}$ and $\tilde{\lambda}$ into the second and third equation in (5.164), then the conditions for the solvability of the resulting equations can be written in the form

$$\begin{pmatrix} \alpha & \beta \\ \beta & \gamma \end{pmatrix} \begin{pmatrix} \delta \\ \kappa \end{pmatrix} = \begin{pmatrix} \psi_0^*(c_2 - T_{yy}^0 \varphi_0 y_1) \\ \psi_0^*(c_3 - T_{yy}^0 s_0 y_1 - T_{y\lambda}^0 y_1) \end{pmatrix}. \tag{5.165}$$

The assumption $\tau < 0$ guarantees that there exists a unique solution (δ_0, κ_0) of this system. This in turn gives the result that

$$(\tilde{y}, \tilde{\lambda}) = (y_1 + \delta_0 \varphi_0 + \kappa_0 s_0, \kappa_0)$$

is uniquely determined. This allows us to write the second and third equation in (5.164) in the form

$$T_y^0 \tilde{\varphi} = \bar{c}_2 \in \mathbb{R}(T_y^0),$$
$$T_y^0 \tilde{s} = \bar{c}_3 \in \mathbb{R}(T_y^0).$$

It follows

$$\tilde{\varphi} = \varphi_1 + d_1 \varphi_0, \quad \varphi_1 \in Y_1, \quad d_1 \in \mathbb{R},$$
$$\tilde{s} = s_1 + d_2 \varphi_0, \quad s_1 \in Y_1, \quad d_2 \in \mathbb{R}.$$

Using the fourth and fifth equation in (5.164), we obtain

$$\varphi_0^* \tilde{\varphi} = \underbrace{\varphi_0^* \varphi_1}_{=0} + d_1 \underbrace{\varphi_0^* \varphi_0}_{=1} = \mu_1, \quad \text{i.e., } d_1 = \mu_1,$$

$$\varphi_0^* \tilde{s} = \underbrace{\varphi_0^* s_1}_{=0} + d_2 \underbrace{\varphi_0^* \varphi_0}_{=1} = \mu_2, \quad \text{i.e., } d_2 = \mu_2.$$

In summary, we have the result that $\tilde{\varphi}$ and $\tilde{s}$ are uniquely determined and the claim of our theorem is proved. ∎

Remark 5.53 The above proof shows that the statement of Theorem 5.52 remains valid, if the condition $\tau < 0$ is replaced by $\tau \neq 0$. □

With respect to the reversal of the statement of Theorem 5.52, we have the following result.

Theorem 5.54 *Suppose* $\tilde{z}_0 \equiv (y_0, \lambda_0, \varphi_0, s_0, 0)$ *is an isolated solution of problem* (5.160), *i.e.,* $\tilde{T}'(\tilde{z}_0)$ *is a linear homeomorphism of* $\tilde{Z}$ *onto* $\tilde{W}$. *Then,* $z_0 \equiv (y_0, \lambda_0)$ *satisfies:*

- $T(z_0) = 0$,
- $\dim \mathcal{N}(T'(z_0)) = 2$, $\quad \mathrm{codim}\, \mathcal{R}(T'(z_0)) = 1$,
- $\tau \equiv \alpha\gamma - \beta^2 \neq 0$, *and*
- $T_\lambda(z_0) \in \mathcal{R}(T_y(z_0))$.

Proof (i) Assume that $\tau = 0$ and $(c_1, c_2)^T$ is a nontrivial solution of the problem

$$\begin{pmatrix} \alpha & \beta \\ \beta & \gamma \end{pmatrix} \begin{pmatrix} c_1 \\ c_2 \end{pmatrix} = \begin{pmatrix} 0 \\ 0 \end{pmatrix}.$$

Moreover, let $w_1, w_2 \in Y_1$ be the unique solutions of the equations

$$T_y^0 w_1 = -\left[c_1 T_{yy}^0 \varphi_0^2 + c_2 \left(T_{yy}^0 \varphi_0 s_0 + T_{y\lambda}^0 \varphi_0 \right) \right],$$
$$T_y^0 w_2 = -\left[c_1 \left(T_{yy}^0 s_0 \varphi_0 + T_{y\lambda}^0 \varphi_0 \right) + c_2 \left(T_{yy}^0 s_0^2 + 2 T_{y\lambda}^0 s_0 + T_{\lambda\lambda}^0 \right) \right].$$

Then,

$$\tilde{T}'(\tilde{z}_0)(c_1\varphi_0 + c_2 s_0, c_2, w_1, w_2, 0)^T$$
$$= \begin{pmatrix} c_2 T_y^0 s_0 + c_2 T_\lambda^0 \\ c_1 T_{yy}^0 \varphi_0^2 + c_2 T_{yy}^0 \varphi_0 s_0 + c_2 T_{y\lambda}^0 \varphi_0 + T_y^0 w_1 \\ c_1 T_{yy}^0 s_0 \varphi_0 + c_2 T_{yy}^0 s_0^2 + c_1 T_{y\lambda}^0 \varphi_0 + 2 c_2 T_{y\lambda}^0 s_0 + c_2 T_{\lambda\lambda}^0 + T_y^0 w_2 \\ \varphi_0^* w_1 \\ \varphi_0^* w_2 \end{pmatrix} = \begin{pmatrix} 0 \\ 0 \\ 0 \\ 0 \\ 0 \end{pmatrix}.$$

However, our assumption is $\mathcal{N}(\tilde{T}'(\tilde{z}_0)) = \{0\}$. Thus we have a contradiction and it must hold $\tau \neq 0$.

(ii) Next we assume that there exists a $w_0 \in Y$, $w_0 \neq 0$, such that

$$T_y^0 w_0 = 0, \quad \varphi_0^* w_0 = 0.$$

But this implies $(0, 0, w_0, 0, 0) \in \mathcal{N}(\tilde{T}'(\tilde{z}_0))$, which is a contradiction to our assumption. Thus, it must hold $w_0 = 0$.

(iii) The remaining conditions are trivially fulfilled. ∎

Corollary 5.55 *If the secondary simple bifurcation points $z^{(i)} = (y^{(i)}, \lambda^{(i)})$ of the original operator equation (5.11) have to be computed, then those isolated solutions $\tilde{z}^{(i)}$ of the extended problem (5.160) must be determined, which are of the form*

$$\tilde{z}^{(i)} = (y^{(i)}, \lambda^{(i)}, \varphi^{(i)}, s^{(i)}, 0).$$

$\square$

The implementation of the extended problem (5.160) for BVPs of the form (5.10), which is based on the functional (5.47), is

$$
\begin{aligned}
y'(x) &= f(x, y; \lambda) - \mu\psi_0(x), & B_a y(a) + B_b y(b) &= 0, \\
\varphi'(x) &= f_y(x, y; \lambda)\varphi(x), & B_a \varphi(a) + B_b \varphi(b) &= 0, \\
s'(x) &= f_y(x, y; \lambda)s(x) + f_\lambda(x, y; \lambda), & B_a s(a) + B_b s(b) &= 0, \\
\xi_1'(x) &= \varphi(x)^T \varphi(x), & \xi_1(a) &= 0, \ \xi_1(b) = 1, \quad (5.166) \\
\xi_2'(x) &= \varphi(x)^T s(x), & \xi_2(a) &= 0, \ \xi_2(b) = 0, \\
\lambda'(x) &= 0, \\
\mu'(x) &= 0.
\end{aligned}
$$

This again is a nonlinear BVP in standard form of dimension $3n + 4$ that can be solved with the known numerical techniques, e.g., with the multiple shooting code presented in Sect. 4.6. It makes sense to use starting vectors, whose last component μ is zero.

Remark 5.56 If the BVP is self-adjoint, the first ODE in (5.166) can be replaced by the equation

$$y'(x) = f(x, y; \lambda) - \mu\varphi_0(x).$$

$\square$

The implementation (5.166) of the extension technique (5.159) is based on the functional (5.47). The number of equations in the BVP (5.166) can be further reduced, if we use instead of (5.47) the functional (5.53), i.e.,

$$\varphi_0^* v(x) \equiv \varphi_0(a)^T v(a),$$

whereas before $\varphi_0(x)$ is the nontrivial solution of the second (block-) equation in (5.166), i.e., the corresponding eigenfunction. In this way, we obtain the following extended BVP of dimension $3n + 2$, which is constructed in a self-explanatory manner:

$$
\begin{aligned}
y'(x) &= f(x, y; \lambda) - \mu\psi_0(x), & B_a y(a) + B_b y(b) &= 0, \\
\varphi'(x) &= f_y(x, y; \lambda)\varphi(x), & B_a\varphi(a) + B_b\varphi(b) &= 0, \\
s'(x) &= f_y(x, y; \lambda)s(x) + f_\lambda(x, y; \lambda), & B_a s(a) + B_b s(b) &= 0, \quad (5.167) \\
\lambda'(x) &= 0, & \varphi(a)^T \varphi(a) &= 1, \\
\mu'(x) &= 0, & \varphi(a)^T s(a) &= 0.
\end{aligned}
$$

An other extension technique for the determination of secondary simple bifurcation points has been presented in [88]. Here, the second (block-) equation in (5.160) is replaced by the corresponding adjoint problem. The extended operator $\hat{T}$ of T is

$$
\hat{T} : \begin{cases}
\hat{Z} \equiv Y \times \mathbb{R} \times Y^* \times \mathbb{R} \to \hat{W} \equiv W \times Y^* \times \mathbb{R} \times \mathbb{R} \\
\hat{z} \equiv (z, \psi^*, \mu) = (y, \lambda, \psi^*, \mu) \to \begin{pmatrix} T(y, \lambda) + \mu\psi_0 \\ T_y^*(y, \lambda)\,\psi^* \\ \psi^* T_\lambda(y, \lambda) \\ \psi^*\psi_0 - 1 \end{pmatrix},
\end{cases} \quad (5.168)
$$

and the corresponding extended problem is

$$
\hat{T}(\hat{z}) = 0. \tag{5.169}
$$

Obviously, if $z_0 \equiv (y_0, \lambda_0) \in Z$ is a secondary simple bifurcation point of the original operator equation (5.11), then $\hat{z}_0 \equiv (y_0, \lambda_0, \psi_0^*, 0) \in \hat{Z}$ is a solution of the extended operator equation (5.169). Moreover, we have the following result.

Theorem 5.57 *Let the hypotheses of Theorem 5.52 be satisfied. Then, the operator $\hat{T}'(\hat{z}_0)$ is a linear homeomorphism of $\hat{Z}$ onto $\hat{W}$, i.e.,*

$$
\mathcal{N}(\hat{T}'(\hat{z}_0)) = \{0\}, \quad \mathcal{R}(\hat{T}'(\hat{z}_0)) = \hat{W}.
$$

Proof See, e.g., [88]. ∎

If the element φ_0 is not required for a subsequent curve tracing, the extended problem (5.169) is to be favorized compared to the problem (5.160) since it has fewer equations.

The appropriate implementation of the extended problem (5.169) for BVPs of the form (5.10), which is based on the functional (5.47), is the following extended BVP of dimension $2n + 4$:

$$y'(x) = f(x, y; \lambda) - \mu\psi(x), \qquad B_a y(a) + B_b y(b) = 0,$$
$$\psi'(x) = -f_y(x, y; \lambda)^T \psi(x), \qquad B_a^* \psi(a) + B_b^* \psi(b) = 0,$$
$$\xi_1'(x) = -\psi(x)^T f_\lambda(x, y; \lambda), \qquad \xi_1(a) = 0, \;\; \xi_1(b) = 0,$$
$$\xi_2'(x) = \psi(x)^T \psi(x), \qquad\qquad \xi_2(a) = 0, \;\; \xi_2(b) = 1, \qquad (5.170)$$
$$\lambda'(x) = 0,$$
$$\mu'(x) = 0.$$

The matrices B_a^* and B_b^* are defined by (see, e.g., [63]):

$$B_a^* B_a^T - B_b^* B_b^T = 0, \quad \text{rank} \left[B_a^* | B_b^* \right] = n.$$

If we use the functional (5.53) instead of the functional (5.47), the following BVP of the dimension $2n + 2$ results:

$$y'(x) = f(x, y; \lambda) - \mu\psi(x), \qquad B_a y(a) + B_b y(b) = 0,$$
$$\psi'(x) = -f_y(x, y; \lambda)^T \psi(x), \qquad B_a^* \psi(a) + B_b^* \psi(b) = 0,$$
$$\lambda'(x) = 0, \qquad\qquad\qquad\qquad \psi(a)^T \psi(a) = 1, \qquad (5.171)$$
$$\mu'(x) = 0, \qquad\qquad\qquad \psi(a)^T f_\lambda(a, y(a); \lambda) = 0.$$

5.5.3 Determination of Solutions in the Neighborhood of a Secondary Simple Bifurcation Point

Let us assume that $z \equiv (y, \lambda) \in Z$ is an isolated solution of the operator equation (5.11), which is located in the direct neighborhood of a secondary simple bifurcation point $z_0 \equiv (y_0, \lambda_0) \in Z$. The following transformation technique for the numerical approximation of z is based on the statements and results presented in the proof of Theorem 5.50. The starting point is the formula (5.150), i.e.,

$$y(\varepsilon) = y_0 + \varepsilon\{\rho(\varepsilon)p_1 + \sigma(\varepsilon)q_1\} + \varepsilon^2 u(\varepsilon),$$
$$\lambda(\varepsilon) = \lambda_0 + \varepsilon\{\rho(\varepsilon)p_2 + \sigma(\varepsilon)q_2\} + \varepsilon^2 \tau(\varepsilon), \qquad (5.172)$$
$$\rho(\varepsilon)^2 + \sigma(\varepsilon)^2 = 1, \quad (u(\varepsilon), \tau(\varepsilon)) \in \bar{Z}.$$

The representation (5.172) of the solution curve is not based on the special form of $p = (p_1, p_2)^T \in Y \times \mathbb{R}$ and $q = (q_1, q_2)^T \in Y \times \mathbb{R}$, which we have introduced in Sect. 5.5.2. Therefore, any known bases of the null space $\mathcal{N}(T'(z_0))$ can be used. However, if the bifurcation point z_0 has been determined by the extension technique (5.159), (5.160), then it makes sense to set

$$p_1 = \varphi_0, \quad p_2 = 0, \quad q_1 = s_0, \quad q_2 = 1.$$

With $v(\varepsilon) \equiv (u(\varepsilon), \tau(\varepsilon))^T$ and $z(\varepsilon)$ according to formula (5.172), we have

$$T'(z_0)v(\varepsilon) = T_y^0 u(\varepsilon) + T_\lambda^0 \tau(\varepsilon)$$

and

$$T''(z_0)[\rho(\varepsilon)p + \sigma(\varepsilon)q + \varepsilon v(\varepsilon)]^2$$

$$= \begin{pmatrix} T_{yy}^0 & T_{y\lambda}^0 \\ T_{y\lambda}^0 & T_{\lambda\lambda}^0 \end{pmatrix} \begin{pmatrix} \rho(\varepsilon)p_1 + \sigma(\varepsilon)q_1 + \varepsilon u(\varepsilon) \\ \rho(\varepsilon)p_2 + \sigma(\varepsilon)q_2 + \varepsilon \tau(\varepsilon) \end{pmatrix}^2,$$

$$= T_{yy}^0[\rho(\varepsilon)p_1 + \sigma(\varepsilon)q_1 + \varepsilon u(\varepsilon)]^2$$
$$+ 2T_{y\lambda}^0[\rho(\varepsilon)p_1 + \sigma(\varepsilon)q_1 + \varepsilon u(\varepsilon)][\rho(\varepsilon)p_2 + \sigma(\varepsilon)q_2 + \varepsilon \tau(\varepsilon)]$$
$$+ T_{\lambda\lambda}^0[\rho(\varepsilon)p_2 + \sigma(\varepsilon)q_2 + \varepsilon \tau(\varepsilon)]^2.$$

Let $p^* = (p_1^*, p_2^*)^T \in Z^*$ and $q^* = (q_1^*, q_2^*)^T \in Z^*$ be two linear functionals, with

$$p^*p = q^*q \neq 0, \quad p^*q = q^*p = 0.$$

Now, the transformed problem (5.152) can be formulated more precisely as

$$\begin{pmatrix} T_y^0 & T_\lambda^0 \\ p_1^* & p_2^* \\ q_1^* & q_2^* \end{pmatrix} \begin{pmatrix} u(\varepsilon) \\ \tau(\varepsilon) \end{pmatrix} = - \begin{pmatrix} w(u(\varepsilon), \tau(\varepsilon), \rho(\varepsilon), \sigma(\varepsilon); \varepsilon) \\ 0 \\ 0 \end{pmatrix}, \tag{5.173}$$

$$\rho(\varepsilon)^2 + \sigma(\varepsilon)^2 = 1,$$

with

$$w(u, \tau, \rho, \sigma; \varepsilon) \equiv \frac{1}{2} T_{yy}^0 (\rho p_1 + \sigma q_1 + \varepsilon u)^2$$
$$+ T_{y\lambda}^0 (\rho p_1 + \sigma q_1 + \varepsilon u)(\rho p_2 + \sigma q_2 + \varepsilon \tau)$$
$$+ T_{\lambda\lambda}^0 (\rho p_2 + \sigma q_2 + \varepsilon \tau)^2 + \frac{1}{\varepsilon^2} R(y(\varepsilon), \lambda(\varepsilon)).$$

A statement about the solvability of the transformed problem (5.173) is given in the next theorem.

Theorem 5.58 *Let $z_0 \equiv (y_0, \lambda_0) \in Z$ be a hyperbolic point of the operator T. Then, there exist real numbers $\varepsilon_1, \varepsilon_2 > 0$ such that for $|\varepsilon| \leq \varepsilon_i$ $(i = 1, 2)$ problem (5.173) has two continuous curves of isolated solutions*

$$\eta^i(\varepsilon) \equiv \left(u^i(\varepsilon), \tau^i(\varepsilon), \rho^i(\varepsilon), \sigma^i(\varepsilon)\right)^T, \quad i = 1, 2.$$

By inserting $\eta^1(\varepsilon)$ and $\eta^2(\varepsilon)$ into the ansatz (5.172), it is possible to construct two continuous curves of solutions of the operator equation (5.11). These curves intersect at z_0.

Proof The statement is an immediate conclusion from Theorem 5.50.   ∎

At the end of this section, we will show how this transformation technique can be applied to BVPs of the form (5.10). If the functionals $p^*, q^* \in Z^*$ are given in the form

$$p^* r = p_1^* r_1 + p_2^* r_2 \equiv \int_a^b p_1(x)^T r_1(x) dx + p_2 r_2,$$

$$q^* r = q_1^* r_1 + q_2^* r_2 \equiv \int_a^b q_1(x)^T r_1(x) dx + q_2 r_2,$$

$$r = (r_1, r_2)^T \in Z,$$

then the second and third (block-) equation of the transformed problem (5.173) may be written in ODE formulation as follows

$$\xi_1'(x; \varepsilon) = p_1(x)^T u(x; \varepsilon) + p_2 \tau(\varepsilon), \quad \xi_1(a; \varepsilon) = \xi_1(b; \varepsilon) = 0 \qquad (5.174)$$

and

$$\xi_2'(x; \varepsilon) = q_1(x)^T u(x; \varepsilon) + q_2 \tau(\varepsilon), \quad \xi_2(a; \varepsilon) = \xi_2(b; \varepsilon) = 0, \qquad (5.175)$$

respectively. Moreover, we add the trivial ODEs $\tau'(x; \varepsilon) = 0$, $\rho'(x; \varepsilon) = 0$ and $\sigma'(x; \varepsilon) = 0$ to the resulting system. In this way, we obtain the following transformed BVP of dimension $n + 5$

$$
\begin{aligned}
u'(x; \varepsilon) &= f_y^0 u(x; \varepsilon) + f_\lambda^0 \tau(x; \varepsilon) \\
&\quad + \tilde{w}(u(x; \varepsilon), \tau(x; \varepsilon), \rho(x; \varepsilon), \sigma(x; \varepsilon); \varepsilon), \\
\xi_1'(x; \varepsilon) &= p_1(x)^T u(x; \varepsilon) + p_2 \tau(\varepsilon), \\
\xi_2'(x; \varepsilon) &= q_1(x)^T u(x; \varepsilon) + q_2 \tau(\varepsilon) \\
\tau'(x; \varepsilon) &= 0, \quad \rho'(x; \varepsilon) = 0, \quad \sigma'(x; \varepsilon) = 0, \\
B_a u(a; \varepsilon) &+ B_b u(b; \varepsilon) = 0, \\
\xi_1(a; \varepsilon) &= \xi_2(a; \varepsilon) = \xi_1(b; \varepsilon) = \xi_2(b; \varepsilon) = 0, \\
\rho(a; \varepsilon)^2 &+ \sigma(x; \varepsilon)^2 - 1 = 0, \qquad |\varepsilon| \le \varepsilon_0
\end{aligned}
\qquad (5.176)
$$

where

$$
\begin{aligned}
w(u, \tau, \rho, \sigma; \varepsilon) &\equiv \frac{1}{2} f_{yy}^0 (\rho p_1 + \sigma q_1 + \varepsilon u)^2 \\
&\quad + f_{y\lambda}^0 (\rho p_1 + \sigma q_1 + \varepsilon u)(\rho p_2 + \sigma q_2 + \varepsilon \tau) \\
&\quad + \frac{1}{2} f_{\lambda\lambda}^0 (\rho p_2 + \sigma q_2 + \varepsilon \tau)^2 + \frac{1}{\varepsilon^2} \tilde{R}(y(x; \varepsilon), \lambda(x; \varepsilon)),
\end{aligned}
$$

and

$$\tilde{R}(y, \lambda) \equiv f(y, \lambda) - \left\{ f^0 + f_y^0(y - y_0) + f_\lambda^0(\lambda - \lambda_0) + \frac{1}{2} f_{yy}^0(y - y_0)^2 \right.$$
$$\left. + f_{y\lambda}^0(\lambda - \lambda_0)(y - y_0) + \frac{1}{2} f_{\lambda\lambda}^0(\lambda - \lambda_0)^2 \right\}.$$

5.6 Perturbed Bifurcation Problems

5.6.1 Nondegenerate Initial Imperfections

All problems discussed so far are of the general form

$$T : Y \times \Lambda \to W, \quad T(y, \lambda) = 0,$$
$$Y, W, \Lambda \text{ - Banach spaces,} \tag{5.177}$$

where we have set $\Lambda \equiv \mathbb{R}$ in (5.11). In this book, we will not consider the case $\Lambda = \mathbb{R}^k$, i.e., multiparameter bifurcation problems.

As we have seen, a typical bifurcation diagram for the equation (5.11) contains bifurcation and limit points. However, in experiments and in real applications, the sharp transitions of bifurcation rarely occur. Small imperfections, impurities, or other inhomogeneities tend to distort these transitions. In Fig. 5.19 an example for the effect of small initial imperfections on the first bifurcation point of the governing equations (5.6) of the Euler–Bernoulli rod is shown. We have to take such initial imperfections into consideration, since the rod is not a perfectly straight line or the cross-section is not always the same.

To analyze mathematically the perturbation of bifurcations caused by imperfections and other impurities, the classical bifurcation theory is modified by introducing an additional small perturbation parameter, which characterizes the magnitudes of these inhomogeneities. Thus, instead of the one-parameter problem (5.177) the following two-parameter problem is studied

$$\bar{T} : Y \times \Lambda_1 \times \Lambda_2 \to W, \quad \bar{T}(y, \lambda, \tau) = 0,$$
$$Y, W, \Lambda_1, \Lambda_2 \text{ - Banach spaces,} \tag{5.178}$$

where Λ_1 is the control parameter space and Λ_2 is the perturbation parameter space. Here, we set $\Lambda_1 = \Lambda_2 = \mathbb{R}$. The connection between problem (5.177) and (5.178) is made by embedding (5.177) into (5.178) such that

$$\bar{T}(y, \lambda, 0) = T(y, \lambda). \tag{5.179}$$

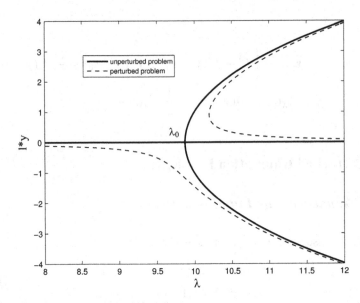

Fig. 5.19 Perturbed primary simple bifurcation

Suppose that the BVP (5.10) enlarged by the perturbation parameter τ is of the form

$$y'(x) = f(x, y; \lambda, \tau), \quad a < x < b,$$
$$B_a y(a) + B_b y(b) = 0,$$

(5.180)

with

$$f : D_f \to \mathbb{R}^n, \quad D_f \subset [a, b] \times \mathbb{R}^n \times \mathbb{R} \times \mathbb{R}, \quad f \in \mathbb{C}^p, \quad p \geq 2,$$
$$\lambda \in D_\lambda \subset \mathbb{R}, \quad \tau \in D_\tau \subset \mathbb{R}, \quad B_a, B_b \in \mathbb{R}^{n \times n}, \quad \text{rank}[B_a | B_b] = n.$$

Here, λ is the control parameter and τ is the perturbation parameter reflecting the *small* initial imperfections in the mathematical model. The corresponding operator form is

$$T(y, \lambda, \tau) = 0, \quad y \in Y, \quad \lambda, \tau \in \mathbb{R},$$

(5.181)

where

$$T : Z \equiv Y \times \mathbb{R} \times \mathbb{R} \to W, \quad T(y, \lambda, \tau) \equiv y' - f(\cdot, y; \lambda, \tau),$$

and

$$Y \equiv B\mathbb{C}^1([a, b], \mathbb{R}^n), \quad W \equiv \mathbb{C}([a, b], \mathbb{R}^n).$$

Assume that the operator T satisfies two additional conditions:

$$T(0, \lambda, 0) = 0 \quad \text{for all } \lambda \in \mathbb{R},$$
$$T(0, \lambda, \tau) \neq 0 \quad \text{for all } \tau \neq 0. \tag{5.182}$$

We start with the description of the unperturbed problem (i.e., the case $\tau = 0$). The first condition of (5.182) implies that the solution field of the unperturbed problem $E(T; \tau = 0)$ contains the trivial solution curve. Let λ_0 be a simple eigenvalue of $T_y(z), z \equiv (0, \lambda, 0) \in Y \times \mathbb{R} \times \mathbb{R}$, i.e., it holds

$$\dim \mathcal{N}(T_y(z_0)) = 1, \quad z_0 \equiv (0, \lambda_0, 0),$$
$$\mathcal{N}(T_y(z_0)) = \text{span}\{\varphi_0\}, \quad \|\varphi_0\| = 1. \tag{5.183}$$

Then, conditions (5.183) say that $T_y(z_0)$ is a Fredholm operator with index zero and the statements (5.17)–(5.21) of the Riesz–Schauder theory are valid. Besides (5.183), we suppose that

$$a_1 \equiv \psi_0^* T_{y\lambda}(z_0)\varphi_0 \neq 0. \tag{5.184}$$

As explained in Sect. 5.4.2, the assumptions (5.183) and (5.184) guarantee that in the neighborhood of $z_0 \equiv (0, \lambda_0, 0)$ the solution field of the unperturbed problem $E(T; \tau = 0)$ consist of exactly two solution curves, which intersect at z_0. One of these curves is the trivial solution curve $\{(0, \lambda, 0) : \lambda \in \mathbb{R}\}$.

We now want to study the effects of nonvanishing initial imperfections (i.e., $\tau \neq 0$) on the operator equation (5.181). The following classification of perturbed problems takes into account the fact that the corresponding solution field $E(T; \tau = 0)$ can be qualitatively changed by small initial imperfections.

Definition 5.59

- The operator equation (5.181) is called a *BP-problem* (bifurcation preserving problem), if in any neighborhood $U \subset Y \times \mathbb{R}$ of $(0, \lambda_0)$, a $\tau(U) > 0$ can be found such that for each fixed $\tau, 0 < |\tau| < \tau(U)$, there exists a bifurcation point in U.
- Otherwise, the operator equation (5.181) is called a *BD-problem* (bifurcation destroying problem). □

The problem (5.181) is always a BD-problem if there exists a neighborhood $V \subset Y \times \mathbb{R} \times \mathbb{R}$ of $(0, \lambda_0, 0)$ for which the following statement is true: $T(y, \lambda, \tau) = 0, (y, \lambda, \tau) \in V$, and $\tau \neq 0$ imply that (y, λ) is not a bifurcation point.

For the next considerations, a special type of perturbations $\tau \neq 0$ is of interest.

Definition 5.60 Let the conditions (5.183) and (5.184) for the unperturbed problem be satisfied. We define

$$b_1 \equiv \psi_0^* T_\tau(z_0). \tag{5.185}$$

The parameter τ is said to represent a *nondegenerate* (*degenerate*) imperfection if $b_1 \neq 0$ ($b_1 = 0$). □

Nondegenerate imperfections play an important role in the bifurcation theory since they have a great influence on the solution field of the unperturbed problem.

Theorem 5.61 *Assume that the conditions* (5.182)–(5.184) *and* $b_1 \neq 0$ *are satisfied. Then, the problem* (5.181) *is a BD-problem.*

Proof See [114]. ∎

Example 5.62 Let us consider once more the BVP (5.6) governing the Euler–Bernoulli rod problem. The only difference now is that we introduce an additional external parameter τ, describing the initial imperfections of the rod, into the right-hand side of the ODE. More precisely, we have the following perturbed BVP

$$y''(x) + \lambda \sin(y(x)) = \tau \cos(\pi x), \quad x \in (0, 1),$$
$$y'(0) = 0, \quad y'(1) = 0. \tag{5.186}$$

As we have seen in Sect. 5.1, the primary simple bifurcation points and the corresponding eigenfunctions of the unperturbed problem are $\lambda_0^{(k)} = (k\pi)^2$ and $\varphi_0^{(k)} = \sqrt{2} \cos(k\pi x)$, respectively. Here, we want to concentrate on the first bifurcation point $\lambda_0 \equiv \lambda_0^{(1)} = \pi^2$ and answer the question, whether a bifurcation point also occurs in the perturbed BVP (5.186). In order to do that, we have to determine the coefficient b_1 defined in (5.185). If we set

$$T(y, \lambda, \tau) \equiv y'' + \lambda \sin(y) - \tau \cos(\pi x),$$

it follows $T_\tau(z_0) = -\cos(\pi x)$ and

$$b_1 = -\sqrt{2} \int_0^1 \cos(\pi x)^2 dx = -\frac{\sqrt{2}}{2} \neq 0.$$

Looking at the assumptions of Theorem 5.61, we see that the BVP (5.186) is a BD-problem. The difference in the structure of the solution field of the unperturbed and the perturbed problem can be seen in Fig. 5.20.

Let us now change the boundary conditions of the BVP (5.186) as follows

$$y''(x) + \lambda \sin(y(x)) = \tau \cos(\pi x), \quad x \in (0, 1),$$
$$y(0) = 0, \quad y(1) = 0. \tag{5.187}$$

Now, for the unperturbed problem we have the following situation. The first primary simple bifurcation point λ_0 is the same as before, but the corresponding eigenfunction changes to $\varphi_0(x) = \sqrt{2} \sin(\pi x)$. Therefore, the coefficient b_1 is calculated as

$$b_1 = -\sqrt{2} \int_0^1 \cos(\pi x) \sin(\pi x) dx = 0.$$

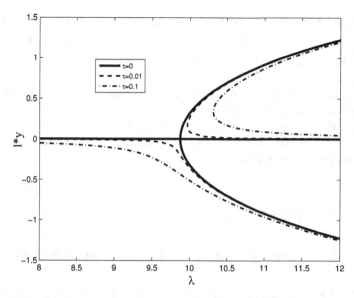

Fig. 5.20 Bifurcation diagram of problem (5.186) for 3 different values of τ

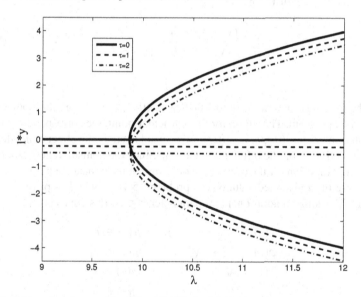

Fig. 5.21 Bifurcation diagram of problem (5.187) for 3 different values of τ

Obviously, Theorem 5.61 provides not enough information to decide whether the BVP (5.187) is a BD- or a BP-problem. But in fact it is a BP-problem as can be seen in Fig. 5.21. □

Our numerical treatment of (5.181), under the condition (5.185), is based on the theory of Keener and Keller [68]. The following steps have to be executed:

1. Determination of a curve of nonisolated solutions of (5.181), which contains the particular point $(y, \lambda, \tau) = (0, \lambda_0, 0)$;
2. Determination of curves of nontrivial isolated solutions of (5.181), which branch off from the curve computed in step 1;
3. Presentation of the solution field $E(T)$ of (5.181) in form of a bifurcation diagram using the solution curves constructed in step 2 and the curve of nonisolated solutions determined in step 1.

5.6.2 Nonisolated Solutions

The nonisolation of solutions of problem (5.181) is explained in the following definition.

Definition 5.63 The quadruple $(y, \varphi, \lambda, \tau) \in Y \times Y \times \mathbb{R} \times \mathbb{R}$ is referred to as nonisolated solution of (5.181) if

$$
\begin{aligned}
T(y, \lambda, \tau) &= 0, \\
T_y(y, \lambda, \tau)\varphi &= 0, \quad \|\varphi\| \neq 0.
\end{aligned}
\tag{5.188}
$$

$\square$

From the previous section, we know that $z_0 \equiv (0, \lambda_0)$ is a bifurcation point of the unperturbed problem. Therefore, for this particular point, the corresponding nonisolated solution is $\tilde{z}_0 \equiv (0, \varphi_0, \lambda_0, 0)$. Now, we want to study the effect of nondegenerate imperfections τ on the bifurcation point z_0. Since the nonisolation of solutions is a necessary condition for the occurrence of bifurcations, it makes sense to determine first a curve of nonisolated solutions of problem (5.181), which is passing through the point $\tilde{z}_0$. Using an additional real parameter $\varepsilon \in \mathbb{R}$, this curve is represented as follows:

$$
\begin{aligned}
y(\varepsilon) &= \varepsilon\varphi_0 + \varepsilon^2 u(\varepsilon), & u(\varepsilon) &\in Y_1, \\
\varphi(\varepsilon) &= \varphi_0 + \varepsilon\chi(\varepsilon), & \chi(\varepsilon) &\in Y_1, \\
\lambda(\varepsilon) &= \lambda_0 + \varepsilon\alpha(\varepsilon), & \alpha(\varepsilon) &\in \mathbb{R}, \\
\tau(\varepsilon) &= \varepsilon^2 \beta(\varepsilon), & \beta(\varepsilon) &\in \mathbb{R}.
\end{aligned}
\tag{5.189}
$$

The Taylor series of T at $\tilde{z}_0$ is

$$
\begin{aligned}
T(y, \lambda, \tau) = {}& T_y^0 y + T_\tau^0 \tau + T_{y\lambda}^0 y(\lambda - \lambda_0) + T_{y\tau}^0 y\tau + \frac{1}{2}T_{yy} y^2 \\
& + T_{\tau\lambda}^0 \tau(\lambda - \lambda_0) + R_1(y, \lambda, \tau).
\end{aligned}
\tag{5.190}
$$

The corresponding Taylor series of T_y is

$$T_y(y, \lambda, \tau) = T_y^0 + T_{y\lambda}^0(\lambda - \lambda_0) + T_{yy}^0 y + R_2(y, \lambda, \tau). \tag{5.191}$$

For the ansatz (5.189), we have

$$R_1 = O(\varepsilon^3) \quad \text{and} \quad R_2 = O(\varepsilon^2).$$

Inserting the ansatz (5.189) into (5.188), we obtain

$$
\begin{aligned}
0 = {} & T_y^0 u(\varepsilon) + T_\tau^0 \beta(\varepsilon) + T_{y\lambda}^0 \alpha(\varepsilon)\varphi_0 + \varepsilon T_{y\lambda}^0 \alpha(\varepsilon)u(\varepsilon) \\
& + \varepsilon T_{y\tau}^0 \beta(\varepsilon)[\varphi_0 + \varepsilon u(\varepsilon)] + \frac{1}{2}T_{yy}^0[\varphi_0 + \varepsilon u(\varepsilon)]^2 \\
& + \varepsilon T_{\tau\lambda}^0 \beta(\varepsilon)\alpha(\varepsilon) + \frac{1}{\varepsilon^2}R_1, \quad \varphi_0^* u(\varepsilon) = 0,
\end{aligned}
\tag{5.192}
$$

and

$$
\begin{aligned}
0 = {} & T_y^0 \chi(\varepsilon) + T_{y\lambda}^0 \alpha(\varepsilon)[\varphi_0 + \varepsilon\chi(\varepsilon)] + T_{yy}^0[\varphi_0 + \varepsilon u(\varepsilon)][\varphi_0 + \varepsilon\chi(\varepsilon)] \\
& + \frac{1}{\varepsilon}R_2[\varphi_0 + \varepsilon\chi(\varepsilon)], \quad \varphi_0^*\chi(\varepsilon) = 0.
\end{aligned}
\tag{5.193}
$$

Equations (5.192) and (5.193) can be written in matrix notation as

$$
\begin{pmatrix}
T_y^0 & 0 & T_{y\lambda}^0\varphi_0 & T_\tau^0 \\
0 & T_y^0 & T_{y\lambda}^0\varphi_0 & 0 \\
\varphi_0^* & 0 & 0 & 0 \\
0 & \varphi_0^* & 0 & 0
\end{pmatrix}
\begin{pmatrix}
u(\varepsilon) \\
\chi(\varepsilon) \\
\alpha(\varepsilon) \\
\beta(\varepsilon)
\end{pmatrix}
= -
\begin{pmatrix}
w_1(u, \alpha, \beta; \varepsilon) \\
w_2(u, \chi, \alpha, \beta; \varepsilon) \\
0 \\
0
\end{pmatrix},
\tag{5.194}
$$

where

$$
\begin{aligned}
w_1(u, \alpha, \beta; \varepsilon) \equiv {} & \frac{1}{2}T_{yy}^0(\varphi_0 + \varepsilon u)^2 + \varepsilon\beta T_{y\tau}^0(\varphi_0 + \varepsilon u) + \varepsilon\alpha T_{y\lambda}^0 u \\
& + \varepsilon\alpha\beta T_{\tau\lambda}^0 + \frac{1}{\varepsilon^2}R_1(y, \lambda, \tau),
\end{aligned}
$$

and

$$
\begin{aligned}
w_2(u, \chi, \alpha, \beta; \varepsilon) \equiv {} & \varepsilon\alpha T_{y\lambda}^0\chi \\
& + \left[T_{yy}^0(\varphi_0 + \varepsilon u) + \frac{1}{\varepsilon}R_2(y, \lambda, \tau)\right](\varphi_0 + \varepsilon\chi).
\end{aligned}
$$

A statement about the solvability of problem (5.194) is given in the next theorem.

Theorem 5.64 *Let the assumptions* (5.182)–(5.184) *and* $b_1 \neq 0$ *be satisfied. Then, there exists a real number* $\varepsilon_0 > 0$ *such that for* $|\varepsilon| \leq \varepsilon_0$ *problem* (5.194) *has a continuous curve of isolated solutions*

$$\eta(\varepsilon) \equiv (u(\varepsilon), \chi(\varepsilon), \alpha(\varepsilon), \beta(\varepsilon))^T .$$

By inserting $\eta(\varepsilon)$ *into the ansatz* (5.189) *it is possible to construct a continuous curve of solutions of the equations* (5.188). *This curve is composed of nonisolated solutions of* (5.181) *and passes through*

$$\tilde{z}_0 \equiv (0, \varphi_0, \lambda_0, 0) \in Y \times Y \times \mathbb{R} \times \mathbb{R}.$$

Proof Consider (5.194) for $\varepsilon \to 0$. We get

$$\mathbb{A}\begin{pmatrix} u(0) \\ \chi(0) \\ \alpha(0) \\ \beta(0) \end{pmatrix} \equiv \begin{pmatrix} T_y^0 & 0 & T_{y\lambda}^0\varphi_0 & T_\tau^0 \\ 0 & T_y^0 & T_{y\lambda}^0\varphi_0 & 0 \\ \varphi_0^* & 0 & 0 & 0 \\ 0 & \varphi_0^* & 0 & 0 \end{pmatrix}\begin{pmatrix} u(0) \\ \chi(0) \\ \alpha(0) \\ \beta(0) \end{pmatrix} = -\begin{pmatrix} \frac{1}{2}T_{yy}^0\varphi_0^2 \\ T_{yy}^0\varphi_0^2 \\ 0 \\ 0 \end{pmatrix}.$$

We set

$$A \equiv \begin{pmatrix} T_y^0 & 0 \\ 0 & T_y^0 \end{pmatrix}, \quad B \equiv \begin{pmatrix} T_{y\lambda}^0\varphi_0 & T_\tau^0 \\ T_{y\lambda}^0\varphi_0 & 0 \end{pmatrix}$$

$$C^* \equiv \begin{pmatrix} \varphi_0^* & 0 \\ 0 & \varphi_0^* \end{pmatrix}, \quad D \equiv \begin{pmatrix} 0 & 0 \\ 0 & 0 \end{pmatrix}, \quad m \equiv 2,$$

and use Lemma 5.25. Obviously, A is not bijective and

$$\mathcal{N}(A) = \text{span}\left\{\begin{pmatrix} \varphi_0 \\ 0 \end{pmatrix}, \begin{pmatrix} 0 \\ \varphi_0 \end{pmatrix}\right\}, \quad \text{i.e., } \dim \mathcal{N}(A) = 2.$$

Thus, the second case applies and we have to verify (c_0)–(c_3).

(c_0): $\mathcal{R}(B) = \text{span}\left\{\begin{pmatrix} T_\tau^0 \\ 0 \end{pmatrix}, \begin{pmatrix} T_{y\lambda}^0\varphi_0 \\ T_{y\lambda}^0\varphi_0 \end{pmatrix}\right\}, \quad \text{i.e., } \dim \mathcal{R}(B) = 2,$

(c_1): It holds: $\begin{pmatrix} v_1 \\ v_2 \end{pmatrix} \in \mathcal{R}(A) \Leftrightarrow \psi_0^* v_1 = 0$ and $\psi_0^* v_2 = 0$.

For the matrix B we have:

$v_1 \underbrace{\psi_0^* T_{y\lambda}^0\varphi_0}_{\neq 0} = 0 \Leftrightarrow v_1 = 0$, which implies

$v_2 \underbrace{\psi_0^* T_\tau^0}_{\neq 0} + v_1 \underbrace{\psi_0^* T_{y\lambda}^0\varphi_0}_{=0} = 0 \Leftrightarrow v_2 = 0,$

i.e., $\mathcal{R}(B) \cap \mathcal{R}(A) = \begin{pmatrix} 0 \\ 0 \end{pmatrix}.$

(c_2) & (c_3) are fulfilled trivially.

Thus, we have shown that operator $\mathbb{A}$ is bijective. The claim of the theorem now follows analogously to the proofs of the transformation techniques presented in the previous sections. ∎

For BVPs of the form (5.180), the equation (5.194) can be implemented as follows

$$
\begin{aligned}
&u'(x; \varepsilon) = f_y^0 u(x; \varepsilon) + \alpha(\varepsilon) f_{y\lambda}^0 \varphi_0(x) + \beta(\varepsilon) f_\tau^0 + \tilde{w}_1(u, \alpha, \beta; \varepsilon), \\
&\chi'(x, \varepsilon) = f_y^0 \chi(x; \varepsilon) + \alpha(\varepsilon) f_{y\lambda}^0 \varphi_0(x) + \tilde{w}_2(u, \chi, \alpha, \beta; \varepsilon), \\
&\xi_1'(x; \varepsilon) = \varphi_0(x)^T u(x; \varepsilon), \\
&\xi_2'(x; \varepsilon) = \varphi_0(x)^T \chi(x; \varepsilon), \\
&\alpha'(\varepsilon) = 0, \\
&\beta'(\varepsilon) = 0, \\
&B_a u(a; \varepsilon) + B_b u(b; \varepsilon) = 0, \quad B_a \chi(a; \varepsilon) + B_b \chi(b; \varepsilon) = 0, \\
&\xi_1(a; \varepsilon) = 0, \quad \xi_1(b; \varepsilon) = 0, \quad \xi_2(a; \varepsilon) = 0, \quad \xi_2(b; \varepsilon) = 0,
\end{aligned}
\tag{5.195}
$$

with

$$
\begin{aligned}
\tilde{w}_1(u, \alpha, \beta; \varepsilon) &\equiv \frac{1}{2} f_{yy}^0 (\varphi_0 + \varepsilon u)^2 + \varepsilon \beta f_{y\tau}^0 (\varphi_0 + \varepsilon u) + \varepsilon \alpha f_{y\lambda}^0 u \\
&\quad + \varepsilon \alpha \beta f_{\tau\lambda}^0 + \frac{1}{\varepsilon^2} \tilde{R}_1, \\
\tilde{w}_2(u, \chi, \alpha, \beta; \varepsilon) &\equiv \varepsilon \alpha f_{y\lambda}^0 \chi + \left[f_{yy}^0 (\varphi_0 + \varepsilon u) + \frac{1}{\varepsilon} \tilde{R}_2 \right] [\varphi_0 + \varepsilon \chi], \\
\tilde{R}_1 &\equiv f(x, y, \lambda, \tau) - \left\{ f_y^0 y + f_\tau^0 \tau + f_{y\lambda}^0 y(\lambda - \lambda_0) + f_{y\tau}^0 y\tau \right. \\
&\quad \left. + \frac{1}{2} f_{yy}^0 y^2 + f_{\tau\lambda}^0 \tau(\lambda - \lambda_0) \right\}, \\
\tilde{R}_2 &\equiv f_y(x, y, \lambda, \tau) - \{ f_y^0 + f_{y\lambda}^0 (\lambda - \lambda_0) + f_{yy}^0 y \}.
\end{aligned}
$$

Now, we want to show that it is also possible to compute the nonisolated solutions in direct dependence on the perturbation parameter τ. As will be seen later, such techniques are very advantageous to generate bifurcation diagrams. They enable to compute simultaneously a nonisolated solution and a curve of nontrivial solutions of (5.181) passing through this solution. Moreover, the use of an additional (external) parameter is not required. The disadvantage of this method, however, is the relatively high analytical and numerical effort caused by the computation of all required partial derivatives of T.

The ansatz for the curve of nonisolated solutions depends on the second bifurcation coefficient a_2 of the unperturbed problem. First, let us consider the case of an asymmetric bifurcation point, i.e., we assume

$$
a_2 \equiv \psi_0^* T_{yy}(0, \lambda_0, 0)\varphi_0^2 \neq 0.
$$

If the last equation in (5.189) is solved for ε, i.e., $\varepsilon = g(\tau)$, the following ansatz for the curve of nonisolated solutions of (5.181) passing through the particular point $\tilde{z}_0 \equiv (0, \varphi_0, \lambda_0, 0)$ is obtained:

$$
\begin{aligned}
y(\xi) &= -\frac{a_1}{a_2}\varphi_0 \xi + \{K(\xi)\varphi_0 + u_1\}\xi^2 + u(\xi)\xi^3, \\
\lambda(\xi) &= \lambda_0 + \xi + \lambda_1(\xi)\xi^2, \\
\varphi(\xi) &= \varphi_0 + \varphi_1 \xi + \psi(\xi)\xi^2, \\
\xi^2 &= \frac{2b_1 a_2}{a_1^2}\tau, \quad \text{sign}(\tau) = \text{sign}(b_1 a_2),
\end{aligned}
\tag{5.196}
$$

where $\lambda_1, K \in \mathbb{R}$ and $v, \psi \in Y_1$. The elements u_1 and φ_1 are the (unique) solutions of the following linear problems

$$
\begin{aligned}
T_y^0 u_1 &= \frac{a_1}{a_2}T_{y\lambda}^0\varphi_0 - \frac{a_1^2}{2b_1 a_2}T_\tau^0 - \frac{1}{2}\left(\frac{a_1}{a_2}\right)^2 T_{yy}^0\varphi_0^2, \\
\varphi_0^* u_1 &= 0,
\end{aligned}
\tag{5.197}
$$

and

$$
\begin{aligned}
T_y^0 \varphi_1 &= -T_{y\lambda}^0\varphi_0 + \frac{a_1}{a_2}T_{yy}^0\varphi_0^2, \\
\varphi_0^* \varphi_1 &= 0.
\end{aligned}
\tag{5.198}
$$

Now, we expand T and T_y into Taylor series:

$$
\begin{aligned}
T(y, \lambda, \tau) = {}& T_y^0 y + T_\tau^0 \tau + T_{y\lambda}^0 y(\lambda - \lambda_0) + T_{y\tau}^0 y\tau + \frac{1}{2}T_{yy}^0 y^2 + T_{\tau\lambda}^0\tau(\lambda - \lambda_0) \\
& + \frac{1}{6}T_{yyy}^0 y^3 + \frac{1}{2}T_{y\lambda\lambda}^0 y(\lambda - \lambda_0)^2 + \frac{1}{2}T_{yy\lambda}^0 y^2(\lambda - \lambda_0) \\
& + R_1(y, \lambda, \tau),
\end{aligned}
\tag{5.199}
$$

and

$$
\begin{aligned}
T_y(y, \lambda, \tau) = {}& T_y^0 + T_{y\lambda}^0(\lambda - \lambda_0) + T_{y\tau}^0\tau + T_{yy}^0 y + T_{yy\lambda}^0 y(\lambda - \lambda_0) \\
& + \frac{1}{2}T_{yyy}^0 y^2 + \frac{1}{2}T_{y\lambda\lambda}^0(\lambda - \lambda_0) + R_2(y, \lambda, \tau).
\end{aligned}
\tag{5.200}
$$

With respect to the ansatz (5.196), the remainders R_1 and R_2 satisfy

$$
R_1 = O(\xi^4) \quad \text{and} \quad R_2 = O(\xi^3).
$$

Inserting (5.196) and (5.199) into the first equation of (5.188), we obtain

$$
\begin{aligned}
0 = {}& \xi T_y^0 \left(-\frac{a_1}{a_2}\varphi_0 + \{K\varphi_0 + u_1\}\xi + u\xi^2 \right) + \xi^2 T_\tau^0 \frac{a_1^2}{2b_1 a_2} \\
&+ \xi^2 T_{y\lambda}^0 \left(-\frac{a_1}{a_2}\varphi_0 + \{K\varphi_0 + u_1\}\xi + u\xi^2 \right)(1 + \lambda_1\xi) \\
&+ \xi^3 T_{y\tau}^0 \left(-\frac{a_1}{a_2}\varphi_0 + \{K\varphi_0 + u_1\}\xi + u\xi^2 \right)\frac{a_1^2}{2b_1 a_2} \\
&+ \xi^2 \frac{1}{2} T_{yy}^0 \left(-\frac{a_1}{a_2}\varphi_0 + \{K\varphi_0 + u_1\}\xi + u\xi^2 \right)^2 \\
&+ \xi^3 T_{\tau\lambda}^0 \frac{a_1^2}{2b_1 a_2}(1 + \lambda_1\xi) &\text{(5.201)} \\
&+ \xi^3 \frac{1}{6} T_{yyy}^0 \left(-\frac{a_1}{a_2}\varphi_0 + \{K\varphi_0 + u_1\}\xi + u\xi^2 \right)^3 \\
&+ \xi^3 \frac{1}{2} T_{y\lambda\lambda}^0 \left(-\frac{a_1}{a_2}\varphi_0 + \{K\varphi_0 + u_1\}\xi + u\xi^2 \right)(1 + \lambda_1\xi)^2 \\
&+ \xi^3 \frac{1}{2} T_{yy\lambda}^0 \left(-\frac{a_1}{a_2}\varphi_0 + \{K\varphi_0 + u_1\}\xi + u\xi^2 \right)^2 (1 + \lambda_1\xi) \\
&+ R_1(y(\xi), \lambda(\xi), \tau(\xi)).
\end{aligned}
$$

Now, we consider the terms in ξ and ξ^2. We obtain:

- Terms in ξ

$$
-\frac{a_1}{a_2} T_y^0 \varphi_0 = 0.
$$

- Terms in ξ^2
  Due to (5.197), it holds

$$
T_y^0 u_1 + \frac{a_1^2}{2b_1 a_2} T_\tau^0 - \frac{a_1}{a_2} T_{y\lambda}^0 \varphi_0 + \frac{1}{2}\left(\frac{a_1}{a_2}\right)^2 T_{yy}^0 \varphi_0^2 = 0.
$$

Thus, the right-hand side of (5.201) is $O(\xi^3)$.

Next, we insert (5.196) and (5.200) into the second equation in (5.188). This yields

$$
\begin{aligned}
0 = {}& \xi T_y^0(\varphi_1 + \psi\xi) + \xi T_{y\lambda}^0(1 + \lambda_1\xi)(\varphi_0 + \varphi_1\xi + \psi\xi^2) \\
&+ \xi^2 T_{y\tau}^0 \frac{a_1^2}{2b_1 a_2}(\varphi_0 + \varphi_1\xi + \psi\xi^2) \\
&+ \xi T_{yy}^0 \left(-\frac{a_1}{a_2}\varphi_0 + \{K\varphi_0 + u_1\}\xi + u\xi^2 \right)(\varphi_0 + \varphi_1\xi + \psi\xi^2) &\text{(5.202)}
\end{aligned}
$$

$$+ \xi^2 T_{yy\lambda}^0 \left(-\frac{a_1}{a_2}\varphi_0 + \{K\varphi_0 + u_1\}\xi + u\xi^2 \right)(1 + \lambda_1\xi)(\varphi_0 + \varphi_1\xi + \psi\xi^2)$$

$$+ \xi^2 \frac{1}{2} T_{yyy}^0 \left(-\frac{a_1}{a_2}\varphi_0 + \{K\varphi_0 + u_1\}\xi + u\xi^2 \right)^2 (\varphi_0 + \varphi_1\xi + \psi\xi^2)$$

$$+ \xi^2 \frac{1}{2} T_{y\lambda\lambda}^0 (1 + \lambda_1\xi)^2 (\varphi_0 + \varphi_1\xi + \psi\xi^2)$$

$$+ R_2(y(\xi), \lambda(\xi), \tau(\xi))(\varphi_0 + \varphi_1\xi + \psi\xi^2)$$

Let us consider the terms in ξ. Due to (5.198), we obtain:

$$T_y^0 \varphi_1 + T_{y\lambda}^0 \varphi_0 - \frac{a_1}{a_2} T_{yy}^0 \varphi_0^2 = 0.$$

Thus, the right-hand side of (5.202) is $O(\xi^2)$. Dividing the equation (5.201) by ξ^3 and the equation (5.202) by ξ^2, we can write the resulting equations in a compact form as

$$\begin{pmatrix} T_y^0 & 0 & -\frac{a_1}{a_2}T_{y\lambda}^0\varphi_0 & -\frac{a_1}{a_2}T_{yy}^0\varphi_0^2 + T_{y\lambda}^0\varphi_0 \\ 0 & T_y^0 & T_{y\lambda}^0\varphi_0 & T_{yy}^0\varphi_0^2 \\ \varphi_0^* & 0 & 0 & 0 \\ 0 & \varphi_0^* & 0 & 0 \end{pmatrix} \begin{pmatrix} u(\xi) \\ \psi(\xi) \\ \lambda_1(\xi) \\ K(\xi) \end{pmatrix} = \begin{pmatrix} w_1 \\ w_2 \\ 0 \\ 0 \end{pmatrix}, \quad (5.203)$$

where

$$w_1(u, \lambda_1, K; \xi) \equiv$$

$$T_{y\lambda}^0 \{u_1 + u\xi + \lambda_1 w_3(u, K; \xi)\} + \frac{a_1^2}{2b_1 a_2} T_{y\tau}^0 w_4(u, K; \xi)$$

$$+ \frac{1}{2} T_{yy}^0 \left\{ -2\frac{1}{2}\varphi_0 u_1 + \left[u_1^2 - \frac{2a_1}{a_2}\varphi_0 u + 2u_1 u\xi + u^2\xi^2 - K^2\varphi_0^2 \right]\xi \right.$$

$$\left. + 2K\varphi_0 w_3(u, K; \xi) \right\}$$

$$+ \frac{a_1^2}{2b_1 a_2} T_{\tau\lambda}^0 \{1 + \lambda_1\xi\} + \frac{1}{6} T_{yyy}^0 w_4(u, K; \xi)^3$$

$$+ \frac{1}{2} T_{y\lambda\lambda}^0 \{1 + \lambda_1\xi\}^2 w_4(u, K; \xi) + \frac{1}{\xi^3} R_1(y, \lambda, \tau),$$

$$w_2(u, \psi, \lambda_1, K; \xi) \equiv$$

$$\left[\frac{a_1^2}{2b_1 a_2} T_{y\tau}^0 + T_{yy}^0 \{u_1 + u\xi\} + T_{yy\lambda}^0 w_4(u, K; \xi) + \lambda_1 T_{yy\lambda}^0 w_4(u, K; \xi)\xi \right.$$

$$\left. + \frac{1}{2} T_{yyy}^0 w_4(u, K; \xi)^2 + \frac{1}{2} T_{y\lambda\lambda}^0 \{1 + \lambda_1\xi\}^2 + \frac{1}{\xi^2} R_2(y, \lambda, \tau) \right] \times$$

$$\times \left[\varphi_0 + \varphi_1\xi + \psi\xi^2 \right] + \left[\lambda_1 T_{y\lambda}^0 + K T_{yy}^0 \varphi_0 \right] \left[\varphi_1\xi + \psi\xi^2 \right]$$

$$+ \left[T_{y\lambda}^0 - \frac{a_1}{a_2} T_{yy}^0 \varphi_0 \right] [\varphi_1 + \psi \xi],$$

$$w_3(u, K; \xi) \equiv (K\varphi_0 + u_1 + u\xi)\xi,$$

$$w_4(u, K; \xi) \equiv -\frac{a_1}{a_2}\varphi_0 + w_3(u, K; \xi).$$

The solvability properties of the equations (5.203) are stated in the next theorem.

Theorem 5.65 *Let the assumptions* (5.182)–(5.184) *and* $a_2 \equiv \psi_0^* T_{yy}^0 \varphi_0^2 \neq 0$ *be satisfied. Then, there exists a real constant* $\tau_0 > 0$ *such that for* $|\tau| \leq \tau_0$ *and* $\mathrm{sign}(\tau) = \mathrm{sign}(b_1 a_2)$ *problem* (5.203) *has a continuous curve of isolated solutions*

$$\eta(\xi) \equiv (u(\xi), \psi(\xi), \lambda_1(\xi), K(\xi))^T.$$

By inserting $\eta(\xi)$ *into the ansatz* (5.196) *it is possible to construct a continuous curve of solutions of the equations* (5.188). *This curve is composed of nonisolated solutions of* (5.181) *and passes through*

$$\tilde{z}_0 \equiv (0, \varphi_0, \lambda_0, 0) \in Y \times Y \times \mathbb{R} \times \mathbb{R}.$$

Proof Consider (5.203) for $\xi \to 0$, i.e., for $\tau \to 0$. We get

$$\mathbb{A} \begin{pmatrix} u(0) \\ \psi(0) \\ \lambda_1(0) \\ K(0) \end{pmatrix} = - \begin{pmatrix} w_1(u, \lambda_1, K; 0) \\ w_2(u, \psi, \lambda_1, K; 0) \\ 0 \\ 0 \end{pmatrix},$$

where $\mathbb{A}$ denotes the system matrix in (5.203). We set

$$A \equiv \begin{pmatrix} T_y^0 & 0 \\ 0 & T_y^0 \end{pmatrix}, \quad B \equiv \begin{pmatrix} -\dfrac{a_1}{a_2} T_{y\lambda}^0 \varphi_0 & -\dfrac{a_1}{a_2} T_{yy}^0 \varphi_0^2 + T_{y\lambda}^0 \varphi_0 \\ T_{y\lambda}^0 \varphi_0 & T_{yy}^0 \varphi_0^2 \end{pmatrix},$$

$$C^* \equiv \begin{pmatrix} \varphi_0^* & 0 \\ 0 & \varphi_0^* \end{pmatrix}, \quad D \equiv \begin{pmatrix} 0 & 0 \\ 0 & 0 \end{pmatrix}, \quad m \equiv 2,$$

and use Lemma 5.25. Obviously, A is not bijective and $\dim \mathcal{N}(A) = 2$, i.e., the second case applies, and we have to verify (c_0)–(c_3).

(c_0): $\dim \mathcal{R}(B) = 2$,

(c_1): For $l_1 \in Y$ and $l_2 \in \mathbb{R}$ we have

$$\begin{pmatrix} -\dfrac{a_1}{a_2} T_{y\lambda}^0 \varphi_0 l_1 + \left(-\dfrac{a_1}{a_2} T_{yy}^0 \varphi_0^2 + T_{y\lambda}^0 \varphi_0 \right) l_2 \\ T_{y\lambda}^0 \varphi_0 l_1 + T_{yy}^0 \varphi_0^2 l_2 \end{pmatrix} \in \mathcal{R}(B).$$

We know that $\begin{pmatrix} v_1 \\ v_2 \end{pmatrix} \in \mathcal{R}(A) \Leftrightarrow \psi_0^* v_1 = 0$ and $\psi_0^* v_2 = 0$.

Let us now consider the matrix B. We set

$$0 \doteq -\frac{a_1}{a_2}\psi_0^* T_{y\lambda}^0 \varphi_0 l_1 + \left(-\frac{a_1}{a_2}\psi_0^* T_{yy}^0 \varphi_0^2 + \psi_0^* T_{y\lambda}^0 \varphi_0\right) l_2,$$

$$0 \doteq \psi_0^* T_{y\lambda}^0 \varphi_0 l_1 + \psi_0^* T_{yy}^0 \varphi_0^2 l_2.$$

These two equations can be written in matrix-vector notation as

$$\begin{pmatrix} -\dfrac{a_1^2}{a_2} & 0 \\ a_1 & a_2 \end{pmatrix} \begin{pmatrix} l_1 \\ l_2 \end{pmatrix} = \begin{pmatrix} 0 \\ 0 \end{pmatrix}.$$

Since the determinant of the system matrix does not vanish, we obtain

$$l_1 = 0 \text{ and } l_2 = 0, \text{ i.e., } \mathcal{R}(A) \cap \mathcal{R}(B) = \begin{pmatrix} 0 \\ 0 \end{pmatrix}.$$

The conditions (c_2) & (c_3) are fulfilled trivially. Thus, we have shown that operator $\mathbb{A}$ is bijective. The claim of the theorem now follows analogously to the proofs of the transformation techniques presented in the previous sections. ∎

Let us now consider the case of a symmetric bifurcation point of the unperturbed problem, i.e., we assume

$$a_2 \equiv \psi_0^* T_{yy}(0, \lambda_0, 0)\varphi_0^2 = 0, \quad a_3 \equiv \psi_0^* \left(6T_{yy}^0 \varphi_0 w_0 + T_{yyy}^0 \varphi_0^3\right) \neq 0.$$

We use the following ansatz for the curve of nonisolated solutions of (5.181) passing through the particular point $\tilde{z}_0 \equiv (0, \varphi_0, \lambda_0, 0)$:

$$\begin{aligned}
y(\xi) &= \sqrt{\tilde{C}_2}\varphi_0\xi + \left\{\tilde{C}_2\tilde{u}_1 + K(\xi)\varphi_0\right\}\xi^2 \\
&\quad + \left\{u_2 + 2\sqrt{\tilde{C}_2}K(\xi)\tilde{u}_1\right\}\xi^3 + \left\{u(\xi) + K(\xi)^2\tilde{u}_1\right\}\xi^4, \\
\lambda(\xi) &= \lambda_0 + \xi^2 + \lambda_1(\xi)\xi^3, \\
\varphi(\xi) &= \varphi_0 + 2\sqrt{\tilde{C}_2}\tilde{u}_1 + \{\varphi_2 + 2K(\xi)\tilde{u}_1\}\xi^2 + \psi(\xi)\xi^3, \\
\xi^3 &= -\frac{3}{2}\frac{b_1}{\sqrt{\tilde{C}_2}a_1}\tau,
\end{aligned} \tag{5.204}$$

where $u, \psi \in Y_1$ and $\lambda_1, K \in \mathbb{R}$. The constant $\tilde{C}_2$ is defined as

$$\tilde{C}_2 \equiv -\frac{2a_1}{a_3}, \tag{5.205}$$

and we suppose that

$$\tilde{C}_2 > 0. \tag{5.206}$$

The elements $\tilde{u}_1$, u_2, and φ_2 are the (unique) solutions of the following linear problems

$$T_y^0 \tilde{u}_1 = -\frac{1}{2} T_{yy}^0 \varphi_0^2, \qquad\qquad \varphi_0^* \tilde{u}_1 = 0, \qquad (5.207)$$

$$T_y^0 u_2 = \frac{2}{3} \frac{\sqrt{\tilde{C}_2} a_1}{b_1} T_\tau^0 - \sqrt{\tilde{C}_2} T_{y\lambda}^0 \varphi_0$$
$$- \frac{1}{6} \tilde{C}_2 \sqrt{\tilde{C}_2} \left(T_{yyy}^0 \varphi_0^3 + 6 T_{yy}^0 \varphi_0 \tilde{u}_1 \right), \qquad \varphi_0^* u_2 = 0, \qquad (5.208)$$

$$T_y^0 \varphi_2 = - T_{y\lambda}^0 \varphi_0 - \frac{1}{2} \tilde{C}_2 \left(T_{yyy}^0 \varphi_0^3 + 6 T_{yy}^0 \varphi_0 \tilde{u}_1 \right), \qquad \varphi_0^* \varphi_2 = 0. \qquad (5.209)$$

Now, we expand T and T_y into Taylor series:

$$T(y, \lambda, \tau) = T_y^0 y + T_\tau^0 \tau + T_{y\lambda}^0 y(\lambda - \lambda_0) + T_{y\tau}^0 y\tau + \frac{1}{2} T_{yy}^0 y^2$$
$$+ \frac{1}{6} T_{yyy}^0 y^3 + \frac{1}{2} T_{yy\lambda}^0 y^2 (\lambda - \lambda_0) + \frac{1}{24} T_{yyyy}^0 y^4 \qquad (5.210)$$
$$+ R_3(y, \lambda, \tau),$$

and

$$T_y(y, \lambda, \tau) = T_y^0 + T_{y\lambda}^0 (\lambda - \lambda_0) + T_{y\tau}^0 \tau + T_{yy}^0 y + T_{yy\lambda}^0 y(\lambda - \lambda_0)$$
$$+ \frac{1}{2} T_{yyy}^0 y^2 + \frac{1}{6} T_{yyyy}^0 y^3 + R_4(y, \lambda, \tau). \qquad (5.211)$$

With respect to the ansatz (5.204), the remainders R_3 and R_4 satisfy

$$R_3 = O(\xi^5) \quad \text{and} \quad R_4 = O(\xi^4).$$

Inserting (5.204) and (5.210) into the first equation in (5.188), we obtain

$$0 = \xi T_y^0 \left(\sqrt{\tilde{C}_2} \varphi_0 + \alpha\xi + \beta\xi^2 + \gamma\xi^3 \right)$$
$$- \xi^3 \frac{2}{3} \frac{\sqrt{\tilde{C}_2} a_1}{b_1} T_\tau^0 + \xi^3 T_{y\lambda}^0 \left(\sqrt{\tilde{C}_2} \varphi_0 + \alpha\xi + \beta\xi^2 + \gamma\xi^3 \right)(1 + \lambda_1 \xi)$$
$$- \xi^4 T_{y\tau}^0 \frac{2}{3} \frac{\sqrt{\tilde{C}_2} a_1}{b_1} \left(\sqrt{\tilde{C}_2} \varphi_0 + \alpha\xi + \beta\xi^2 + \gamma\xi^3 \right)$$
$$+ \xi^2 \frac{1}{2} T_{yy}^0 \left(\sqrt{\tilde{C}_2} \varphi_0 + \alpha\xi + \beta\xi^2 + \gamma\xi^3 \right)^2 \qquad (5.212)$$
$$+ \xi^3 \frac{1}{6} T_{yyy}^0 \left(\sqrt{\tilde{C}_2} \varphi_0 + \alpha\xi + \beta\xi^2 + \gamma\xi^3 \right)^3$$

$$+ \xi^4 \frac{1}{2} T_{yy\lambda}^0 \left(\sqrt{\tilde{C}_2} \varphi_0 + \alpha \xi + \beta \xi^2 + \gamma \xi^3 \right)^2 (1 + \lambda_1 \xi)$$

$$+ \xi^4 \frac{1}{24} \left(\sqrt{\tilde{C}_2} \varphi_0 + \alpha \xi + \beta \xi^2 + \gamma \xi^3 \right)^4 + R_3,$$

where we have used the abbreviations

$$\alpha \equiv \tilde{C}_2 \tilde{u}_1 + K(\xi) \varphi_0, \quad \beta \equiv u_2 + 2\sqrt{\tilde{C}_2} K(\xi) \tilde{u}_1,$$
$$\gamma \equiv u(\xi) + K(\xi)^2 \tilde{u}_1.$$

Now, we consider the terms in ξ, ξ^2 and ξ^3. We obtain:

- Terms in ξ

$$\sqrt{\tilde{C}_2} T_y^0 \varphi_0 = 0.$$

- Terms in ξ^2
  Due to (5.207), it holds

$$\tilde{C}_2 \left(T_y^0 \tilde{u}_1 + \frac{1}{2} T_{yy}^0 \varphi_0^2 \right) = 0.$$

- Terms in ξ^3
  Due to (5.207) and (5.208), we get

$$T_y^0 u_2 + 2\sqrt{\tilde{C}_2} K T_y^0 \tilde{u}_1 - \frac{2}{3} \frac{\sqrt{\tilde{C}_2} a_1}{b_1} T_\tau^0 + T_{y\lambda}^0 \sqrt{\tilde{C}_2} \varphi_0$$

$$+ \frac{1}{2} T_{yy}^0 \left(2\sqrt{\tilde{C}_2} \varphi_0 \{\tilde{C}_2 \tilde{u}_1 + K \varphi_0\} \right) + \sqrt{\tilde{C}_2} \tilde{C}_2 \frac{1}{6} T_{yyy}^0 \varphi_0^3$$

$$= 2\sqrt{\tilde{C}_2} K \left(T_y^0 \tilde{u}_1 + \frac{1}{2} T_{yy}^0 \varphi_0^2 \right) + T_y^0 u_2 - \frac{2}{3} \frac{\sqrt{\tilde{C}_2} a_1}{b_1} T_\tau^0$$

$$+ \sqrt{\tilde{C}_2} T_{y\lambda}^0 \varphi_0 + \frac{1}{6} \sqrt{\tilde{C}_2} \tilde{C}_2 \left(T_{yyy}^0 \varphi_0^3 + 6 T_{yy}^0 \varphi_0 \tilde{u}_1 \right)$$

$$= 0.$$

Thus, the right-hand side of (5.212) is $O(\xi^4)$.

Next, we insert (5.204) and (5.211) into the second equation of (5.188). This yields

$$0 = T_y^0 \left(\varphi_0 + 2\sqrt{\tilde{C}_2}\tilde{u}_1\xi + \delta\xi^2 + \psi\xi^3 \right)$$

$$+ \xi^2 T_{y\lambda}^0 (1 + \lambda_1\xi) \left(\varphi_0 + 2\sqrt{\tilde{C}_2}\tilde{u}_1\xi + \delta\xi^2 + \psi\xi^3 \right)$$

$$- \xi^3 \frac{2}{3}\frac{\sqrt{\tilde{C}_2}a_1}{b_1} T_{y\tau}^0 \left(\varphi_0 + 2\sqrt{\tilde{C}_2}\tilde{u}_1\xi + \delta\xi^2 + \psi\xi^3 \right)$$

$$+ \xi T_{yy}^0 \left(\sqrt{\tilde{C}_2}\varphi_0 + \alpha\xi + \beta\xi^2 + \gamma\xi^3 \right) \left(\varphi_0 + 2\sqrt{\tilde{C}_2}\tilde{u}_1\xi + \delta\xi^2 + \psi\xi^3 \right)$$

$$+ \xi^3 (1 + \lambda_1\xi) \left(\sqrt{\tilde{C}_2}\varphi_0 + \alpha\xi + \beta\xi^2 + \gamma\xi^3 \right) \times \tag{5.213}$$

$$\times \left(\varphi_0 + 2\sqrt{\tilde{C}_2}\tilde{u}_1\xi + \delta\xi^2 + \psi\xi^3 \right)$$

$$+ \xi^2 \frac{1}{2} T_{yyy}^0 \left(\sqrt{\tilde{C}_2}\varphi_0 + \alpha\xi + \beta\xi^2 + \gamma\xi^3 \right)^2 \times$$

$$\times \left(\varphi_0 + 2\sqrt{\tilde{C}_2}\tilde{u}_1\xi + \delta\xi^2 + \psi\xi^3 \right)$$

$$+ \xi^3 \frac{1}{6} T_{yyyy}^0 \left(\sqrt{\tilde{C}_2}\varphi_0 + \alpha\xi + \beta\xi^2 + \gamma\xi^3 \right)^3 \times$$

$$\times \left(\varphi_0 + 2\sqrt{\tilde{C}_2}\tilde{u}_1\xi + \delta\xi^2 + \psi\xi^3 \right)$$

$$+ R_5 \left(\varphi_0 + 2\sqrt{\tilde{C}_2}\tilde{u}_1\xi + \delta\xi^2 + \psi\xi^3 \right),$$

where in addition we have used the abbreviation

$$\delta \equiv \varphi_2 + 2K(\xi)\tilde{u}_1.$$

As before, we consider the terms in ξ and ξ^2. We obtain:

- Terms in ξ
  Due to (5.207), it holds

$$2\sqrt{\tilde{C}_2} \left(T_y^0 \tilde{u}_1 + \frac{1}{2} T_{yy}^0 \varphi_0^2 \right) = 0.$$

- Terms in ξ^2
  Due to (5.207) and (5.209), we get

$$T_y^0 \varphi_2 + 2K T_y^0 \tilde{u}_1 + T_{y\lambda}^0 \varphi_0 + T_{yy}^0 \left(\tilde{C}_2 \tilde{u}_1 \varphi_0 + K \varphi_0^2 \right)$$

$$+ T_{yy}^0 \left(2\sqrt{\tilde{C}_2} \varphi_0 \sqrt{\tilde{C}_2} \tilde{u}_1 \right) + \frac{1}{2} T_{yyy}^0 \sqrt{\tilde{C}_2} \varphi_0^3$$

$$= T_y^0 \varphi_2 + 2K \left(T_y^0 \tilde{u}_1 + \frac{1}{2} T_{yy}^0 \varphi_0^2 \right) + T_{y\lambda}^0 \varphi_0$$

$$+ \frac{1}{2} \tilde{C}_2 \left(T_{yyy}^0 \varphi_0^3 + 6 T_{yy}^0 \tilde{u}_1 \varphi_0 \right)$$

$$= 0.$$

Thus, the right-hand side of (5.213) is $O(\xi^3)$. Dividing the equation (5.212) by ξ^4 and the equation (5.213) by ξ^3, we can write the resulting equations in a compact form as

$$\mathbb{A} \begin{pmatrix} v(\xi) \\ \psi(\xi) \\ \lambda_1(\xi) \\ K(\xi) \end{pmatrix} = - \begin{pmatrix} w_1(v, \lambda_1, K; \xi) \\ w_2(v, \psi, \lambda_1, K; \xi) \\ 0 \\ 0 \end{pmatrix}, \tag{5.214}$$

where

$$\mathbb{A} \equiv \begin{pmatrix} T_y^0 & 0 & \sqrt{\tilde{C}_2} T_{y\lambda}^0 \varphi_0 & \frac{1}{2} \left(2 T_{y\lambda}^0 \varphi_0 + 6 \tilde{C}_2 T_{yy}^0 \varphi_0 \tilde{u}_1 + \tilde{C}_2 T_{yyy}^0 \varphi_0^3 \right) \\ 0 & T_y^0 & T_{y\lambda}^0 \varphi_0 & \sqrt{\tilde{C}_2} \left(6 T_{yy}^0 \varphi_0 \tilde{u}_1 + T_{yyy}^0 \varphi_0^3 \right) \\ \varphi_0^* & 0 & 0 & 0 \\ 0 & \varphi_0^* & 0 & 0 \end{pmatrix},$$

and

$$w_1(v, \lambda_1, K; \xi) = T_{y\lambda}^0 \left(w_3(v, K; \xi) + \lambda_1 w_7(v, K; \xi)\xi \right)$$

$$+ T_{yy}^0 \left(\frac{1}{2} w_3(v, K; \xi)^3 + \sqrt{\tilde{C}_2} \varphi_0 (u_2 + \{v + K^2 \tilde{u}_1\}\xi) \right.$$

$$\left. + K \varphi_0 \{ w_3(v, K; \xi) - \tilde{C}_2 \tilde{u}_1 \} \right)$$

$$+ \frac{1}{6} T_{yyy}^0 \left(3 \tilde{C}_2 \varphi_0^2 w_3(v, K; \xi) + 3\sqrt{\tilde{C}_2} \varphi_0 w_7(v, K; \xi)^2 \xi \right.$$

$$\left. + w_7(v, K; \xi)^3 \xi^2 \right)$$

$$- \frac{2}{3} \frac{\sqrt{\tilde{C}_2} a_1}{b_1} T_{y\tau}^0 w_4(v, K; \xi)$$

$$+ \frac{1}{2} T_{yy\lambda}^0 \{ 1 + \lambda_1 \xi \} w_4(v, K; \xi)^2$$

$$+ \frac{1}{24} T^0_{yyyy} w_4(v, K; \xi)^4 + \frac{1}{\xi^4} R_3(y, \lambda, \tau),$$

$$
\begin{aligned}
w_2(v, \psi, \lambda_1, K; \xi) = \Bigg[&-\frac{2}{3} \frac{\sqrt{\tilde{C}_2} a_1}{b_1} T^0_{y\tau} + T^0_{yy}(u_2 + \{v + K^2 \tilde{u}_1\} \xi) \\
&+ T^0_{yy\lambda}\{1 + \lambda_1 \xi\} w_4(v, K; \xi) \\
&+ \frac{1}{2} T^0_{yyy}\left(2\sqrt{\tilde{C}_2}\varphi_0 w_3(v, K; \xi) + w_7(v, K; \xi)^2 \xi \right) \\
&+ \frac{1}{6} T^0_{yyyy} w_4(v, K; \xi)^3 + \frac{1}{\xi^3} R_4(y, \lambda, \tau) \Bigg] w_6(\psi, K; \xi) \\
&+ \Bigg[T^0_{y\lambda}\{1 + \lambda_1 \xi\} + T^0_{yy}\tilde{C}_2 \tilde{u}_1 \\
&+ \frac{1}{2} T^0_{yyy}\left(\tilde{C}_2 \varphi_0^2 + 2\sqrt{\tilde{C}_2} K \varphi_0^2 \xi \right) \Bigg] w_5(\psi, K; \xi) \\
&+ T^0_{yy}\Bigg(\sqrt{\tilde{C}_2}\varphi_0\{\varphi_2 + \psi\xi\} \\
&\qquad\qquad + K\varphi_0\{\varphi_2 + 2K\tilde{u}_1 + \psi\xi\}\xi \Bigg),
\end{aligned}
$$

$$w_3(v, K; \xi) \equiv \tilde{C}_2 \tilde{u}_1 + \left\{ u_2 + 2\sqrt{\tilde{C}_2} K \tilde{u}_1 \right\} \xi + \{v + K^2 \tilde{u}_1\}\xi^2,$$

$$w_4(v, K; \xi) \equiv \sqrt{\tilde{C}_2}\varphi_0 + K\varphi_0\xi + w_3(v, K; \xi)\xi,$$

$$w_5(\psi, K; \xi) \equiv 2\sqrt{\tilde{C}_2}\tilde{u}_1 + \{\varphi_2 + 2K\tilde{u}_1\}\xi + \psi\xi^2,$$

$$w_6(\psi, K; \xi) \equiv \varphi_0 + w_5(\psi, K; \xi)\xi,$$

$$w_7(v, K; \xi) \equiv K\varphi_0 + w_3(v, K; \xi).$$

The solvability properties of the equations (5.214) are stated in the next theorem.

Theorem 5.66 *Let the assumptions (5.182)–(5.184), $a_2 = 0$, $a_3 \neq 0$ and $\tilde{C}_2 > 0$ be satisfied. Then, there exists a real constant $\tau_0 > 0$ such that for $|\tau| \leq \tau_0$ problem (5.214) has a continuous curve of isolated solutions*

$$\eta(\xi) \equiv (u(\xi), \psi(\xi), \lambda_1(\xi), K(\xi))^T.$$

By inserting $\eta(\xi)$ into the ansatz (5.204) it is possible to construct a continuous curve of solutions of the equations (5.188). This curve is composed of nonisolated solutions of (5.181) and passes through

$$\tilde{z}_0 \equiv (0, \varphi_0, \lambda_0, 0) \in Y \times Y \times \mathbb{R} \times \mathbb{R}.$$

Proof Consider (5.214) for $\xi \to 0$, i.e., for $\tau \to 0$. We obtain

$$A \begin{pmatrix} u(0) \\ \psi(0) \\ \lambda_1(0) \\ K(0) \end{pmatrix} = - \begin{pmatrix} w_1(u, \lambda_1, K; 0) \\ w_2(u, \psi, \lambda_1, K; 0) \\ 0 \\ 0 \end{pmatrix}.$$

We set

$$A \equiv \begin{pmatrix} T_y^0 & 0 \\ 0 & T_y^0 \end{pmatrix},$$

$$B \equiv \begin{pmatrix} \sqrt{\tilde{C}_2} T_{y\lambda}^0 \varphi_0 & \frac{1}{2}\left(2T_{y\lambda}^0 \varphi_0 + 6\tilde{C}_2 T_{yy}^0 \varphi_0 \tilde{u}_1 + \tilde{C}_2 T_{yyy}^0 \varphi_0^3\right) \\ T_{y\lambda}^0 \varphi_0 & \sqrt{\tilde{C}_2}\left(6T_{yy}^0 \varphi_0 \tilde{u}_1 + T_{yyy}^0 \varphi_0^3\right) \end{pmatrix},$$

$$C^* \equiv \begin{pmatrix} \varphi_0^* & 0 \\ 0 & \varphi_0^* \end{pmatrix}, \quad D \equiv \begin{pmatrix} 0 & 0 \\ 0 & 0 \end{pmatrix}, \quad m = 2,$$

and use Lemma 5.25. Obviously, A is not bijective and $\dim \mathcal{N}(A) = 2$, i.e., the second case applies, and we have to verify (c_0)–(c_3).

(c_0): $\dim \mathcal{R}(B) = 2$,

(c_1):   For $l_1 \in Y$ and $l_2 \in \mathbb{R}$, the vector

$$\begin{pmatrix} \sqrt{\tilde{C}_2} T_{y\lambda}^0 \varphi_0 l_1 + \frac{1}{2}\left(2T_{y\lambda}^0 \varphi_0 + 6\tilde{C}_2 T_{yy}^0 \varphi_0 \tilde{u}_1 + \tilde{C}_2 T_{yyy}^0 \varphi_0^3\right) l_2 \\ T_{y\lambda}^0 \varphi_0 l_1 + \sqrt{\tilde{C}_2}\left(6T_{yy}^0 \varphi_0 \tilde{u}_1 + T_{yyy}^0 \varphi_0^3\right) l_2 \end{pmatrix}$$

is in $\mathcal{R}(B)$.

We know that $\begin{pmatrix} v_1 \\ v_2 \end{pmatrix} \in \mathcal{R}(A) \Leftrightarrow \psi_0^* v_1 = 0$ and $\psi_0^* v_2 = 0$.

Let us now consider the matrix B. We set

$$0 \doteq \sqrt{\tilde{C}_2} \psi_0^* T_{y\lambda}^0 \varphi_0 l_1$$
$$+ \frac{1}{2}\left(2\psi_0^* T_{y\lambda}^0 \varphi_0 + 6\tilde{C}_2 \psi_0^* T_{yy}^0 \varphi_0 \tilde{u}_1 + \tilde{C}_2 \psi_0^* T_{yyy}^0 \varphi_0^3\right) l_2,$$

$$0 \doteq \psi_0^* T_{y\lambda}^0 \varphi_0 l_1 + \sqrt{\tilde{C}_2}\left(6\psi_0^* T_{yy}^0 \varphi_0 \tilde{u}_1 + \psi_0^* T_{yyy}^0 \varphi_0^3\right) l_2.$$

These two equations can be written in matrix-vector notation as

$$\begin{pmatrix} \sqrt{\tilde{C}_2} a_1 & \frac{1}{2}\left(2a_1 + \tilde{C}_2 a_3\right) \\ a_1 & \sqrt{\tilde{C}_2} a_3 \end{pmatrix} \begin{pmatrix} l_1 \\ l_2 \end{pmatrix} = \begin{pmatrix} 0 \\ 0 \end{pmatrix}.$$

Since the determinant of the system matrix does not vanish, we obtain

$$l_1 = 0 \text{ and } l_2 = 0, \text{ i.e., } \mathcal{R}(A) \cap \mathcal{R}(B) = \begin{pmatrix} 0 \\ 0 \end{pmatrix}.$$

The conditions (c_2) & (c_3) are fulfilled trivially. Thus, we have shown that operator A is bijective. The claim of the theorem now follows analogously to the proofs of the transformation techniques presented in the previous sections. ∎

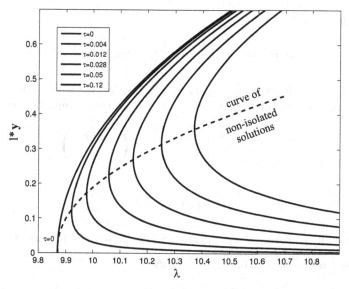

Fig. 5.22 Curves of nontrivial solutions passing through nonisolated solutions of the perturbed problem

5.6.3 Solution Curves Through Nonisolated Solutions

Let $\bar{\tau}$ be a fixed value of the perturbation parameter. It is now our goal to determine a curve of nontrivial solutions of the operator equation (5.181), which passes through the corresponding nonisolated solution $(\bar{y}, \bar{\varphi}, \bar{\lambda}, \bar{\tau})$, see Fig. 5.22.

For this solution curve, we use the following ansatz

$$
\begin{aligned}
y(\delta) &= \bar{y} + \delta\bar{\varphi} + \delta^2 w(\delta), \quad w \in \bar{X}_1, \\
\lambda(\delta) &= \bar{\lambda} + \delta^2 \rho(\delta), \quad \rho \in \mathbb{R}.
\end{aligned}
\tag{5.215}
$$

At $\bar{z} \equiv (\bar{y}, \bar{\lambda}, \bar{\tau})$ we expand T into a Taylor series:

$$
\begin{aligned}
T(y, \lambda, \bar{\tau}) = {}& T(\bar{z}) + T_y(\bar{z})(y - \bar{y}) + T_\lambda(\bar{z})(\lambda - \bar{\lambda}) \\
& + \frac{1}{2} T_{yy}(\bar{z})(y - \bar{y})^2 + R_5(y, \lambda, \bar{\tau}).
\end{aligned}
\tag{5.216}
$$

With respect to the ansatz (5.215), the remainder R_5 satisfies $R_5 = O(\delta^3)$.

Inserting (5.215) and (5.216) into the equation

$$
T(y, \lambda, \bar{\tau}) = 0,
\tag{5.217}
$$

we obtain

$$0 = \delta T_y(\bar{z})(\bar{\varphi} + \delta w) + T_\lambda(\bar{z})\delta^2\rho + \frac{1}{2}T_{yy}(\bar{z})(\bar{\varphi} + \delta w)^2 + R_5(y, \lambda, \bar{\tau}).$$

Since $T_y(\bar{z})\bar{\varphi} = 0$, the right-hand side is $O(\delta^2)$. Dividing the equation by δ^2, we can write the resulting equation and the equation $\bar{\varphi}^*w = 0$ in a compact notation as

$$\begin{pmatrix} T_y(\bar{z}) & T_\lambda(\bar{z}) \\ \bar{\varphi}^* & 0 \end{pmatrix} \begin{pmatrix} w(\delta) \\ \rho(\delta) \end{pmatrix} = - \begin{pmatrix} g(w, \rho; \delta) \\ 0 \end{pmatrix}, \qquad (5.218)$$

where

$$g(w, \rho; \delta) \equiv \frac{1}{2}T_{yy}(\bar{z})(\bar{\varphi} + \delta w)^2 + \frac{1}{\delta^2}R_5.$$

The solvability properties of the equation (5.218) are stated in the next theorem.

Theorem 5.67 *Let the operator $T_y(\bar{z})$ satisfy*

$$\dim \mathcal{N}(T_y(\bar{z})) = 1, \quad \mathcal{N}(T_y(\bar{z})) = \text{span}\{\bar{\varphi}\}, \quad \|\bar{\varphi}\| = 1.$$

Furthermore, assume that

$$b_2 \equiv \bar{\psi}^*T_\lambda(\bar{z}) \neq 0, \qquad (5.219)$$

where $\bar{\psi}^*$ is defined by

$$\mathcal{N}(T_y^*(\bar{z})) = \text{span}\{\bar{\psi}^*\}, \quad \|\bar{\psi}^*\| = 1.$$

Then, there exists a real constant $\delta_0 > 0$ such that for $|\delta| \leq \delta_0$ problem (5.218) has a continuous curve of isolated solutions

$$\eta(\delta) \equiv (w(\delta), \rho(\delta))^T.$$

By inserting $\eta(\delta)$ into the ansatz (5.215) it is possible to construct a continuous curve of solutions of the equation (5.217), which contains the nonisolated solution $\bar{z}$.

Proof Consider (5.218) for $\delta \to 0$. We obtain

$$\mathbb{A}\begin{pmatrix} w(0) \\ \rho(0) \end{pmatrix} \equiv \begin{pmatrix} T_y(\bar{z}) & T_\lambda(\bar{z}) \\ \bar{\varphi}^* & 0 \end{pmatrix} \begin{pmatrix} w(0) \\ \rho(0) \end{pmatrix} = - \begin{pmatrix} \frac{1}{2}T_{yy}(\bar{z})\bar{\varphi}^2 \\ 0 \end{pmatrix}.$$

We set

$$A \equiv T_y(\bar{z}), \quad B \equiv T_\lambda(\bar{z}),$$
$$C^* \equiv \bar{\varphi}^*, \quad D \equiv 0, \quad m = 1,$$

and use Lemma 5.25. Obviously, A is not bijective and $\dim \mathcal{N}(A) = 1$, i.e., the second case applies, and we have to verify (c_0)–(c_3).

(c_0): $\dim \mathcal{R}(B) = 1$,
(c_1): It holds $v \in \mathcal{R}(A) \Leftrightarrow \bar{\psi}^* v = 0$.
    We have $T_\lambda(\bar{z})\alpha \in \mathcal{R}(B)$, and because of the assumption (5.219),
    $\bar{\psi}^* T_\lambda(\bar{z})\alpha$ vanishes if and only if $\alpha = 0$.
    Therefore, $\mathcal{R}(A) \cap \mathcal{R}(B) = \{0\}$.
(c_2) & (c_3) are fulfilled trivially.

Thus, we have shown that the operator $\mathbb{A}$ is bijective. The claim of the theorem now follows analogously to the proofs of the transformation techniques presented in the previous sections. ∎

5.7 Path-Following Methods for Simple Solution Curves

5.7.1 Tangent Predictor Methods

In the previous sections, we have presented the analytical and numerical methods for operator equations of the form (5.11). The reason for this general approach is that the extension and transformation techniques are not restricted to ODEs. If the operator T is defined appropriately, these general techniques can also be used to solve, e.g., parametrized nonlinear algebraic equations, parametrized nonlinear integral equations and parametrized nonlinear partial differential equations.

However, in this section we want to deal exclusively with BVPs of the form (5.10). As before, let $E(f)$ be the solution field of the BVP.

Assume that a solution $z^{(1)} \equiv (y^{(1)}, \lambda^{(1)})$ of the BVP (5.10) has been determined. Now, the *path-following problem* consists in the computation of further solutions

$$(y^{(2)}, \lambda^{(2)}), (y^{(3)}, \lambda^{(3)}), \ldots$$

on the corresponding solution branch until a certain endpoint $\lambda = \lambda_b$ is reached.

The kth step of the path-following method starts with a known solution $(y^{(k)}, \lambda^{(k)}) \in E(f)$, and for the next value $\lambda^{(k+1)}$ of the parameter λ it is attempted to determine a solution $(y^{(k+1)}, \lambda^{(k+1)}) \in E(f)$:

$$(y^{(k)}, \lambda^{(k)}) \quad \rightarrow \quad (y^{(k+1)}, \lambda^{(k+1)}).$$

When the so-called *predictor–corrector methods* are used, the step $k \to k + 1$ is realized in two partial steps (see Fig. 5.23):

$$(y^{(k)}, \lambda^{(k)}) \xrightarrow{\text{predictor step}} (\bar{y}^{(k+1)}, \bar{\lambda}^{(k+1)}) \xrightarrow{\text{corrector step}} (y^{(k+1)}, \lambda^{(k+1)}).$$

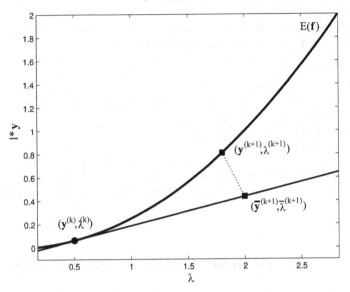

Fig. 5.23 Predictor-corrector method with a tangent predictor

In general, the result $(\bar{y}, \bar{\lambda})$ of the predictor is not a solution of (5.10). Rather, the predictor step provides sufficiently accurate starting values for the subsequent *corrector iteration*, so that this iteration converges. The special predictor-corrector methods can be subdivided into two classes. For the first class, the most work is spend to determine a predictor, which is near to the solution branch. Thus, the corrector requires only a few iteration steps to generate a sufficiently exact approximation of the solution. The members of the other class use an almost globally convergent corrector iteration. Therefore, the starting values can be quite inaccurate, i.e., the result of the predictor is generated with a reduced amount of work.

In the following, the distance between two successive solutions $(y^{(k)}, \lambda^{(k)})$ and $(y^{(k+1)}, \lambda^{(k+1)})$ is referred to as *step length*. As we will see later, a relation must be added to the BVP (5.10), which determines the position of a solution on the solution branch. This relation depends on the special strategy of the parametrization, which is used to follow the branch.

The path-following methods can be distinguished, among other things, by the following key elements:

1. the predictor,
2. the corrector,
3. the strategy of parametrization, and
4. the control of the step length.

Elements 1–3 can be varied independently of each other. However, the control of the step length must be consistent with the predictor, the corrector, and the parametrization.

In the following, one of the simplest path-following methods is described, which is based on the *tangent predictor*. Let us assume that a solution $(y^{(k)}, \lambda^{(k)})$ has already been calculated. The next step consists in the determination of the tangent $(v^{(k)}, \eta^{(k)})$ to the solution curve at $(y^{(k)}, \lambda^{(k)})$. This can be done by solving the linearization of the ODE (5.11). To simplify the representation, we refer to the operator form (5.11) of the BVP and write the tangent equation as

$$T_y^{(k)} v^{(k)} + T_\lambda^{(k)} \eta^{(k)} = 0, \tag{5.220}$$

with $\|(v^{(k)}, \eta^{(k)})\| = 1$. On this tangent, a predictor point

$$(y^{(k+1,0)}, \lambda^{(k+1,0)}) \equiv (y^{(k)}, \lambda^{(k)}) + \tau^{(k)} (v^{(k)}, \eta^{(k)}) \tag{5.221}$$

is determined using the step length $\tau^{(k)}$. This predictor point is used as a starting approximation for the subsequent corrector iteration, which is applied to the nonlinear system of equations

$$\begin{aligned}
T(y, \lambda) &= 0, \\
l^{(k)}(y, \lambda) - l^{(k)}(y^{(k+1,0)}, \lambda^{(k+1,0)}) &= 0
\end{aligned} \tag{5.222}$$

to compute the next solution $(y^{(k+1)}, \lambda^{(k+1)}) \in E(f)$. Here, $l^{(k)}$ is a suitably chosen functional that characterizes the type of parametrization.

In Algorithm 2.1 the above path-following technique is described in more detail.

Algorithm 2.1 Path-following with a tangent predictor

Step 1: Choose $\tau_{\max} > 0$, $\tau^{(1)} \in (0, \tau_{\max})$, $\alpha \in (0, 1)$, $\eta^{(0)} \in (-1, 1)$;
Give $(y^{(1)}, \lambda^{(1)}) \in E(f)$, $l_0 \in \{\pm l^{(0)}, \pm l_a^{(1)}, \ldots, \pm l_a^{(n)}\}$,
with $l^{(0)} \equiv \eta^{(0)}$, $l_a^{(i)} \equiv v_i(a)$, $i = 1, \ldots, n$;
Set $k := 1$;

Step 2: If $l_{k-1} = \pm l^{(0)}$ then determine $v^{(k)}$ as the solution of the following BVP
$$v'(x) = f_y(x, y^{(k)}; \lambda^{(k)}) v(x) + \eta^{(k-1)} f_\lambda(x, y^{(k)}; \lambda^{(k)}),$$
$$B_a v(a) + B_b v(b) = 0,$$
and set $\eta^{(k)} := \eta^{(k-1)}$;
If $l_{k-1} = \pm l_a^{(i)}$, $i \in \{1, \ldots, n\}$, then determine $(v^{(k)}, \eta^{(k)})$ as the solution of the following BVP
$$v'(x) = f_y(x, y^{(k)}; \lambda^{(k)}) v(x) + \eta(x) f_\lambda(x, y^{(k)}; \lambda^{(k)}),$$
$$\eta'(x) = 0,$$
$$B_a v(a) + B_b v(b) = 0,$$
$$v_i(a) = \pm 1;$$

Step 3 : Determine l_k according to the following procedure:

    (a) $kmax := \max\{|v_1(a)|, \ldots, |v_n(a)|, |\eta|\}$,

    (b) If $kmax = |\eta|$ then $l_k := \text{sign}(\eta)l^{(0)}$,

    (c) If $kmax = |v_i(a)|$ then $l_k := \text{sign}(v_i(a))l_a^{(i)}$,

    (d) If $l_k \neq l_{k-1}$ then $(\boldsymbol{v}^{(k)}, \eta^{(k)}) := \dfrac{1}{kmax}(\boldsymbol{v}^{(k)}, \eta^{(k)})$;

Step 4 : Determine the starting approximations for the computation of a
new point on the solution curve:

$$\boldsymbol{y}^{(k+1,0)} := \boldsymbol{y}^{(k)} + \tau^{(k)}\boldsymbol{v}^{(k)},$$
$$\lambda^{(k+1,0)} := \lambda^{(k)} + \tau^{(k)}\eta^{(k)};$$

Step 5 : If $l_k = \pm l^{(0)}$ then determine $\boldsymbol{y}^{(k+1)}$ as the solution of the
following BVP

$$\boldsymbol{y}'(x) = \boldsymbol{f}(x, \boldsymbol{y}; \lambda^{(k+1,0)}),$$
$$B_a \boldsymbol{y}(a) + B_b \boldsymbol{y}(b) = \boldsymbol{0},$$

and set $\lambda^{(k+1)} := \lambda^{(k+1,0)}$;

If $l_k = \pm l_a^{(i)}$ then determine $(\boldsymbol{y}^{(k+1)}, \lambda^{(k+1)})$ as the solution
of the following BVP

$$\boldsymbol{y}'(x) = \boldsymbol{f}(x, \boldsymbol{y}; \lambda),$$
$$\lambda'(x) = 0,$$
$$B_a \boldsymbol{y}(a) + B_b \boldsymbol{y}(b) = \boldsymbol{0},$$
$$y_i(a) = y_i^{(k+1,0)}(a);$$

Step 6 : If the new point $(\boldsymbol{y}^{(k+1)}, \lambda^{(k+1)})$ could not be determined
in Step 5 then set

$$\tau^{(k)} := \alpha\tau^{(k)},$$

and go to Step 2.

If the new point has been computed in Step 5 then choose

$$\tau^{(k+1)} \in [\tau^{(k)}, \tau_{\max}],$$

set

$$k := k + 1,$$

and go to Step 2.

Possible termination criteria for this algorithm are:

- a boundary of the interesting parameter interval $[\lambda_a, \lambda_e]$ has been reached,
- the maximum number of points on the solution curve has been computed,

- if the BVPs are solved by a shooting method, the condition of the associated systems of linear algebraic equations is too bad,
- the step length $\tau^{(k)}$ is too small.

5.7.2 Arclength Continuation

Now, we come to another variant of the path-following method, the so-called *arclength continuation*. The idea behind this path-following method is that a simple solution curve can be parametrized by the arclength, i.e.,

$$y = y(s), \quad \lambda = \lambda(s). \tag{5.223}$$

Let us assume that for a given $s^{(k)}$ the corresponding solution $(y^{(k)}, \lambda^{(k)}) \in E(f)$ is known. Then,

$$(\dot{y}^{(k)}, \dot{\lambda}^{(k)}) \equiv \left(\frac{dy}{ds}(s^{(k)}), \frac{d\lambda}{ds}(s^{(k)}) \right)$$

represents the direction of the solution curve at $(y^{(k)}, \lambda^{(k)})$.

To improve the clarity of the presentation, we come back to the operator form (5.11) of the BVP, and insert the parametrization (5.223) into the operator equation. This yields

$$T(y(s), \lambda(s)) = 0.$$

Differentiating the above identity with respect to s, we obtain

$$T_y(y(s), \lambda(s)) \dot{y}(s) + T_\lambda(y(s), \lambda(s)) \dot{\lambda}(s) = 0. \tag{5.224}$$

Setting $s = s^{(k)}$, it follows

$$0 = T_y(y^{(k)}, \lambda^{(k)}) \dot{y}^{(k)} + \dot{\lambda}^{(k)} T_\lambda(y^{(k)}, \lambda^{(k)}). \tag{5.225}$$

Adding the normalization condition

$$\| \dot{y}^{(k)} \|^2 + \left(\dot{\lambda}^{(k)} \right)^2 = 1, \tag{5.226}$$

the point $(\dot{y}^{(k)}, \dot{\lambda}^{(k)})$ is uniquely determined up to sign by (5.225) and (5.226).

Comparing (5.225) with (5.220), we see that we have computed the tangent on the solution curve.

Now, the new point $(y^{(k+1)}, \lambda^{(k+1)}) = (y(s^{(k+1)}), \lambda(s^{(k+1)}))$ on the solution curve is determined as the solution of the following system of equations

Fig. 5.24 The plane, which
is defined by (5.228)

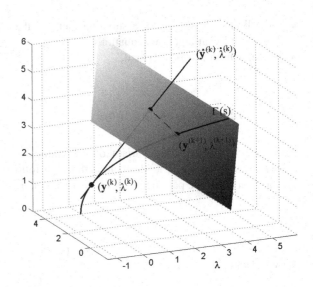

$$T(y(s), \lambda(s)) = 0,$$
$$N(y(s), \lambda(s), s) = 0,$$
(5.227)

where

$$N(y(s), \lambda(s), s) \equiv \varphi_k^*(y(s) - y^{(k)}) + \dot{\lambda}^{(k)}(\lambda(s) - \lambda^{(k)}) - (s^{(k+1)} - s^{(k)}). \quad (5.228)$$

Here, $\varphi_k^* \in Y^*$ is a linear functional that satisfies the conditions

$$\|\varphi_k^*\| = \|\dot{y}^{(k)}\|, \quad \varphi_k^*(\dot{y}^{(k)}) = \|\dot{y}^{(k)}\|^2.$$

Formula (5.228) is the equation of a plane that is perpendicular to the tangent $(\dot{y}^{(k)}, \dot{\lambda}^{(k)})$, and is located at a distance $(s^{(k+1)} - s^{(k)})$ from $(y^{(k)}, \lambda^{(k)})$ (see Fig. 5.24). This plane will intersect the curve Γ if $(s^{(k+1)} - s^{(k)})$ and the curvature of Γ are not too large.

The control of the step-size $\triangle s \equiv s^{(k+1)} - s^{(k)}$ should be based on the convergence behavior of the corrector iteration. In the monograph [70] it is shown that the arclength continuation can be realized for simple solution curves.

The question is still open, how the tangent can be determined from (5.225) and (5.226). Since we are dealing with the tracing of a *simple* solution curve, we must distinguish between the two cases:

- $(y^{(k)}, \lambda^{(k)})$ is an isolated solution of the BVP (5.10).

The first step is to compute w from the Eq. (5.225), i.e.,

$$T_y^{(k)} w = -T_\lambda^{(k)}.$$

Then, we set

$$\dot{y}^{(k)} \equiv a\boldsymbol{w}, \quad \dot{\lambda}^{(k)} \equiv a,$$

and determine a from (5.226) as

$$a = \frac{\pm 1}{\sqrt{1 + \|\boldsymbol{w}\|^2}}.$$

The sign of a is chosen such that the orientation of the path is preserved. More precisely, if $(\dot{y}^{(k-1)}, \dot{\lambda}^{(k-1)})$ is the preceding tangent vector, then we require

$$\left(\dot{y}^{(k-1)}\right)^T \dot{y}^{(k)} + \dot{\lambda}^{(k-1)} \dot{\lambda}^{(k)} > 0.$$

Thus, the sign of a is chosen such that

$$a\left(\left(\dot{y}^{(k-1)}\right)^T \boldsymbol{w} + \dot{\lambda}^{(k-1)}\right) > 0.$$

- $(y^{(k)}, \lambda^{(k)})$ is a simple turning point of the BVP (5.10).
  First, we consider the equation (5.225), i.e.,

$$T_y^{(k)} \dot{y}^{(k)} = -\dot{\lambda}^{(k)} T_\lambda^{(k)}.$$

This equation is solvable if

$$\dot{\lambda}^{(k)} \underbrace{\psi_0^* T_\lambda^{(k)}}_{\neq 0} = 0,$$

and we obtain $\dot{\lambda}^{(k)} = 0$. Thus $\dot{y}^{(k)} = c\,\varphi_0$, where $\mathrm{span}\{\varphi_0\} = \mathcal{N}(T_y^{(k)})$, $\|\varphi_0\| = 1$. Now, we insert this representation of $\dot{y}^{(k)}$ into the equation (5.226). It follows

$$c^2 \underbrace{\|\varphi_0\|^2}_{=1} = 1, \quad \text{i.e., } c = \pm 1.$$

5.7.3 Local Parametrization

To conclude the description of path-following methods, we will discuss a third strategy, which is often used and relatively simple. It is called *local parametrization*. Here, each component of the vector $z \equiv (y, \lambda)$ can be admitted as homotopy parameter. Let us assume that two consecutive solutions $z^{(i-1)} \equiv (y^{(i-1)}, \lambda^{(i-1)})$ and $z^{(i)} \equiv (y^{(i)}, \lambda^{(i)})$ of the BVP (5.10) are known and located on the same simple solution curve. Then, that component of z is selected as the local homotopy parameter, for which the relative change

$$\delta z_p^{(i)} \equiv \left| \frac{z_p^{(i)} - z_p^{(i-1)}}{z_p^{(i)}} \right|, \quad p \in \{1, \ldots, n+1\},$$

was greatest in the ith continuation step. Then, the new point $z^{(i+1)}$ on the solution curve is determined as the solution of the following system of equations

$$\tilde{T}(z) \equiv \begin{pmatrix} T(z) \\ z_p - z_p^{(i+1)} \end{pmatrix} = \mathbf{0}, \tag{5.229}$$

where

$$z_p^{(i+1)} \equiv z_p^{(i)} + s^{(i+1)}, \tag{5.230}$$

and

$$s^{(i+1)} \equiv \tau^{(i)} \left(z_p^{(i)} - z_p^{(i-1)} \right) = \tau^{(i)} s^{(i)}. \tag{5.231}$$

The step-size $s^{(i+1)}$ is controlled on the basis of the number α_i of iteration steps, which were necessary to approximate the system of equations (5.229) in the ith continuation step with a prescribed accuracy. More precisely, the parameter $\tau^{(i)}$ is computed according to the formula

$$\tau^{(i)} = \frac{\beta}{\alpha_i}, \tag{5.232}$$

where the constant β is the desired number of iteration steps for the respective BVP-solver. Thus, $s^{(i+1)}$ is increased if $\alpha_i < \beta$ and decreased if $\alpha_i > \beta$.

If the iteration method does not converge in the $(i+1)$th continuation step, i.e., the new point $z^{(i+1)}$ cannot be determined, then the step-size $s^{(i+1)}$ is reduced by choosing

$$\tau^{(i+1)} := 0.2 \, \tau^{(i+1)}, \tag{5.233}$$

and the $(i+1)$th step is repeated.

Of particular importance is the generation of good starting vectors $z^{(i+1,0)}$. There are at least three strategies:

- $z^{(i+1,0)} := z^{(i)}$,
- $z^{(i+1,0)} := z^{(i)} + \tau^{(i)} w^{(i)}$, where $w^{(i)}$ is the tangent at $z^{(i)}$, and
- $z^{(i+1,0)} := z^{(i)} + \tau^{(i)} \tilde{w}^{(i)}$, where $\tilde{w}^{(i)}$ is the secant line through the points $z^{(i)}$ and $z^{(i-1)}$.

Let us now present some details for the implementation of this continuation method for the BVP (5.10). In (5.229), the equation $z_p - z_p^{(i+1)} = 0$ is realized by an additional boundary condition

$$z_p(a) - \varkappa = 0. \tag{5.234}$$

With the strategy described above, the homotopy index p is searched among the components of the vector

$$z(a) = (z_1(a), \ldots, z_n(a), z_{n+1}(a))^T, \quad z_{n+1}(a) = \lambda.$$

To ensure, that the number of ODEs matches the number of boundary conditions, the trivial ODE $\lambda'(x) = 0$ is added. Thus, the $(n+1)$-dimensional BVP results

$$
\begin{aligned}
&y'(x) = f(x, y(x); \lambda), \\
&\lambda'(x) = 0, \\
&B_a y(a) + B_b y(b) = 0, \\
&y_p(a) = \varkappa \ (\text{if } p \in \{1, \ldots, n\}), \ \text{or } \lambda = \varkappa \ (\text{if } p = n+1).
\end{aligned}
\tag{5.235}
$$

With

$$\Phi(x, z) \equiv (f(x, z), 0)^T$$

and

$$r(z(a), z(b)) \equiv (B_a y(a) + B_b y(b), z_p(a) - \varkappa)^T$$

we write (5.235) in the form

$$\tilde{T}(z) \equiv \begin{pmatrix} z'(x) - \Phi(x, z) \\ r(z(a), z(b)) \end{pmatrix} = 0, \tag{5.236}$$

where

$$
\begin{aligned}
&\tilde{T} : \tilde{Z} \to \tilde{W}, \quad \tilde{Z} \equiv \mathbb{C}^1([a, b], \mathbb{R}^{n+1}), \\
&\tilde{W} \equiv \mathbb{C}([a, b], \mathbb{R}^n \times \{0\}) \times \mathbb{R}^{n+1}.
\end{aligned}
$$

Now, we come to the characterization of problem (5.236).

Theorem 5.68 *Assume that $z_0 \equiv (y_0, \lambda_0) \in E(f)$ is a simple solution of the BVP (5.10), and $p = p_0$ is the actual homotopy index. Then*

$$\tilde{T}'(z_0) : \tilde{Z} \to \tilde{W}$$

is bijective.

Proof (a) If $\tilde{T}'(z_0)$ is applied to an element $w \equiv (v, \eta) \in Y \times \mathbb{R}$, the following linear BVP results

$$\tilde{T}'(z_0)w = \begin{pmatrix} v' - f_y^0 v - \eta f_\lambda^0 \\ \eta' \\ B_a v(a) + B_b v(b) \\ w_{p_0}(a) \end{pmatrix} = 0. \tag{5.237}$$

Now, it must be shown that (5.237) has only the trivial solution $w(x) \equiv 0$.
(b) Let $z_0 \equiv (y_0, \lambda_0)$ be an *isolated solution* of (5.10). In the neighborhood of z_0, the solution curve passing through z_0 can be parametrized by λ, i.e., we can suppose that $p_0 = n + 1$. The second and fourth equation in (5.237) imply $\eta(x) \equiv 0$. Therefore,

$$
\begin{aligned}
v'(x) - f_y^0 v(x) &= 0, \\
B_a v(a) + B_b v(b) &= 0
\end{aligned}
\quad \Leftrightarrow \quad T_y^0 v = 0.
$$

Since $\mathcal{N}(T_y^0) = \{0\}$, it follows $v(x) \equiv 0$. Thus, we obtain $w(x) = 0$.
(c) Let $z_0 \equiv (y_0, \lambda_0)$ be a *simple turning point* of (5.10), i.e., it holds

$$
T_\lambda^0 \notin \mathcal{R}(T_y^0).
$$

From the first equation in (5.237), we obtain $\eta(x) \equiv 0$. This in turn implies $p_0 \neq n + 1$. Thus, $(v, 0) \in \mathcal{N}(\tilde{T}'(z_0))$ is the solution of the following BVP

$$
\begin{aligned}
v'(x) - f_y^0 v(x) &= 0, \\
B_a v(a) + B_b v(b) &= 0, \\
v_{p_0} &= 0.
\end{aligned}
\tag{5.238}
$$

Assume that $v(x) \not\equiv 0$. Then, there exists an index $\hat{p}$ such that $v_{\hat{p}}(a) \neq 0$. For the homotopy index p_0 it holds

$$
|v_{p_0}(a)| = \max\{|v_1(a)|, \ldots, |v_n(a)|\} \neq 0,
$$

which is in contradiction to the third equation in (5.238).
(d) Now, the surjectivity of $\tilde{T}'(z_0)$ must be shown. We leave this part of the proof to the reader for self-study. ∎

In Algorithm 2.2 the local parametrization method is described in more detail.

Algorithm 2.2 Path-following with the local parametrization method

Step 1: (jumping on a solution curve)

Choose:

$s \in [s_{\min}, s_{\max}] \doteq$ initial step-size,

$\lambda_{\text{start}} \doteq$ initial value for λ,

$redconst \in (0, 1) \doteq$ constant that is used to reduce the step-size,

$p_E \in \{1, 2, \ldots, n + 1\}$, z_{p_E}, $dir \in \{-1, 1\} \doteq$ are used to

terminate the algorithm if $z_{p_E}^{(i)}(a)$ exceeds z_{p_E} from the left

($dir = 1$), or from the right ($dir = -1$),

$\beta \doteq$ desired number of iteration steps for the determination of a solution of the BVP (5.235)

itmax $\doteq$ maximum number of path-following steps;

Set $i := 0$, $p := n + 1$, $\varkappa := \lambda_{\text{start}}$, $q := 0$;

Compute a solution $z^{(i)}$ of the nonlinear BVP (5.235);

Step 2 : (initial homotopy step)

Set $i := 1$, $p_1 := p$, $\varkappa := z_{p_1}^{(0)} + s$,

Compute a solution $z^{(1)}$ of the nonlinear BVP (5.235),

Set $\alpha_1 :=$ number of iteration steps required to compute $z^{(1)}$;

Step 3 : (ith homotopy step)

Set $i := i + 1$,

(a) Computation of the homotopy index p_i

$$\text{Compute } \delta_l := \frac{\left| z_l^{(i-1)}(a) - z_l^{(i-2)}(a) \right|}{\left| z_l^{(i-1)}(a) \right| + 0.1}, \quad l = 1, \ldots, n+1,$$

Set $\delta_{\text{max}} := \max_l \delta_l$ and $p_i := p_{\text{max}}$, with $\delta_{p_{\text{max}}} = \delta_{\text{max}}$;

(b) Computation of the step-size $s^{(i)}$

$$\text{Compute } \tau^{(i-1)} := \frac{\beta}{\alpha_{i-1}},$$

Set $s^{(i)} := \tau^{(i-1)} \left(z_{p_i}^{(i-1)}(a) - z_{p_i}^{(i-2)}(a) \right)$;

If $s^{(i)} < s_{\text{min}}$ then $\tau^{(i)} = \dfrac{\tau^{(i-1)} s_{\text{min}}}{s^{(i)}}$;

If $s^{(i)} > s_{\text{max}}$ then $\tau^{(i)} = \dfrac{\tau^{(i-1)} s_{\text{max}}}{s^{(i)}}$;

(c) Computation of starting values for the corrector iteration

(i) without extrapolation: $z^{(i,0)} := z^{(i-1)}$,

(ii) with extrapolation: $z^{(i,0)} := \left(1 + \tau^{(i-1)} \right) z^{(i-1)} - \tau^{(i-1)} z^{(i-2)}$;

(d) Computation of the ith point on the solution curve

Set $\varkappa := z_{p_i}^{(i,0)}(a)$ and compute the solution $z^{(i)}$

of the BVP (5.235);

Set $\alpha_i :=$ number of iteration steps required to compute $z^{(i)}$;

(e) Test whether a final value has been exceeded

If $z_{p_E}^{(i)}(a) \cdot dir > z_{p_E} \cdot dir$ and $\text{sign}\left(z_{p_E}^{(i)}(a) - z_{p_E}^{(i-1)}(a) \right) = dir$ then

set $q := 4$;

Step 4 : (Evaluation)

If $i = itmax$ then $q := 4$;

If $z^{(i)}$ could be determined and $q \neq 4$, then goto Step 3,

else goto Step 5;

Step 5: Set $q := q + 1$,

(a) Reducing the step-size

If $q < 3$ then

set $\tau^{(i)} := redconst \cdot \tau^{(i)}$ and goto Step 3(b);

(b) Selecting a new homotopy index

If $q = 3$ then

compute an index $\hat{p}_i$ with $\delta_{\hat{p}_i} \equiv \max_{l \neq p_i}\{\delta_l\}$, and

set $p_i := \hat{p}_i$, goto Step 3(c);

(c) End of path-following

If $q > 3$ then Stop.

In Fig. 5.25, Algorithm 2.2 is visualized.

5.7.4 Detection of Singular Points

An important part of path-following algorithms is the detection of bifurcation and turning points. This is realized by so-called *test functions* $\vartheta(z)$, which are monitored during the path-following. A singular point is indicated by a zero of ϑ.

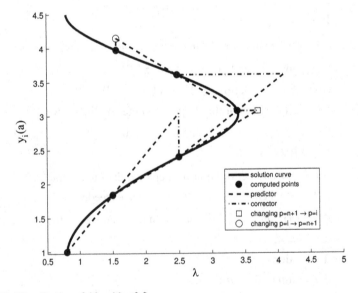

Fig. 5.25 Visualization of Algorithm 2.2

Definition 5.69 Let $z^{(0)} \equiv (y^{(0)}, \lambda^{(0)})$ be a singular point (turning or bifurcation point) of the BVP (5.10). Then $\vartheta(z)$ is called a test function for $z^{(0)}$ if

- ϑ is a continuous function in z,
- $\vartheta(z^{(0)}) = 0$,
- $\vartheta(z)$ changes the sign at $z^{(0)}$. $\qquad\qquad\square$

For each computed point $z^{(i)}$ on the solution curve, the value $\vartheta^{(i)} \equiv \vartheta(z^{(i)})$ is determined. If ϑ changes the sign during the transition from the ith to the $(i + 1)$th point on the curve, i.e., if it holds

$$\vartheta^{(i)} \vartheta^{(i+1)} < 0,$$

then a singular point must be on the curve between $z^{(i)}$ and $z^{(i+1)}$.

However, qualitatively different singular points can only be detected by different test functions. Therefore, let us begin with the case of a simple turning point. An appropriate test function is

$$\vartheta(z) \equiv \frac{d\lambda}{dz_p(a)}(z). \tag{5.239}$$

If $z^{(0)} \equiv (y^{(0)}, \lambda_0)$ is a simple turning point, then we have $\vartheta(z^{(0)}) = 0$. The continuity of ϑ with respect to z follows from the continuity of the solution manifold of the BVP (5.10).

When the ith homotopy step has been executed and the corresponding solution $z^{(i)}$ is determined, the three parameter values $\lambda^{(i-2)}, \lambda^{(i-1)}$ and $\lambda^{(i)}$ are known. Moreover, the sign of $\vartheta(z)$ depends only on the change of the λ-values (using a sufficiently small homotopy step-size). Therefore, by the expression

$$\vartheta^{(i)} \equiv \vartheta(z^{(i)}) = (\lambda^{(i)} - \lambda^{(i-1)}), \tag{5.240}$$

an appropriate test function is also defined. Indeed, if it holds

$$\vartheta^{(i-1)} \vartheta^{(i)} < 0,$$

then a simple turning point has been crossed. But for the position of the turning point with respect to the three computed points on the curve, there are two possibilities (see Fig. 5.26).

In the case (a) the turning point will be crossed in the $(i - 1)$th homotopy step, whereas in the case (b) the turning point is crossed in the ith step. In both cases the crossing of a simple turning point can only be detected with the test function (5.240) in the ith homotopy step.

Let us now come to bifurcation points. We will show how a test function can be defined on the basis of the shooting method, which is used to solve the BVP (5.235). It is sufficient to consider only the simple shooting method (see Sect. 4.2).

If the simple shooting method is applied to solve the BVP (5.10), this BVP is transformed into the finite-dimensional problem

Fig. 5.26 Position of the turning point w.r.t. the three computed points

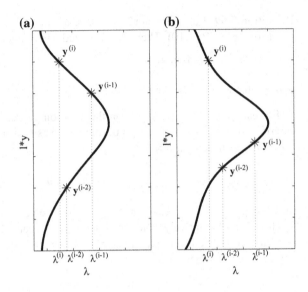

$$F(s) \equiv B_a s + B_b u(b, s; \lambda) = 0, \tag{5.241}$$

where $u(x) \equiv u(x, s; \lambda)$ denotes the solution of the IVP

$$u'(x) = f(x, u(x); \lambda), \quad u(a) = s. \tag{5.242}$$

As we have explained in Chap. 4, the system of nonlinear algebraic equations $F(s) = 0$ is usually solved by a damped and regularized Newton method. This numerical method is based on the Jacobian of F, i.e., we obtain the Jacobian as an by-product of the equation solver. Moreover, the BVP (5.10) possesses an isolated solution $y^{(0)}(x)$ if and only if the Jacobian

$$M(s) \equiv \frac{\partial F(s)}{\partial s} = B_a + B_b \frac{\partial u(b, s; \lambda)}{\partial s} \tag{5.243}$$

is nonsingular at $s = s^{(0)}$, with $s^{(0)} \equiv y^{(0)}(a)$. Thus, if $z^{(0)}$ is a singular point of the operator T in (5.11), with $\dim \mathcal{N}(T'(z^{(0)})) > 0$, then $\det(M(s^{(0)})) = 0$.

It is now obvious to define a test function for the detection of bifurcation points as follows

$$\vartheta(z) \equiv \det(M(s)), \tag{5.244}$$

where $s = z(a)$ establishes the connection between $z(x)$ and s.

We are now facing with the following problem. In the path-following methods, the enlarged BVP (5.235) is used instead of the BVP (5.10). Thus, the question is where we can find the Jacobian $M(s)$ of the given BVP (5.10). In order to find an answer, we write (5.235) in the form

$$\tilde{y}'(x) = \tilde{f}(x, \tilde{y}(x)),$$
$$\tilde{B}_a \tilde{y}(a) + \tilde{B}_b \tilde{y}(b) = \beta,$$

(5.245)

with

$$\tilde{y} \equiv \begin{pmatrix} y \\ \lambda \end{pmatrix}, \quad \tilde{f} \equiv \begin{pmatrix} f \\ 0 \end{pmatrix}, \quad \beta \equiv \begin{pmatrix} \mathbf{0}_n \\ \varkappa \end{pmatrix},$$

$$\tilde{B}_a \equiv \begin{pmatrix} B_a & \mathbf{0}_n \\ (e^{(p)})^T & \end{pmatrix}, \quad \tilde{B}_b \equiv \begin{pmatrix} B_b & \mathbf{0}_n \\ \mathbf{0}_n^T & 0 \end{pmatrix},$$

and $e^{(p)} \in \mathbb{R}^{n+1}$ is the pth unit vector. The associated IVP is

$$\tilde{u}'(x) = \tilde{f}(x, \tilde{u}(x)), \quad \tilde{u}(a) = \tilde{s} \in \mathbb{R}^{n+1},$$

(5.246)

with $\tilde{u} \equiv (u, \lambda)^T$. The simple shooting method transforms the BVP (5.245) into the finite-dimensional problem

$$\tilde{F}(\tilde{s}) \equiv \tilde{B}_a \tilde{s} + \tilde{B}_b \tilde{u}(b; \tilde{s}) - \beta = 0.$$

(5.247)

If the Newton method is used to solve (5.247), the Jacobian of $\tilde{F}$ must be determined. It holds

$$\tilde{M}(\tilde{s}) = \frac{\partial \tilde{F}(\tilde{s})}{\partial \tilde{s}} = \tilde{B}_a + \tilde{B}_b \frac{\partial \tilde{u}(b; \tilde{s})}{\partial \tilde{s}}$$

$$= \tilde{B}_a + \tilde{B}_b \begin{pmatrix} \left. \dfrac{\partial u_j(b; s_1, \ldots, s_n, s_{n+1})}{\partial s_i} \right|_{i,j=1,\ldots,n} & v(b; s_1, \ldots, s_{n+1}) \\ w(b; s_1, \ldots, s_{n+1}) & 1 \end{pmatrix}$$

$$= \tilde{B}_a + \tilde{B}_b \begin{pmatrix} X^e & v(b; s_1, \ldots, s_{n+1}) \\ w(b; s_1, \ldots, s_{n+1}) & 1 \end{pmatrix}$$

$$= \begin{pmatrix} B_a + B_b X^e & B_b v(b; s_1, \ldots, s_{n+1}) \\ e_p^T & \end{pmatrix}$$

$$= \begin{pmatrix} M(s) & B_b v(b; s_1, \ldots, s_{n+1}) \\ e_p^T & \end{pmatrix},$$

where

$$v(b; s_1, \ldots, s_{n+1}) \equiv \left. \frac{\partial u_j(b; s_1, \ldots, s_n, s_{n+1})}{\partial s_{n+1}} \right|_{j=1,\ldots,n},$$

$$w(b; s_1, \ldots, s_{n+1}) \equiv \left. \frac{\partial u_{n+1}(b; s_1, \ldots, s_n, s_{n+1})}{\partial s_i} \right|_{i=1,\ldots,n},$$

and X^e has been defined in Sect. 4.2.

Thus, the Jacobian $M(s)$, which is required to compute the test function $\vartheta(z)$ (see formula (5.244)), can be obtained from the Jacobian $\tilde{M}(\tilde{s})$ by deleting the last

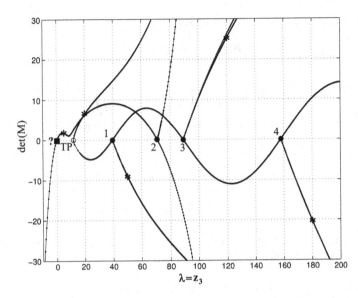

Fig. 5.27 Plot of the test function $\det(M)$ versus λ for problem (5.186) with $\tau = 1$; TP—turning point; $1, \ldots, 4$—bifurcation points; $*$—points where the path-following was started; ?—unknown singularity

row and column. In case that the multiple shooting method is used to solve the nonlinear equations (5.247), we have to proceed block-wise, i.e., we have to delete in each $(n + 1) \times (n + 1)$ block of the Jacobian $\tilde{M}$ the last row and column.

Using (5.244), for each computed point $z^{(i)}$ on the solution curve the value $\vartheta^{(i)} \equiv \vartheta(z^{(i)})$ is determined. If ϑ changes the sign during the transition from the ith to the $(i + 1)$th point on the curve, a singular point must be on the curve between $z^{(i)}$ and $z^{(i+1)}$. Moreover, based on the plot of the test function, it is often possible to recognize that the determinant is zero and there must exist a singular point between $z^{(i)}$ and $z^{(i+1)}$. In Figs. 5.27 and 5.28 the plot of the test function (5.244) and the corresponding bifurcation diagram are given for the perturbed BVP (5.204) with $\tau = 1$.

The unknown singularity detected in Fig. 5.27 (see black square) must be studied separately. It can be seen that for $\lambda = 0$ problem (5.186) is reduced to a singular *linear* problem and the entire straight line L is a solution.

During path-following, it is also possible to determine rough approximations of the critical parameters by interpolation. To simplify the notation, we abbreviate the initial values $y_p(a)$ and $z_p(a)$ with y_p and z_p, respectively. Let us begin with simple turning points. Here, we use the obvious ansatz

$$\lambda = c_1(z_p - c_2)^2 + c_3, \quad p \neq n + 1. \tag{5.248}$$

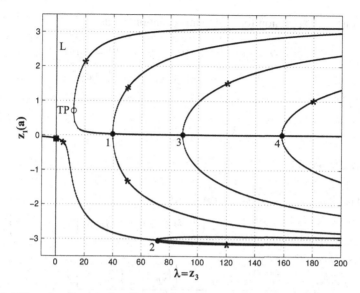

Fig. 5.28 Bifurcation diagram of problem (5.186); *—points where the path-following was started

In the ith homotopy step, the free parameters $c_1, c_2, c_3 \in \mathbb{R}$ can be determined by inserting the computed solutions $z^{(i-2)}$, $z^{(i-1)}$, and $z^{(i)}$ into the ansatz (5.248) as follows:

$$\lambda^{(i)} = c_1(z_p^{(i)} - c_2)^2 + c_3,$$
$$\lambda^{(i-1)} = c_1(z_p^{(i-1)} - c_2)^2 + c_3,$$
$$\lambda^{(i-2)} = c_1(z_p^{(i-2)} - c_2)^2 + c_3.$$

The first equation yields

$$c_3 = \lambda^{(i)} - c_1(z_p^{(i)} - c_2)^2.$$

Subtracting the second equation from the first one gives

$$\lambda^{(i)} - \lambda^{(i-1)} = c_1(z_p^{(i)} - c_2)^2 - c_1(z_p^{(i-1)} - c_2)^2,$$

and we obtain

$$c_1 = \frac{\lambda^{(i)} - \lambda^{(i-1)}}{(z_p^{(i)} - c_2)^2 - (z_p^{(i-1)} - c_2)^2}.$$

Now we substitute the above expression for c_3 into the third equation, and get

$$\lambda^{(i-2)} - \lambda^{(i)} = c_1 \left[(z_p^{(i-2)} - c_2)^2 - (z_p^{(i)} - c_2)^2 \right].$$

Substituting the expression for c_1 into this equation yields

$$\frac{\lambda^{(i-2)} - \lambda^{(i)}}{\lambda^{(i)} - \lambda^{(i-1)}} = \frac{(z_p^{(i-2)} - c_2)^2 - (z_p^{(i)} - c_2)^2}{(z_p^{(i)} - c_2)^2 - (z_p^{(i-1)} - c_2)^2}.$$

Let us set

$$w^{(i)} \equiv \frac{\lambda^{(i-2)} - \lambda^{(i)}}{\lambda^{(i)} - \lambda^{(i-1)}}.$$

Then, we can write

$$\left[\left(z_p^{(i)}\right)^2 - \left(z_p^{(i-1)}\right)^2 \right] w^{(i)} + 2c_2 w^{(i)} \left[z_p^{(i-1)} - z_p^{(i)} \right]$$
$$= \left(z_p^{(i-2)}\right)^2 - \left(z_p^{(i)}\right)^2 + 2c_2 \left[z_p^{(i)} - z_p^{(i-2)} \right].$$

It follows

$$c_2 = \frac{\left(z_p^{(i-2)}\right)^2 - \left(z_p^{(i)}\right)^2 - \left[\left(z_p^{(i)}\right)^2 - \left(z_p^{(i-1)}\right)^2 \right] w^{(i)}}{2 \left(w^{(i)} \left[z_p^{(i-1)} - z_p^{(i)} \right] - \left[z_p^{(i)} - z_p^{(i-2)} \right] \right)}.$$

Differentiating the ansatz (5.248) gives

$$\frac{d\lambda}{dz_p} = 2c_1(z_p - c_2).$$

This derivative must vanish in the turning point, i.e., it holds

$$2c_1(z_p - c_2) \doteq 0.$$

Thus

$$\hat{z}_p^{\text{tp}} = c_2 \text{ and } \hat{\lambda}^{\text{tp}} = c_3 \tag{5.249}$$

are approximations for the critical value z_p^{tp} and the critical parameter λ^{tp}, respectively.

Next, we come to simple bifurcation points. Let us assume that $a_2 \neq 0$ and the test function (5.244) is integrated into the path-following algorithm. If $\vartheta(z)$ changes the sign during the transition from the $(i-1)$th to the ith point of the traced solution curve Γ_1, the following quadratic ansatz can be used for the approximation of these bifurcation points

$$\vartheta = c_1 z_p^2 + c_2 z_p + c_3. \tag{5.250}$$

Inserting three successive value pairs

$$(\vartheta^{(i-2)}, z_p^{(i-2)}), \quad (\vartheta^{(i-1)}, z_p^{(i-1)}), \quad (\vartheta^{(i)}, z_p^{(i)})$$

into the ansatz (5.250) yields

$$\vartheta^{(i)} = c_1 \left(z_p^{(i)}\right)^2 + c_2 z_p^{(i)} + c_3,$$
$$\vartheta^{(i-1)} = c_1 \left(z_p^{(i-1)}\right)^2 + c_2 z_p^{(i-1)} + c_3,$$
$$\vartheta^{(i-2)} = c_1 \left(z_p^{(i-2)}\right)^2 + c_2 z_p^{(i-2)} + c_3.$$

From the third equation, we have

$$c_3 = \vartheta^{(i-2)} - c_1 \left(z_p^{(i-2)}\right)^2 - c_2 z_p^{(i-2)}. \tag{5.251}$$

Subtracting the third equation from the second one gives

$$\vartheta^{(i-1)} - \vartheta^{(i-2)} = c_1 \left[\left(z_p^{(i-1)}\right)^2 - \left(z_p^{(i-2)}\right)^2 \right] + c_2 (z_p^{(i-1)} - z_p^{(i-2)}),$$

and we obtain

$$c_2 = \frac{\vartheta^{(i-1)} - \vartheta^{(i-2)}}{z_p^{(i-1)} - z_p^{(i-2)}} - c_1 (z_p^{(i-1)} + z_p^{(i-2)}). \tag{5.252}$$

Now, we substitute the above expressions for c_2 and c_3 into the first equation, and get

$$\vartheta^{(i)} = c_1 \left(z_p^{(i)}\right)^2 + \frac{\vartheta^{(i-1)} - \vartheta^{(i-2)}}{z_p^{(i-1)} - z_p^{(i-2)}} z_p^{(i)} - c_1 (z_p^{(i-1)} + z_p^{(i-2)}) z_p^{(i)}$$
$$+ \vartheta^{(i-2)} - c_1 \left(z_p^{(i-2)}\right)^2 - \frac{\vartheta^{(i-1)} - \vartheta^{(i-2)}}{z_p^{(i-1)} - z_p^{(i-2)}} z_p^{(i-2)}$$
$$+ c_1 (z_p^{(i-1)} + z_p^{(i-2)}) z_p^{(i-2)}.$$

It follows

$$c_1 = \frac{1}{z_p^{(i)} - z_p^{(i-1)}} \left(\frac{\vartheta^{(i)} - \vartheta^{(i-2)}}{z_p^{(i)} - z_p^{(i-2)}} - \frac{\vartheta^{(i-1)} - \vartheta^{(i-2)}}{z_p^{(i-1)} - z_p^{(i-2)}} \right). \tag{5.253}$$

The solution of the equation

$$c_1 z_p^2 + c_2 z_p + c_3 = 0,$$

where c_1, c_2 and c_3 are given in (5.251)–(5.253), can be used as an approximation $\hat{z}_p^b$ for the critical value z_p^b, i.e.,

$$\hat{z}_p^b = -\frac{c_2}{2c_1} \pm \sqrt{\frac{c_2^2}{4c_1^2} - \frac{c_3}{c_1}}. \tag{5.254}$$

Since $\hat{z}_p^{\mathrm{b}}$ must be located between $z_p^{(i)}$ and $z_p^{(i-1)}$, the sign in (5.254) has to be chosen such that

$$\left(z_p^{(i)} - \hat{z}_p^{\mathrm{b}}\right)\left(\hat{z}_p^{\mathrm{b}} - z_p^{(i-1)}\right) > 0. \tag{5.255}$$

Thus, if it holds $p = n + 1$, we have determined an approximation $\hat{\lambda}^{\mathrm{b}} = \hat{z}_p^{\mathrm{b}}$ for the critical parameter λ^{b}.

If $a_2 = 0$ and the test function $\vartheta(z)$ does not change the sign during the path-following of a solution curve Γ_1 but is nearly zero at $\lambda = \tilde{\lambda}$ (see Fig. 5.28), the other solution curve Γ_2 should be traced in the neighborhood of $\tilde{\lambda}$. In the case that the corresponding test function changes its sign near $\tilde{\lambda}$, there must be a bifurcation point. Now, using the ansatz (5.250) with $p = n + 1$ for the solution curve Γ_2, the critical parameter λ^b can be approximated as described above.

5.8 Parametrized Nonlinear BVPs from the Applications

5.8.1 Buckling of Thin-Walled Spherical Shells

In this section, we present some examples of parametrized nonlinear two-point BVPs from the applications where we have used the numerical techniques discussed in the previous chapters to determine the corresponding solution manifold.

The study of the buckling behavior of thin-walled spherical shells under an external pressure constitutes still today a challenge to mathematicians and engineers. It seems to be hopeless to handle all possible deformations in a systematic way, because in the solution manifold of the corresponding nonlinear, parametrized equations singularities of very high order can occur. Therefore, many authors simplify the buckling problem by presuming that only axisymmetric deformations of the shell are possible. As explained in [44], this restriction is not as academic as it might appear at the first moment, because the production process of spherical shells frequently favors axisymmetric imperfections, which create a preference for axisymmetric buckling patterns.

The shell equations presented in [19, 121] belong to the most-studied models describing the axisymmetric buckling of spherical shells. Both models are based on the assumptions:

- the material of the shell is elastic, isotropic and homogeneous,
- the thickness of the shell is small and constant,
- only axisymmetric deformations of the spherical shell are possible,
- the uniform external pressure keeps always the direction pointing towards the origin of the spherical shell.

The governing equations in [19] constitute a parametrized nonlinear BVP for a system of four first-order differential equations

$$y_1'(x) = (\nu - 1)\cot(x)y_1(x) + y_2(x) + \left(k\cot^2(x) - \lambda\right)y_4(x)$$
$$\qquad + \cot(x)y_2(x)y_4(x),$$
$$y_2'(x) = y_3(x),$$
$$y_3'(x) = \left(\cot^2(x) - \nu\right)y_2(x) - \cot(x)y_3(x) - y_4(x) - 0.5\cot(x)y_4^2(x),$$
$$y_4'(x) = \frac{1 - \nu^2}{k}y_1(x) - \nu\cot(x)y_4(x),$$
$$y_2(0) = y_4(0) = y_2(\pi) = y_4(\pi) = 0. \tag{5.256}$$

The components y_1, y_2, y_3, and y_4 are proportional, respectively, to the radial bending moment, the transversal shear, the circumferential membrane stress, and the angle of rotation of a tangent to the meridian. The quantity ν is Poisson's ratio. The radii of the inner and outer surfaces of the spherical shell are given by $R \mp h/2$, where R is the radius of the midsurface. Finally,

$$\lambda \equiv \frac{pR}{2Eh}, \quad k \equiv \frac{1}{12}\left(\frac{h}{R}\right)^2, \tag{5.257}$$

where E denotes Young's modulus (see, e.g., [11]) and p is the external pressure. The underlying geometry of the shell is sketched in Fig. 5.29,(a).

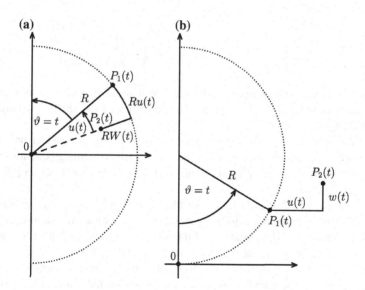

Fig. 5.29 Geometry of the spherical shell

The model equations formulated in [121], Appendix K, represent a BVP for a system of two second-order differential equations:

$$\beta''(x) = -\cot(x)\beta'(x) - \cos(x - \beta(x))\frac{\sin(x - \beta(x)) - \sin(x)}{\sin^2(x)}$$

$$+ \nu\frac{\cos(x - \beta(x)) - \cos(x)}{\sin(x)} - \frac{1}{\delta}\frac{\sin(x - \beta(x))}{\sin(x)}\psi^*(x)$$

$$- 2\frac{\lambda}{\delta}\sin(x)\cos(x - \beta(x)),$$

$$\psi^{*''}(x) = -\cot(x)\psi^{*'}(x)$$

$$+ \left(\frac{\cos^2(x - \beta(x))}{\sin^2(x)} - \nu\frac{\sin(x - \beta(x))}{\sin(x)}[1 - \beta'(x)]\right)\psi^*(x)$$

$$+ \frac{1}{\delta}\frac{\cos(x - \beta(x)) - \cos(x)}{\sin(x)} - 2\lambda\sin(x - \beta(x))\cos(x - \beta(x))$$

$$- 2\lambda\nu\sin(x)\cos(x - \beta(x))[1 - \beta'(x)] + 4\lambda\nu\sin(\beta(x))$$

$$+ 12\lambda\sin(x)\cos(x),$$

$$\beta(0) = \psi^*(0) = \beta(\pi) = \psi^*(\pi) = 0.$$

$$(5.258)$$

Here, β describes the angle enclosed by the tangents to the deformed and undeformed meridian. Moreover, ψ^* is a stress function depending on the horizontal stress resultant H,

$$\psi^*(\cdot) \equiv \frac{4H\sin(\cdot)}{R\,p_c}, \qquad (5.259)$$

where

$$p_c \equiv 4E\delta^2\sqrt{12(1 - \nu^2)}, \quad \delta^2 \equiv \left(\frac{h}{R}\right)^2\frac{1}{12(1 - \nu^2)},$$

and h, R, ν, and E are defined as before. The loading parameter λ is related to the external pressure p by $\lambda \equiv p/p_c$. The geometry of the shell is visualized in Fig. 5.29,(b).

Because of the assumptions stated above, it is sufficient to consider only an arc of the midsurface, which is indicated by a dotted line. The defining equations for the displacements u, W (Fig. 5.29,(a)) and u, w (Fig. 5.29,(b)) are given in [19, 121], respectively.

The shell Eqs. (5.256) and (5.258) have been studied by many authors (see, e.g., [44, 64, 72, 109]). In the paper [62], a general class of parametrized nonlinear BVPs is defined, which includes equations of the type (5.256) and (5.258). For this class the occurring bifurcation phenomena have been analyzed in detail, also with the analytical and numerical methods described in this book. Here, we present the results of some experiments, which were based on the following values of the internal parameters: $\nu = 0.32$ (steel), $k = 10^{-5}$ and $\delta = 0.00333779$ (thin sphere). We have solved the corresponding BVPs with the multiple shooting code RWPM [59].

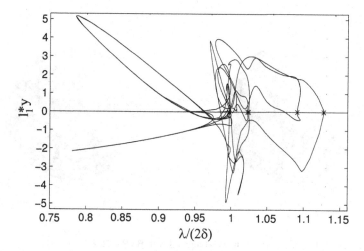

Fig. 5.30 Bifurcation diagram of the BVP (5.256)

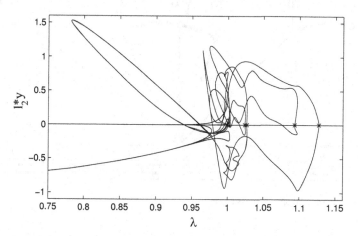

Fig. 5.31 Bifurcation diagram of the BVP (5.258)

Moreover, the code RWPKV [65] was used to realize the tracing of solution curves. In Figs. 5.30 and 5.31 parts of the solution manifold of the BVPs (5.256) and (5.258), respectively, are shown. For the definition of the functionals l_1^* and l_2^* see [62]. Stars mark the primary bifurcation points.

It can be seen that there are only a few differences in the solution fields of the problems (5.256) and (5.258), which differ significantly in the degree of the nonlinearity. This was an unexpected result. Moreover, in Fig. 5.32 two deformed shells are presented.

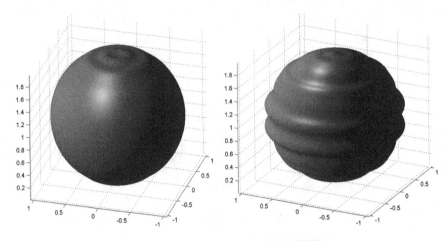

Fig. 5.32 Examples for deformed shells (computed with BVP (5.258))

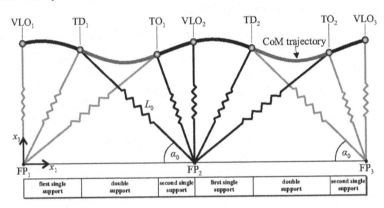

Fig. 5.33 The bipedal spring-mass model for walking

5.8.2 A Bipedal Spring-Mass Model

One of the most successful templates for the simulation of human walking and running is the spring-mass model. The model for running and hopping consists of a mass point representing a center of mass of the human body riding on a linear leg spring [22]. For the study of walking gaits, the planar bipedal model with two leg springs was introduced in [41]. It consists of two massless leg springs supporting the point mass m (see Fig. 5.33).

Both leg springs have the same stiffness k_0 and rest length L_0. For a fixed time t the location and velocity of the center of mass in the real plane $\mathbb{R}^2$ are given by $(x(t), y(t))^T$ and $(\dot{x}(t), \dot{y}(t))^T$, respectively. Here, the dot denotes the derivative w.r.t. the time t. Any walking gait is completely characterized by four fundamental system parameters (leg stiffness k_0, angle of attack α_0, rest length L_0, system energy E_0)

and the four-dimensional vector of initial conditions $(x(t_0), \dot{x}(t_0), y(t_0), \dot{y}(t_0))^T = (x_0, \dot{x}_0, y_0, \dot{y}_0)^T$.

A walking step comprises a single-support phase and a double-support phase (see Fig. 5.33). Events of touch-down and take-off are transitions between the phases. The trajectory of the center of mass in each phase is the solution of an IVP. The calculation of a walking step starts at time $t = t_0$ during the single-support phase at the instant of the vertical leg orientation VLO (see Fig. 5.33 and [105]). Using VLO as the initial point of a step (i.e., as the Poincaré section), allows to reduce the dimension of the return map [84]. Here, the center of mass is located exactly over the foot point of the supporting leg spring.

The initial values for the first IVP are $(x_0, \dot{x}_0, y_0, \dot{y}_0)^T$. The initial vector of each subsequent phase is the last point of the corresponding previous phase. The motion of the center of mass during the single-support phase is described by the equations

$$m\ddot{x}(t) = k_0 (L_0 - L_1(t)) \frac{x(t) - x_{FP_1}}{L_1(t)},$$
$$m\ddot{y}(t) = k_0 (L_0 - L_1(t)) \frac{y(t)}{L_1(t)} - mg,$$

(5.260)

where $L_1(t) \equiv \sqrt{(x(t) - x_{FP_1})^2 + y^2(t)}$ is the length of the compressed leg spring during stance. The position of the footpoint FP$_1$ is given by $(x_{FP_1}, 0)$. The transition (touch-down) from single-support phase to double-support phase happens, when the landing condition

$$y(t_1) = L_0 \sin(\alpha_0)$$

is fulfilled (see Fig. 5.33). Here, $t = t_1$ is the time when the touch-down occurs.

The equations of the motion during the double-support phase are given by

$$m\ddot{x}(t) = k_0 (L_0 - L_1(t)) \frac{x(t) - x_{FP_1}}{L_1(t)} + k_0 (L_0 - L_2(t)) \frac{x(t) - x_{FP_2}}{L_2(t)},$$
$$m\ddot{y}(t) = k_0 (L_0 - L_1(t)) \frac{y(t)}{L_1(t)} + k_0 (L_0 - L_2(t)) \frac{y(t)}{L_2(t)} - mg,$$

(5.261)

where $L_2(t) \equiv \sqrt{(x(t) - x_{FP_2})^2 + y^2(t)}$ is the length of the second compressed leg spring. The positions of both footpoints FP$_1$ and FP$_2$ are given by $(x_{FP_1}, 0)$ and $(x_{FP_2}, 0)$, respectively. The transition (take-off) from the double-support phase to the single-support phase occurs when the extending length L_1 of the first leg spring reaches the rest length L_0, i.e., when the take-off condition

$$(x(t_2) - x_{FP_1})^2 + y^2(t_2) = L_0^2$$

is fulfilled. Here, $t = t_2$ is the time when the take-off occurs.

The walking step ends at the next VLO at time $t = t_3$, when the condition $x(t_3) = x_{FP_2}$ is fulfilled. The system is energy-conservative, i.e., the system energy E_0 remains constant during the whole step.

Periodic solutions of the model often correspond to continuous locomotion patterns. The investigations of stable periodic solutions is of particular interest. The mathematical definition of stability is considered as the property of the system to absorb small perturbations. Periodic walking solutions are found and analyzed using the well-known Poincaré return map [95]. Their stability is determined by the value of the corresponding Floquet multipliers [39]. The detailed mathematical description can be seen, e.g., in [45].

Since the end of each phase of a running or walking step is defined by an event, the times t_1, t_2, and t_3 are not known in advance. Therefore, instead of solving the IVPs successively it is more appropriate to scale each phase to the unit interval $[0, 1]$ and to solve all three phases simultaneously by a parametrized nonlinear BVP of the form

$$y'(x) = f(y(x)), \quad 0 < x < 1,$$
$$r(y(0), y(1); \lambda) = 0. \tag{5.262}$$

A further advantage of this strategy is that the numerical methods described in the previous sections can be applied directly to the BVP (5.262) to determine the corresponding solution manifold. But notice that operator T in the operator equation (5.11) has to be defined now as (see formulas (5.13) and (5.14))

$$T(y, \lambda) \equiv \begin{pmatrix} y' - f(y) \\ r(y(0), y(1); \lambda) \end{pmatrix}, \tag{5.263}$$

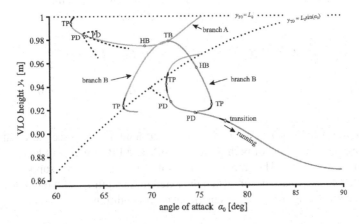

Fig. 5.34 Initial VLO height y_0 of periodic walking patterns versus the angle of attack α_0. The dots represent singular points: *TP* turning point, *TB* secondary bifurcation point, *HB* Hopf bifurcation point, *PD* perioddoubling bifurcation point

where the associated functional spaces must be chosen adequately. A special feature of this BVP is that the bifurcation parameter λ occurs only in the boundary function r. More information about this model and its extensions can be found in [81, 82, 83].

In Fig. 5.34, a part of the solution manifold of the bipedal spring-mass model is shown. Here, the system energy is $E_0 = 820\,J$ corresponding to the average horizontal velocity $v_x \approx 1.1$ m/s. Furthermore, we have set $k_0 = 16\,kN/m$, $L_0 = 1\,m$, $m = 80\,kg$, and $g = 9.81$ m/s^2.

5.9 Exercises

Exercise 5.1 Draw the bifurcation diagrams for the following scalar functions ($l \in \mathbb{R}$):

1. $l\,u = \lambda\,u$,
2. $l\,u + c\,u^2 = \lambda\,u$, $c \neq 0$,
3. $l\,u + c\,u^3 = \lambda\,u$, $c \neq 0$,
4. $a(u) = \lambda\,u$, with

$$a(u) \equiv \begin{cases} u, & |u| < 1 \\ (u^2 + 1)/2, & u \geq 1, \\ -(u^2 + 1)/2, & u \leq -1. \end{cases}$$

Exercise 5.2 Analyze the solution field of the following equation in $\mathbb{R}^2$ and draw the corresponding bifurcation diagram:

$$A\,u = \lambda\,u,$$

where

$$A\,u \equiv L\,u + C\,u,$$

$$L\,u \equiv \begin{bmatrix} \beta_1 & 0 \\ 0 & \beta_2 \end{bmatrix} \begin{bmatrix} u_1 \\ u_2 \end{bmatrix} = \begin{bmatrix} \beta_1 u_1 \\ \beta_2 u_2 \end{bmatrix}, \quad \beta_2 > \beta_1 > 0,$$

$$C\,u \equiv \begin{bmatrix} \gamma_1 u_1 (u_1^2 + u_2^2) \\ \gamma_2 u_2 (u_1^2 + u_2^2) \end{bmatrix}, \quad \gamma_1 > \gamma_2 > 0.$$

HINT: It holds $C(\alpha u) = \alpha^3 C\,u$.

Exercise 5.3 On the Internet at http://www-history.mcs.st-and.ac.uk, under the menu item *Famous Curves Index*, you will find the curve of the *trifolium*. The corresponding formula is

$$(u^2 + v^2)(v^2 + u(u + a)) = 4auv^2.$$

1. Set $a = 1$, $u = \lambda$, and $v = y$.
2. Write this formula as an operator equation $T(y, \lambda) = 0$.

3. Formulate the corresponding extended system for the computation of simple turning points and determine the solutions of this extended system (in order to do that you can use the symbolic tool of the MATLAB).
4. Draw the solution curve (using the MATLAB functions contour and meshgrid) and mark the turning points.
5. Set $a = 1$, $u = y$ and $v = \lambda$ and repeat the steps 2–4.

Exercise 5.4 On the website mentioned in Exercise 5.3, you will also find the curve of the *limacon of Pascal*. The corresponding formula is

$$(u^2 + v^2 - 2au)^2 = b^2(u^2 + v^2).$$

Set $a = b = 1$ and proceed as described in Exercise 5.3.

Exercise 5.5 On the website mentioned in Exercise 5.3, you will also find the *eight curve*. The corresponding formula is

$$u^4 = a^2(u^2 - v^2).$$

Set $a = 1$ and proceed as described in Exercise 5.3.

Exercise 5.6 Prove Corollary 5.17:
The limit point $z_0 \in Z$ is a double turning point if and only if

$$a_2 \equiv \psi_0^* T_{yy}^0 \varphi_0^2 = 0 \quad \text{and} \quad a_3 \equiv \psi_0^*(6T_{yy}^0 \varphi_0 w_0 + T_{yyy}^0 \varphi_0^3) \neq 0,$$

where $w_0 \in Y_1$ is the unique solution of the equation $T_y^0 w_0 = -\dfrac{1}{2}T_{yy}^0 \varphi_0^2$.

Exercise 5.7 On the basis of the multiple shooting method (see Algorithm 4.2), implement the path-following technique (see Algorithm 2.2) into a MATLAB m-file rwpmaink.m. Use rwpmaink to trace the solution curve of Bratu's problem (5.53). Determine the existing simple turning point with the appropriate extended system. Draw the bifurcation diagram $y'(0)$ versus λ and mark the turning point.

Exercise 5.8 Use rwpmaink (see Exercise 5.7) to trace the solution curve of the following BVP

$$y''(x) = \lambda\, y(x) \sin(\lambda\, y(x)), \quad y(0) = y(1) = 1.$$

Draw the bifurcation diagram $y'(0)$ versus λ.
Hint: Start with $\lambda = 0$ and determine the solution field for $\lambda \in [0, 25]$.

Exercise 5.9 Use rwpmaink (see Exercise 5.7) to trace the solution curve of the following BVP (*dance with a ribbon*, see [85])

$$y''(x) = \sin\left((x - x^2)y(x)^2\right) - \lambda^2 + \lambda^3, \quad y(0) = y(1), \quad y'(0) = y'(1).$$

Determine the existing simple turning points with the appropriate extended system. Draw the bifurcation diagram $y(0)$ versus λ and mark the turning points. Hint: Start with $\lambda = 0$ and determine the solution field for $\lambda \in [-0.8, 1.2]$.

Exercise 5.10 Consider the BVP

$$y''(x) = \lambda\, y(x)\omega(y), \quad y(1) = 1, \quad y'(0) = 0,$$

where

$$\omega(y) \equiv \exp\left(\frac{\gamma\beta(1 - y)}{1 + \beta(1 - y)}\right).$$

Set $\gamma = 20$, $\beta = 0.4$, and determine the solution field for $\lambda \in [0.01, 0.19]$. Approximate the singular points. Draw the bifurcation diagram $y(0)$ versus λ and mark the computed singular points.

Exercise 5.11 Use rwpmaink to trace the solution curve of the following BVP (*iron*, see [85])

$$y''(x) = \left(\sin(x^2 - x)y(x)^2 + \lambda\right)^2 + \lambda, \quad y(0) = y(1), \quad y'(0) = y'(1).$$

Draw the bifurcation diagram $y(0)$ versus λ and mark the singular points. Hint: Start with $\lambda = -0.4$ and determine the solution field for $\lambda \in [-1, 0]$.

Exercise 5.12 Use rwpmaink to trace the solution curve of the following BVP (*boomerang*, see [85])

$$y''(x) = \left(\sin(x^2 - x)y(x)^2 + \lambda\right)^2 - \lambda, \quad y(0) = y(1), \quad y'(0) = y'(1).$$

Draw the bifurcation diagram $y(0)$ versus λ and mark the singular points. Hint: Start with $\lambda = 4$ and determine the solution field for $\lambda \in [0, 7]$.

Exercise 5.13 Use rwpmaink to trace the solution curve of the following BVP (*loop*, see [85])

$$y''(x) = -(\lambda + \pi^2)\sin(y(x)) - \lambda^4, \quad y(0) = 0, \quad y(1) = 0.$$

Draw the bifurcation diagram $y'(0)$ versus λ and mark the singular points. Hint: Start with $\lambda = -1$ and determine the solution field for $\lambda \in [-1, 1]$.

Exercise 5.14 Use rwpmaink to trace the solution curve of the following BVP (*raindrop*, see [85])

$$y''(x) = \left(x^3 - \frac{1}{4}\right)\sin\left(y(x)^2\right) + \lambda^3 + \lambda^4, \quad y(0) = y(1), \quad y'(1) = 0.$$

Draw the bifurcation diagram $y(0)$ versus λ and mark the singular points. Hint: Start with $\lambda = -0.7$ and determine the solution field for $\lambda \in [-1, 0]$.

Exercise 5.15 Use `rwpmaink` to trace the solution curve of the following BVP (*molar*, see [85])

$$y''(x) = \left(x^3 - \frac{1}{4}\right) \sin\left(y(x)^2\right) + \lambda^3 - \lambda^4, \quad y(0) = y(1), \quad y'(1) = 0.$$

Draw the bifurcation diagram $y(0)$ versus λ and mark the singular points.
Hint: Start with $\lambda = 0.2$ and determine the solution field for $\lambda \in [-0.5, 1.2]$.

Exercise 5.16 Use `rwpmaink` to trace the solution curve of the following BVP (*Mordillo U*, see [85])

$$y''(x) = \sin\left((x^2 - x)y(x)^4 - \lambda\right) y(x) + \lambda + \lambda^2, \quad y(0) = y(1), \quad y'(0) = y'(1).$$

Draw the bifurcation diagram $y(0)$ versus λ and mark the singular points.
Hint: Start with $\lambda = -6$ and determine the solution field for $\lambda \in [-10, 7]$.

Exercise 5.17 Use `rwpmaink` to trace the solution curve of the following BVP (*between 8 and* ∞, see [85])

$$y''(x) = \left(\sin\left((x - x^2)y(x)^3\right) + \lambda\right) y(x) + \lambda^4, \quad y(0) = y(1), \quad y'(0) = y'(1).$$

Draw the bifurcation diagram $y(0)$ versus λ and mark the singular points.
Hint: Start with $\lambda = -0.5$ and determine the solution field for $\lambda \in [-1, 1]$.

Exercise 5.18 Use `rwpmaink` to trace the solution curve of the following BVP (*speech bubble*, see [85])

$$y''(x) = \sin\left((x - x^2)y(x)^2\right) y(x) - \lambda^2, \quad y(0) = y(1), \quad y'(0) = y'(1).$$

Draw the bifurcation diagram $y(0)$ versus λ and mark the singular points.
Hint: Start with $\lambda = -1$ and determine the solution field for $\lambda \in [-1.8, 1.8]$.

Exercise 5.19 Use `rwpmaink` to trace the solution curve of the following BVP (*bow tie*, see [85])

$$y''(x) = \sin\left((x - x^2)y(x)^2\right) + \left(x^3 - \frac{1}{4}\right) \sin\left(y(x)^2\right)$$
$$- 2\lambda^4 + y(x)^5 + \lambda^6 + \lambda^8,$$
$$y(0) = y(1), \quad 2y'(0) = y'(1).$$

Draw the bifurcation diagram $y(0)$ versus λ and mark the singular points.
Hint: Start with $\lambda = -1$ and determine the solution field for $\lambda \in [-1.5, 1.5]$.

Exercise 5.20 Use `rwpmaink` to trace the solution curve of the following BVP
(*Janus*, see [85])

$$y''(x) = \left(x^3 - \frac{1}{4}\right) \sin\left(y(x)^2\right) y(x) + \lambda \sin(y(x)),$$
$$y(0) = y(1), \quad 2y'(0) = y'(1).$$

Draw the bifurcation diagram $y(0)$ versus λ and mark the singular points.
Hint: Start with $\lambda = 0.5$ and determine the solution field for $\lambda \in [-1, 1]$.

References

1. Abbaoui, K., Cherruault, Y.: Convergence of Adomian's method applied to differential equations. Comput. Math. Appl. **28**(5), 103–109 (1994)
2. Abbaoui, K., Cherruault, Y.: Convergence of Adomian's method applied to nonlinear equations. Mathl. Comput. Modell. **20**(9), 69–73 (1994)
3. Abbasbandy, S.: The application of homotopy analysis method to nonlinear equations arising in heat transfer. Phys. Lett. A **360**(1), 109–113 (2006)
4. Abbasbandy, S., Shivanian, E.: Prediction of multiplicity of solutions of nonlinear boundary value problems: novel application of homotopy analysis method. Commun. Nonlinear Sci. Numer. Simul. **15**, 3830–3846 (2010)
5. Abdelkader, M.A.: Sequences of nonlinear differential equations with related solutions. Annali di Matematica Pura ed Applicata **81**(1), 249–258 (1969)
6. Abramowitz, M., Stegun, I.A.: Handbook of Mathematical Functions. Dover Publications, New York (1972)
7. Adomian, G.: Nonlinear Stochastic Operator Equations. Academic Press Inc., Orlando, FL (1986)
8. Adomian, G.: A review of the decomposition method in applied mathematics. J. Math. Anal. Appl. **135**, 501–544 (1988)
9. Adomian, G.: Solving Frontier Problems in Physics: The Decomposition Method. Kluwer, Dordrecht (2004)
10. Adomian, G., Rach, R.: Noise terms in decomposition series solution. Comput. Math. Appl. **24**(11), 61–64 (1992)
11. Allaby, M.: A Dictionary of Earth Sciences, 3rd edn. Oxford University Press, Oxford (2008)
12. Almazmumy, M., Hendi, F.A., Bakodah, H.O., Alzumi, H.: Recent modifications of Adomian decomposition method for initial value problem in ordinary differential equations. Am. J. Comput. Math. **2**(3) (2012). doi:10.4236/ajcm.2012.23,030
13. Ascher, U.M., Mattheij, R.M.M., Russell, R.D.: Numerical Solution of Boundary Value Problems for Ordinary Differential Equations. Prentice Hall Series in Computational Mathematics. Prentice-Hall Inc., Englewood Cliffs (1988)
14. Askari, H., Younesian, D., Saadatnia, Z.: Nonlinear oscillations analysis of the elevator cable in a drum drive elevator system. Adv. Appl. Math. Mech. **7**(1), 43–57 (2015)
15. Askeland, D.R., Phulé, P.P.: The Science and Engineering of Materials, 5th edn. Cengage Learning (2006)
16. Azreg-Aïnou, M.: A developed new algorithm for evaluating Adomian polynomials. Comput. Model. Eng. Sci. **42**(1), 1–18 (2009)

© Springer India 2016
M. Hermann and M. Saravi, *Nonlinear Ordinary Differential Equations*,
DOI 10.1007/978-81-322-2812-7

17. Bader, G., Deuflhard, P.: A semi-implicit midpoint rule for stiff systems of ODEs. Numerische Mathematik **41**, 373–398 (1983)
18. Batiha, B.: Numerical solution of Bratu-type equations by the variational iteration method. Hacettepe J. Math. Stat. **39**(1), 23–29 (2010)
19. Bauer, L., Reiss, E.L., Keller, H.B.: Axisymmetric buckling of hollow spheres and hemispheres. Commun. Pure Appl. Math. **23**, 529–568 (1970)
20. Biazar, J., Shafiof, S.M.: A simple algorithm for calculating Adomian polynomials. Int. J. Contemp. Math. Sci. **2**(20), 975–982 (2007)
21. Bickley, W.G.: The plane jet. Philos. Mag. **28**, 727 (1937)
22. Blickhan, R.: The spring-mass model for running and hopping. J. Biomech. **22**(11–12), 1217–1227 (1989)
23. Boresi, A.P., Schmidt, R.J., Sidebottom, O.M.: Advanced Mechanics of Materials. Wiley, New York (1993)
24. Brezzi, F., Rappaz, J., Raviart, P.A.: Finite dimensional approximation of nonlinear problems. Part III: Simple bifurcation points. Numer. Math. **38**, 1–30 (1981)
25. Bucciarelli, L.L.: Engineering Mechanics for Structures. Dover Publications, Dover Civil and Mechanical Engineering (2009)
26. Chandrasekhar, S.: Introduction to the Study of Stellar Structure. Dover Publications, New York (1967)
27. Chow, S.N., Hale, J.K.: Methods of Bifurcation Theory. Springer, New York (1982)
28. Crandall, M.G., Rabinowitz, P.H.: Bifurcation from simple eigenvalues. J. Funct. Anal. **8**, 321–340 (1971)
29. Dangelmayr, G.: Katastrophentheorie nichtlinearer Euler-Lagrange-Gleichungen und Feynman'scher Wegintegrale. Ph.D. thesis, Universität Tübingen (1979)
30. Davis, H.T.: Introduction to Nonlinear Differential and Integral Equations. Dover Publications (1960)
31. Decker, D.W., Keller, H.B.: Multiple limit point bifurcation. J. Math. Anal. Appl. **75**, 417–430 (1980)
32. Dennis, J.E., Schnabel, R.B.: Numerical Methods for Unconstrained Optimization and Nonlinear Equations. Prentice-Hall Inc., Englewood Cliffs, NJ (1983)
33. Deuflhard, P.: Recent advances in multiple shooting techniques. In: Gladwell/Sayers (ed.) Computational Techniques for Ordinary Differential Equations, pp. 217–272. Academic Press, London, New York (1980)
34. Deuflhard, P., Bader, G.: Multiple shooting techniques revisited. In: Deuflhard, P., Hairer, E. (eds.) Numerical Treatment of Inverse Problems in Differential and Integral Equations, pp. 74–94. Birkhäuser Verlag, Boston, Basel, Stuttgart (1983)
35. Durmaz, S., Kaya, M.O.: High-order energy balance method to nonlinear oscillators. J. Appl. Math. **2012**(ID 518684), 1–7 (2012)
36. Fardi, M., Kazemi, E., Ezzati, R., Ghasemi, M.: Periodic solution for strongly nonlinear vibration systems by using the homotopy analysis method. Math. Sci. **6**(65), 1–5 (2012)
37. Feng, Z.: On explicit exact solutions to the compound Burgers-KdV equation. Phys. Lett. A **293**, 57–66 (2002)
38. Finlayson, B.: The Method of Weighted Residuals and Variational Principles. Academic Press Inc., New York (1972)
39. Floquet, G.: Sur les équations différentielles linéaires à coefficients périodiques. Annales scientifiques de l'École Normale Supérieure **12**, 47–88 (1883)
40. Froese, B.: Homotopy analysis method for axisymmetric flow of a power fluid past a stretching sheet. Technical report, Trinity Western University (2007)
41. Geyer, H., Seyfarth, A., Blickhan, R.: Compliant leg behaviour explains basic dynamics of walking and running. Proc. R. Soc. B: Biol. Sci. **273**(1603), 2861–2867 (2006)
42. Golub, G.H., Van Loan, C.F.: Matrix Computations. The John Hopkins University Press, Baltimore and London (1996)
43. Golubitsky, M., Schaeffer, D.: Singularities and Groups in Bifurcation Theory, vol. I. Springer Verlag, New York (1984)

44. Gräff, M., Scheidl, R., Troger, H., Weinmüller, E.: An investigation of the complete post-buckling behavior of axisymmetric spherical shells. J. Appl. Math. Phys. **36**, 803–821 (1985)
45. Guckenheimer, J., Holmes, P.: Nonlinear Oscillations, Dynamical Systems, and Bifurcations of Vector Fields. Applied Mathematical Sciences, vol. 42. Springer (1983)
46. Hartman, P.: Ordinary Differential Equations. Birkhäuser Verlag, Boston, Basel, Stuttgart (1982)
47. Hassan, H.N., Semary, M.S.: Analytic approximate solution for the Bratu's problem by optimal homotopy analysis method. Commun. Numer. Anal. **2013**, 1–14 (2013)
48. He, H.J.: Semi-inverse method of establishing generalized variational principles for fluid mechanics with emphasis on turbomachinery aerodynamics. Int. J. Turbo Jet-Engines **14**(1), 23–28 (1997)
49. He, J.H.: Approximate analytical solution for seepage flow with fractional derivatives in porous media. Comput. Meth. Appl. Mech. Eng. **167**(1–2), 57–68 (1998)
50. He, J.H.: Variational iteration method–a kind of non-linear analytical technique: some examples. Int. J. Non-Linear Mech. **34**(4), 699–708 (1999)
51. He, J.H.: Variational iteration method—some recent results and new interpretations. J. Comput. Appl. Math. **207**(1) (2007)
52. He, J.H.: Hamiltonian approach to nonlinear oscillators. Phys. Lett. A **374**(23), 2312–2314 (2010)
53. He, J.H.: Hamiltonian approach to solitary solutions. Egypt.-Chin. J. Comput. Appl. Math. **1**(1), 6–9 (2012)
54. He, J.H., Wu, X.H.: Variational iteration method: new development and applications. Comput. Math. Appl. **54**(7–8), 881–894 (2007)
55. Hermann, M.: Ein ALGOL-60-Programm zur Diagnose numerischer Instabilität bei Verfahren der linearen Algebra. Wiss. Ztschr. HAB Weimar **20**, 325–330 (1975)
56. Hermann, M.: Shooting methods for two-point boundary value problems–a survey. In: Hermann, M. (ed.) Numerische Behandlung von Differentialgleichungen, Wissenschaftliche Beiträge der FSU Jena, pp. 23–52. Friedrich-Schiller-Universität, Jena (1983)
57. Hermann, M.: Numerik gewöhnlicher Differentialgleichungen - Anfangs- und Randwertprobleme. Oldenbourg Verlag, München und Wien (2004)
58. Hermann, M.: Numerische Mathematik, 3rd edn. Oldenbourg Verlag, München (2011)
59. Hermann, M., Kaiser, D.: RWPM: a software package of shooting methods for nonlinear two-point boundary value problems. Appl. Numer. Math. **13**, 103–108 (1993)
60. Hermann, M., Kaiser, D.: Shooting methods for two-point BVPs with partially separated endconditions. ZAMM **75**, 651–668 (1995)
61. Hermann, M., Kaiser, D.: Numerical methods for parametrized two-point boundary value problems—a survey. In: Alt, W., Hermann, M. (eds.) Berichte des IZWR, vol. Math/Inf/06/03, pp. 23–38. Friedrich-Schiller-Universität Jena, Jenaer Schriften zur Mathematik und Informatik (2003)
62. Hermann, M., Kaiser, D., Schröder, M.: Bifurcation analysis of a class of parametrized two-point boundary value problems. J. Nonlinear Sci. **10**, 507–531 (2000)
63. Hermann, M., Saravi, M.: A First Course in Ordinary Differential Equations: Analytical and Numerical Methods. Springer (2014)
64. Hermann, M., Ullmann, T., Ullrich, K.: The nonlinear buckling problem of a spherical shell: bifurcation phenomena in a BVP with a regular singularity. Technische Mechanik **12**, 177–184 (1991)
65. Hermann, M., Ullrich, K.: RWPKV: a software package for continuation and bifurcation problems in two-point boundary value problems. Appl. Math. Lett. **5**, 57–62 (1992)
66. Hussels, H.G.: Schrittweitensteuerung bei der Integration gewöhnlicher Differentialgleichungen mit Extrapolationsverfahren. Master's thesis, Universität Köln (1973)
67. Jordan, D.W., Smith, P.: Nonlinear Ordinary Differential Equations: An Introduction for Scientists and Engineers. Oxford Texts in Applied and Engineering Mathematics. Oxford University Press Inc., New York (2007)

68. Keener, J.P., Keller, H.B.: Perturbed bifurcation theory. Arch. Rat. Mech. Anal. **50**, 159–175 (1973)
69. Keller, H.B.: Shooting and embedding for two-point boundary value problems. J. Math. Anal. Appl. **36**, 598–610 (1971)
70. Keller, H.B.: Lectures on Numerical Methods in Bifurcation Problems. Springer, Heidelberg, New York (1987)
71. Khuri, S.A.: A new approach to Bratu's problem. Appl. Math. Comput. **147**(1), 131–136 (2004)
72. Lange, C.G., Kriegsmann, G.A.: The axisymmetric branching behavior of complete spherical shells. Q. Appl. Math. **2**, 145–178 (1981)
73. Lesnic, D.: The decomposition method for forward and backward time-dependent problems. J. Comp. Appl. Math. **147**(1), 27–39 (2002)
74. Levi-Città, T.: Détermination rigoureuse des ondes permanentes d'ampleur finie. Math. Ann. **93**, 264–314 (1925)
75. Liao, S.: The proposed homotopy analysis technique for the solution of nonlinear problems. Ph.D. thesis, Shanghai Jiao Tong University (1992)
76. Liao, S.: An explicit, totally analytic approximation of Blasius' viscous flow problems. Int. J. Non-Linear Mech. **34**(4), 759–778 (1999)
77. Liao, S.: Beyond Perturbation: Introduction to the Homotopy Analysis Method. Chapman & Hall/CRC Press, Boca Raton (2003)
78. Liao, S.: Notes on the homotopy analysis method: some definitions and theorems. Commun. Nonlinear Sci. Numer. Simul. **14**, 983–997 (2009)
79. Liao, S.: Advances in the Homotopy Analysis Method. World Scientific Publishing Co. Pte. Ltd., New Jersey et al. (2014)
80. Soliman, M.A.: Rational approximation for the one-dimensional Bratu equation. Int. J. Eng. Technol. **13**(5), 54–61 (2013)
81. Merker, A.: Numerical bifurcation analysis of the bipedal spring-mass model. Ph.D. thesis, Friedrich-Schiller-Universität Jena (2014)
82. Merker, A., Kaiser, D., Hermann, M.: Numerical bifurcation analysis of the bipedal spring-mass model. Physica D: Nonlinear Phenomena **291**, 21–30 (2014)
83. Merker, A., Kaiser, D., Seyfarth, A., Hermann, M.: Stable running with asymmetric legs: a bifurcation approach. Int. J. Bifurcat. Chaos **25**(11), 1–13 (2015)
84. Merker, A., Rummel, J., Seyfarth, A.: Stable walking with asymmetric legs. Bioinspiration Biomimetics **6**(4), 045,004 (2011)
85. Middelmann, W.: Konstruktion erweiterter Systeme für Bifurkationsphänomene mit Anwendung auf Randwertprobleme. Ph.D. thesis, Friedrich Schiller Universität Jena (1998)
86. Mistry, P.R., Pradhan, V.H.: Exact solutions of non-linear equations by variational iteration method. Int. J. Appl. Math. Mech. **10**(10), 1–8 (2014)
87. Mohsen, A.: A simple solution of the Bratu problem. Comput. Math. Appl. **67**, 26–33 (2014)
88. Moore, G.: The numerical treatment of non-trivial bifurcation points. Numer. Funct. Anal. Optim. **2**, 441–472 (1980)
89. Nayfeh, A.H.: Introduction to Perturbation Techniques. Wiley, New York (1981)
90. Nekrassov, A.I.: Über Wellen vom permanenten Typ i. Polyt. Inst. I. Wosnenski, pp. 52–65 (1921) (in russian language)
91. Nekrassov, A.I.: Über Wellen vom permanenten Typ ii. Polyt. Inst. I. Wosnenski, pp. 155–171 (1922)
92. Nierenberg, L.: Topics in Nonlinear Fuctional Analysis. Courant Lecture Notes in Mathematics 6. American Mathematical Society, Providence, Rhode Island (2001)
93. Ortega, J.M., Rheinboldt, W.C.: Iterative Solution of Nonlinear Equations in Several Variables. Academic Press, New York (1970)
94. Pashaei, H., Ganji, D.D., Akbarzade, M.: Application of the energy balance method for strongly nonlinear oscillators. Prog. Electromagnet. Res. M **2**, 47–56 (2008)
95. Poincaré, H.: Sur le problème des trois corps et les équations de la dynamique. Acta Mathematica **13**(1), A3–A270 (1890)

96. Polyanin, A.D., Zaitsev, V.Z.: Handbook of Exact Solutions for Ordinary Differential Equations, 2nd edn. Chapman & Hall/CRC, Boca Raton, Florida (2002)
97. Poston, T., Stewart, I.: Catastrophe Theory and Its Applications. Pitman, San Francisco (1978)
98. Prince, F.J., Dormand, J.R.: High order embedded Runge-Kutta formulas. J. Comput. Appl. Math. 7, 67–75 (1981)
99. Rafei, M., Ganji, D.D., Daniali, H., Pashaei, H.: The variational iteration method for nonlinear oscillators with discontinuities. J. Sound Vib. 305, 614–620 (2007)
100. Rashidinia, J., Maleknejad, K., Taheri, N.: Sinc-Galerkin method for numerical solution of the Bratu's problems. Numer. Algorithms 62, 1–11 (2012)
101. Rheinboldt, W.C.: Methods for Solving Systems of Nonlinear Equations. SIAM, Philadelphia (1998)
102. Roberts, S.M., Shipman, J.S.: Continuation in shooting methods for two-point boundary value problems. J. Math. Anal. Appl. pp. 45–58 (1967)
103. Ronto, M., Samoilenko, A.M.: Numerical-Analytic Methods in the Theory of Boundary-Value Problems. World Scientific Publishing Co., Inc., Singapore (2000)
104. Roose, D., Piessens, R.: Numerical computation of turning points and cusps. Numer. Math. 46, 189–211 (1985)
105. Rummel, J., Blum, Y., Seyfarth, A.: Robust and efficient walking with spring-like legs. Bioinspiration Biomimetics 5(4), 046,004 (13 pp.) (2010)
106. Sachdev, P.L.: Nonlinear Ordinary Differential Equations and Their Applications. No. 142 in Chapman & Hall Pure and Applied Mathematics. CRC Press (1990)
107. Saravi, M.: Pseudo-first integral method for Benjamin-Bona-Mahony, Gardner and foam drainage equations. J. Basic. Appl. Sci. Res. 7(1), 521–526 (2011)
108. Saravi, M., Hermann, M., Khah, E.H.: The comparison of homotopy perturbation method with finite difference method for determination of maximum beam deflection. J. Theor. Appl. Phys. (2013). doi:10.1186/2251-7235-7-8
109. Scheidl, R.: On the axisymmetric buckling of thin spherical shells. Int. Ser. Numer. Math. 70, 441–451 (1984)
110. Schwetlick, H.: Numerische Lösung nichtlinearer Gleichungen. VEB Deutscher Verlag der Wissenschaften, Berlin (1979)
111. Scott, M.R., Watts, H.A.: A systemalized collection of codes for solving two-point boundary value problems. In: Aziz, A.K. (ed.) Numerical Methods for Differential Systems, pp. 197–227. Academic Press, New York and London (1976)
112. Seydel, R.: Numerical computation of branch points in nonlinear equations. Numer. Math. 33, 339–352 (1979)
113. Shampine, L.F., Baca, C.: Fixed vs. variable order Runge-Kutta. Technical report 84-1410, Sandia National Laboratories, Albuquerque (1984)
114. Shearer, M.: One-parameter perturbations of bifurcation from a simple eigenvalue. Math. Proc. Camb. Phil. Soc. 88, 111–123 (1980)
115. Skeel, R.D.: Iterative refinement implies numerical stability for Gaussian elimination. Math. Comput. 35, 817–832 (1980)
116. Spence, A., Werner, B.: Non-simple turning points and cusps. IMA J. Numer. Anal. 2, 413–427 (1982)
117. Stoer, J., Bulirsch, R.: Introduction to Numerical Analysis. Springer, New York, Berlin, Heidelberg (2002)
118. Su, W.P., Wu, B.S., Lim, C.W.: Approximate analytical solutions for oscillation of a mass attached to a streched elastic wire. J. Sound Vib. 300, 1042–1047 (2007)
119. Teschl, G.: Ordinary Differential Equations and Dynamical Systems, Graduate Studies in Mathematics, vol. 140. AMS, Providence, RI (2012)
120. Troesch, B.A.: A simple approach to a sensitive two-point boundary value problem. J. Comput. Phys. 21, 279–290 (1976)
121. Troger, H., Steindl, A.: Nonlinear Stability and Bifurcation Theory. Springer, Wien, New York (1991)

122. Vazquez-Leal, H., Khan, Y., Fernandez-Anaya, G., Herrera-May, A., Sarmiento-Reyes, A., Filobello-Nino, U., Jimenez-Fernandez, V.M., Pereyra-Diaz, D.: A general solution for Troesch's problem. Mathematical Problems in Engineering (2012). doi:10.1155/2012/208375

123. Wallisch, W., Hermann, M.: Numerische Behandlung von Fortsetzungs- und Bifurkationsproblemen bei Randwertaufgaben. Teubner-Texte zur Mathematik, Bd. 102. Teubner Verlag, Leipzig (1987)

124. Wazwaz, A.M.: A reliable modification of Adomian decomposition method. Appl. Math. Comput. **102**(1), 77–86 (1999)

125. Wazwaz, A.M.: Adomian decomposition method for a reliable treatment of the Emden-Fowler equation. Appl. Math. Comput. **61**, 543–560 (2005)

126. Wazwaz, A.M.: Partial Differential Equations and Soltary Waves Theory. Springer Publisher (2009)

127. Wazwaz, A.M.: The variational iteration method for analytic treatment for linear and nonlinear ODEs. Appl. Math. Comput. **212**(1), 120–134 (2009)

128. Wazwaz, A.M.: A reliable study for extensions of the Bratu problem with boundary conditions. Math. Meth. Appl. Sci. **35**(7), 845–856 (2012)

129. Weber, H.: Numerische Behandlung von Verzweigungsproblemen bei Randwertaufgaben gewöhnlicher Differentialgleichungen. Ph.D. thesis, Gutenberg-Universität Mainz (1978)

130. Weber, H.: Numerische Behandlung von Verzweigungsproblemen bei gewöhnlichen Differentialgleichungen. Numer. Math. **32**, 17–29 (1979)

131. Weber, H.: On the numerical approximation of secondary bifurcation problems. In: Allgower, K.G.E.L., Peitgen, H.O. (eds.) Numerical Solution of Nonlinear Equations, pp. 407–425. Springer, Berlin, Heidelberg, New York (1980)

132. Werner, D.: Funktionalanalysis. Springer, Heidelberg et al (2011)

133. Zeeman, E.C.: Catastrophe Theory. Addison-Wesley Publishing Co. Inc., Reading (1977)

134. Zeidler, E.: Nonlinear Functional Analysis and Its Applications I, Fixed Point Theorems. Berlin, et al, Springer, New York (1986)

135. Zhang, H.L.: Periodic solutions for some strongly nonlinear oscillations by He's energy balance method. Comput. Math. Appl. **58**(11–12), 2480–2485 (2009)

Index

A
Abel's equation, 48
Adomian decomposition method (ADM), 44, 93
Adomian polynomial, 46
Arclength continuation, 273

B
Ballistic trajectory, 130
BD-problem, *see* Bifurcation destroying problem
Beam, 65
Bending, 65
Bernoulli equation, 9, 62
Bifurcation
    primary, 194
    secondary, 194
Bifurcation coefficient, 227
    first, 196
    second, 179
    third, 179
Bifurcation destroying problem, 249
Bifurcation diagram, 171
Bifurcation point, 171
    asymmetric, 199, 224, 255
    primary simple, 172
    secondary multiple, 172
    secondary simple, 172
    simple, 196
    symmetric, 199, 218, 260
Bifurcation point equation, 196
Bifurcation preserving problem, 249
Block elimination technique, 150
Block Gaussian elimination, 139
Boomerang, 297
Bordered matrix, 139

Boundary conditions
    completely separated, 150
    partially separated, 131, 146
Bow tie, 298
BP-problem, *see* Bifurcation preserving problem
Bratu's equation, 24
Bratu's problem, 107, 162, 183
    first extension, 120
Bratu-Gelfand equation, 20
Buckling, 122
Burgers' equation, 28
BVP, *see* Two-point boundary value problem

C
Cannon, 130
Cannoneer, 130
Canonical form, 187
Cauchy-Euler equation, 22
Characteristic coefficient, 176, 195
Clairaut equation, 15
Collocation method, 72, 123
Compactification, 139
Complementary functions, 134
Continuation method, 124
Contour, 296
Correction functional, 34
Corrector iteration, 270
Cusp catastrophe, 187
Cusp curve, 187
Cusp point, 187

D
Damping strategy, 142
Dance with a ribbon, 296

© Springer India 2016
M. Hermann and M. Saravi, *Nonlinear Ordinary Differential Equations*,
DOI 10.1007/978-81-322-2812-7

Printed in the United States
By Bookmasters